Environmental History and Climate Dynamics in Tropical Africa

The African continent will be hit hardest by climate change, despite its people contributing the least to global warming. Indeed, African rural populations have a long tradition of adapting their needs to the soil, vegetation and climate. This tradition is understudied yet is significant for adapting to climate change. This book provides insight into how these communities have lived and used their land in relation to climate and vegetation.

This 36th volume of Palaeoecology of Africa presents an interdisciplinary perspective on the historical relationship between humans and their environment, as well as the evolution of the climate in sub-Saharan and tropical Africa. It also explores the challenges posed by climate change and its consequences. It covers a wide geographical area extending from west to east across the center of the African continent. Furthermore, this book covers a broad timeframe, from Pleistocene and Holocene epochs, through the Iron Age, to the last five millennia, up to the present day. Using data and archaeological findings, this book sheds light on Africa's environmental history and climate dynamics. Natural resource exploitation, changes and degradation in vegetation, climate change and the relationship between humans and their environment are introduced and discussed through numerous case studies with the aim to stimulate the international climate debate and future sustainable management practices. Junior and established scientists from Germany, Switzerland, South Africa and Madagascar have strengthened their cooperation and networking in sub-Saharan Africa, a region rich in natural resources and home to a diversity of people, vegetation and animal species.

This Open Access book establishes a strong foundation for future research into the climate and vegetation of sub-Saharan Africa, as well as the relationship between humans and natural resources. It enables us to understand the dynamics of climate change and how humans adapt to their natural environment. It will be of interest to anyone concerned with climate evolution, vegetation dynamics and the transformation of the natural environment by humans, particularly in the sub-Saharan and tropical regions.

Palaeoecology of Africa: International Yearbook of Landscape Evolution and Palaeoenvironments

Series editor: Jürgen Runge

The series publishes interdisciplinary scientific papers on landscape evolution and on former environments of the African continent, e.g. changes in climate and in vegetation cover interconnected to all kinds of environmental dynamics form the Cainozoic up to the present, including since the Late Quaternary the growing influence of humans in many of the areas studied.

Dynamics of Forest Ecosystems in Central Africa During the Holocene
Past – Present – Future
Volume 28
Edited by Jürgen Runge

Holocene Palaeoenvironmental History of the Central Sahara
Volume 29
Edited by Roland Baumhauer and Jürgen Runge

African Palaeoenvironments and Geomorphic Landscape Evolution
Volume 30
Edited by Jürgen Runge

Landscape Evolution, Neotectonics and Quaternary Environmental Change in Southern Cameroon
Volume 31
Edited by Jürgen Runge

New Studies on Former and Recent Landscape Changes in Africa
Volume 32
Edited by Jürgen Runge

Changing Climates, Ecosystems and Environments within Arid Southern Africa and Adjoining Regions
Volume 33
Edited by Jürgen Runge

The African Neogene
Climate, Environments and People
Volume 34
Edited by Jürgen Runge

Quaternary Vegetation Dynamics
The African Pollen Database
Volume 35
Edited by Jürgen Runge, William Gosling, Anne-Marie Lézine, and Louis Scott

Environmental History and Climate Dynamics in Tropical Africa
Landscapes, Humans and Change
Volume 36
Edited by Jürgen Runge

For more information about this series, please visit: https://www.routledge.com/Palaeoecology-of-Africa/book-series/POA

Environmental History and Climate Dynamics in Tropical Africa

Landscapes, Humans and Change

Volume 36

Edited by
Jürgen Runge

CRC Press
Taylor & Francis Group
Boca Raton London New York

CRC Press is an imprint of the
Taylor & Francis Group, an **informa** business

Designed cover image: Landscape in Yarda, Northern Borku, Chad, granaries, settlements, pastoral farming, and aerodynamic relief (yardangs) in the background (see chapter 4, photo by T. Musch, 2023)

First edition published 2027
by CRC Press
4 Park Square, Milton Park, Abingdon, Oxon, OX14 4RN

and by CRC Press
2385 NW Executive Center Drive, Suite 320, Boca Raton FL 33431

CRC Press is an imprint of informa UK Limited

Funding body: Goethe University and the Franziska and Georg Speyer University Foundation
(Speyer'sche Hochschulstiftung)

British Library Cataloguing-in-Publication Data
A catalogue record for this book is available from the British Library

ISBN: 9781041215400 (hbk)
ISBN: 9781041215431 (pbk)
ISBN: 9781003727781 (ebk)

DOI: 10.1201/9781003727781

Typeset in Times
by Deanta Global Publishing Services, Chennai, India

Contents

Preface

Climate change remains one of the major threats to African countries. Environmental modifications of ecosystems are important for the future of all African states. As a researcher, looking back into the past, it is often misunderstood as an exotic endeavor, even though climate changes during the Holocene and before illustrate how early societies must adapt to new environmental conditions. This is much like what we must do today because of rapid changes in a sensitive environment.

The ten scientific papers presented in the 36th volume of Palaeoecology of Africa (PoA) focus on the topics of environmental history and climate dynamics in tropical Africa. The volume explores both the environmental dynamics of today's arid and humid regions and the challenges they pose for human economic activity. Methodologically, this is shown using rock carvings, interpretation of radiocarbon dates, types of water harvesting, aspects of food security, biomass, medicine and economic circumstances of pastoralists and smallholder farmers. The diversity of African landscapes and their variability are clearly evident in all contributions.

An obituary at the beginning of this book pays tribute to the longstanding editor of PoA, Professor Klaus Heine (1940–2023). This volume is dedicated to him.

The peer review process was thankfully supported by Dr Andoniana Rahamiralala, Lena Grieger, S. Vital Cédric Yao and numerous anonymous reviewers. Like volume 35 of PoA, this edition is published as a contemporary Open Access publication (OA) and is also available free of charge online.

I would like to thank the Open Access Fund of Goethe University and the Franziska and Georg Speyer University Foundation (Speyer'sche Hochschulstiftung) for their financial support.

Finally, I take this opportunity to express my gratitude to the Publishing House Routledge/Taylor & Francis/CRC Press, and in particular to Mrs. Anahita Nanda and Mrs. Radika Gupta as publishers in New Delhi, India, for their constant support and advice.

Jürgen Runge
Frankfurt am Main
March 2026

Editor's Biography

Jürgen Runge is a Professor of Physical Geography and Geoecology at Goethe University Frankfurt, Germany. He has worked for many years on the evolution of tropical landscapes and former and recent climate changes in lower latitudes. He is the editor of the series "Palaeo" and a member of several scientific editorial boards.

Contributors

Aboubacar Adamou
Département de Géographie
Université Abdou Moumouni de Niamey
Niamey, Republic of Niger

Sehinde Akinbiola
First Technical University
Ibadan, Oyo State, Nigeria

Julie Aleman
CNRS, Aix-Marseille Université
IRD, INRA, CEREGE
Aix-en-Provence, France

Olusegun Awotoye
First Technical University
Ibadan, Oyo State, Nigeria

Kate Bailey-Neale
University of Applied Sciences Northwestern Switzerland (FHNW) and Academy of Art and Design
Münchenstein, Switzerland

Marion Bamford
Evolutionary Studies Institute
University of the Witwatersrand
Johannesburg, South Africa

Roland Baumhauer
Institut für Geographie und Geologie
Julius-Maximilians-Universität Würzburg
Würzburg, Germany

David Braun
George Washington University
Washington, DC, USA

Johnny Panga Brushworth
Global Geostrategies Network
Seronera, Tanzania

Colin J. Courtney-Mustaphi
Knowledge Core LLC
Basel, Switzerland

Abraham Nqabutho Dabengwa
Evolutionary Studies Institute
University of the Witwatersrand
Johannesburg, South Africa

Benjamin Davies
School of Arts and Sciences and Environmental Studies Program
Tufts University
Boston, USA

Matthew J. Douglass
College of Agriculture and Natural Resources
University of Nebraska
Lincoln, USA

Jana Eger-Karberg
Institute of Egyptology and Coptology
University of Muenster
Münster, Germany

Esther N. Githumbi
Institute of Soil Science and Site Ecology
Technische Universität Dresden
Dresden, Germany

Sani Ibrahim
Département de Géographie
Université Abdou Moumouni de Niamey
Niamey, Republic of Niger

Makarius Peter Itambu
Department of Archaeology and Heritage Studies
University of Dar es Salaam
Dar es Salaam, Tanzania

Tim Karberg
Institute of Egyptology and Coptology and HTW Dresden
University of Muenster
Münster, Germany

Rebecca Kariuki
School of Sustainability
Arizona State University
Tempe, USA

Husna Mashaka Katambo
School of Human Evolution and Social Change
Institute of Human Origins
Arizona State University
Tempe, USA

Rahab N. Kinyanjui
Earth Sciences Department
National Museum of Kenya
Nairobi, Kenya

Paul J. Lane
Department of Archaeology
University of Cambridge
Cambridge, United Kingdom

Pierre Lapellegerie
Department of Environmental Sciences, Geoecology
University of Basel
Basel, Switzerland

Robert Marchant
Department of Environment and Geography
Institute for Tropical Ecosystems
University of York
York, United Kingdom

Alexes Mes
Department of Archaeology
University of Cambridge
Cambridge, United Kingdom

Linus Munishi
School of Life Sciences and Bio-Engineering
The Nelson Mandela African Institution of Science and Technology
Tengeru, Tanzania

Venanzio Munyaka
Department of Geosciences
Baylor University
Waco, USA

Tilman Musch
Institute of African Studies (IAS)
Bayreuth University
Bayreuth, Germany

Emmanuel E. Ndiema
Earth Sciences Department
National Museum of Kenya
Nairobi, Kenya

Ben N. Nyanchoga
Department of History and Archaeology
University of Nairobi
Nairobi, Kenya

Chantal Kabonyi Nzabandora
Department of Biology
Université Officielle de Bukavu
Bukavu, Democratic Republic of Congo

Ayomide Olubaju
First Technical University
Ibadan, Oyo State, Nigeria

Emile Roche†
Department of Biology
Université Officielle de Bukavu
Bukavu, Democratic Republic of Congo

Jürgen Runge
Institute of Physical Geography
Goethe University and Centre of Interdisciplinary African Studies (ZIAF)
Frankfurt am Main, Germany

Ayobami Salami
First Technical University
Ibadan, Oyo State, Nigeria

Erhard Schulz
Institut für Geographie und Geologie
Julius-Maximilians-Universität Würzburg
Würzburg, Germany

Anna Shoemaker
Terra Archaeology Limited
Victoria, Canada

Barbara Sponholz
Institut für Geographie und Geologie
Julius-Maximilians-Universität Würzburg
Würzburg, Germany

Karl Heinz Striedter
Frobenius Institut
Goethe University
Frankfurt am Main, Germany

Ephraim W. Wahome
Department of History and Archaeology
University of Nairobi
Nairobi, Kenya

Sylvia N. Wemanya
Department of Anthropology
Rice University
Houston, USA

Editorial Board

Obituary Klaus Heine (1940–2023) 1

Jürgen Runge

It is with great sadness that we take farewell to the long-time editor of the interdisciplinary and international book series *Palaeoecology of Africa* (PoA) (volumes 20–27), Professor Dr rer. nat. Klaus Heine, who passed away on 27th July 2023 at the age of 83. Klaus Heine was an outstanding geographer and geoscientist whose life and work were characterised by a deep passion for the exploration of the Earth and its natural phenomena, especially of the palaeoclimate of the planet (Figure 1.1).

Born in Wuppertal (North Rhine-Westphalia, Germany) on 7th May 1940, Heine spent his childhood in Einbeck, Lower Saxony, where his family had moved due to World War II. His school career began at Einbeck primary school and continued at Goethe-Gymnasium. In the 1960s, he went to the universities of Munich, Göttingen, Marburg and Bonn, where he studied geography, geology, soil science and meteorology.

In 1962, while still a student, Heine undertook a four-month research trip through South Africa, Southwest Africa (now Namibia), Betschuanaland (now Botswana) and the Federation of Rhodesia and Nyasaland (now Zimbabwe and Zambia). The impressive African landscapes made a deep impression on him and awakened his interest in geographical research. In 1968, he completed his doctorate thesis at the University of Bonn with an outstanding dissertation entitled "Valley history of the Lahn River in the Marburg area" (Fluss- und Talgeschichte der Lahn im Raum Marburg). He then worked as a research assistant, initially in Marburg.

Heine continued his academic career as a research assistant and later as a professor. In 1973, he habilitated with a ground-breaking thesis on the late Quaternary glacial morphology of Mexican volcanoes, which made an important contribution to climate development. In the years that followed, he worked at renowned universities, including the University of Bonn, where he was appointed Professor of Special and Applied Physical Geography in 1973.

Prof. Heine's close relationship with Africa began in 1977. His scientific work was characterised by a total of twelve research trips to southern Africa, including Namibia and South Africa. He also visited Ecuador, Argentina, Chile, Mexico and the USA. His work on the late Quaternary climate and landscape history of southern Africa, in which he reconstructed the climate of past millennia using terrestrial climate witnesses, is particularly noteworthy. This work was recognised worldwide and brought him into international collaborations, including with the University of Kyoto (Japan), UNAM in Mexico and the University of Bloemfontein (South Africa).

In addition to his scientific work, Prof. Heine was also a committed university lecturer. In 1980, he was appointed to the Chair of Physical Geography at Saarland University and in 1983, he accepted an appointment at the University of Regensburg, where he worked as a professor until his retirement in

DOI: 10.1201/9781003727781-1

FIGURE 1.1 Klaus Heine (1940–2023) in Cape Town (private picture).

January 2009. During this time, he was Dean of the Faculty of Philosophy III from 1987 to 1989 and was particularly characterised by his commitment to research into soil erosion, risk analysis of natural hazards and climate research. He published almost 200 scientific papers (see selected bibliography). From 1988 to 2004, he was the editor in chief of the interdisciplinary, international book series *Palaeoecology of Africa* and the journal *Relief-Boden-Paläoklima* of the Bavarian Academy of Sciences and Humanities.

Klaus Heine's outstanding life's work was the German Springer Spektrum book (754 pp.) on *the Quaternary in the tropics* (*Das Quartär in den Tropen*, 2019, Figure 1.2), in which he succeeded in impressively summarising and evaluating decades of international research in the lower latitudes, in South America, Africa, Asia and Australia. He always combined glacial geomorphology, desert research, fluvial processes and soil science in a multidisciplinary way. He discussed his research findings in the context of national and international seminars, projects and conferences, as well as numerous field trips on five continents. He had an unrivalled overview of the regional development of international global change research and was usually able to mediate controversial views on climate change.

During his retirement, Heine remained active and was a visiting professor at the University of British Columbia in Canada (1994) and Kyoto University in Japan (2005–2006). He also worked for many years as a peer reviewer for international journals and made a significant contribution to scientific development in his field.

Professor Heine leaves behind a significant academic legacy and will be remembered both as a researcher and teacher by his colleagues and his many students. Geography has lost a visionary thinker and passionate scientist. In Professor Dr Klaus Heine, *Palaeoecology of Africa* has lost an exceptionally dedicated academic. We will always honour his memory.

FIGURE 1.2 Cover of the Springer Spektrum book (2019) by Klaus Heine, *Das Quartär in den Tropen – eine Rekonstruktion des Paläoklimas* (754 pp.). In 2024, it was also published as an English textbook, *The Quaternary in the Tropics*.

SELECTED BIBLIOGRAPHY OF PROFESSOR HEINE

(Publications in *Palaeoecology of Africa* in bold)

Heine, K. (1970): Fluss- und Talgeschichte im Raum Marburg. *Bonner Geogr. Abh.* 42: 1–195, Bonn.

Heine, K. (1971): Fossile Bodenbildungen auf quartären Flussschottern an der Mittellahn und ihre Bedeutung für die Terrassenstratigraphie. *Eiszeitalter u. Gegenwart* 22: 17–22, Öhringen.

Golte, W. and Heine, K. (1971): Zur Deutung eiszeitlicher Ablagerungen und Formen am Niederrhein. *Rhein. Heimatpflege, N.F.* III: 180–187, Düsseldorf.

Hard, G. and Heine, K. (1971): Reliefgenerationen, Bodenarten und Pflanzengesellschaften an einer pleistozänen Bruchstufe ("Swistsprung") bei Bonn. *Decheniana* 123: 235–247, Bonn.

Heine, K. (1972): Die Bedeutung pedologischer Untersuchungen bei der Trennung von Reliefgenerationen. *Z. Geomorph. Suppl. Bd.* 14: 113–137, Berlin-Stuttgart.

Heine, K. (1973): Die jungpleistozänen und holozänen Gletschervorstöße am Malinche-Vulkan, Mexiko. *Eiszeitalter u. Gegenwart* 23: 46–62, Öhringen.

Heine, K. and Schönhals, E. (1973): Entstehung und Alter der "toba"-Sedimente in Mexiko. *Eiszeitalter u. Gegenwart* 23/24: 201–215, Öhringen.

Heine, K. (1973): Zur Glazialmorphologie und präkeramischen Archäologie des mexikanischen Hochlandes während des Spätglazials (Wisconsin) und Holozäns. *Erdkunde* XXVII: 161–180, Bonn.

Heine, K. and Heide-Weise, H. (1973): Jungquartäre Förderfolgen des Malinche-Vulkans und des Popocatépetl (Sierra Nevada, Mexiko) und ihre Bedeutung für die Glazialgeologie, Paläoklimatologie und Archäologie. *Münster. Forsch. Geol. Paläont.* 31/32: 303–322, Münster.

Golte, W. and Heine, K. (1974): Fossile Riesen-Eiskeilnetze am Niederrhein. *Eiszeitalter u. Gegenwart* 25: 132–140, Öhringen.

Heine, K. (1975): *Studien zur jungquartären Glazialmorphologie mexikanischer Vulkane – mit einem Ausblick auf die Klimaentwicklung*. Das Mexiko-Projekt der Deutschen Forschungsgemeinschaft VII: 178pp., Steiner, Wiesbaden.

Heine; K. (1975): Permafrost am Pico de Orizaba/Mexiko. *Eiszeitalter u. Gegenwart* 26: 212–217, Öhringen.

Heine, K. and Ohngemach, D. (1976): Die Pleistozän-Holozän-Grenze in Mexiko. *Münster. Forsch. Geol. Paläont.* 38/39: 229–251, Münster.

Heine, K. (1976): Auf den Spuren der Eiszeit in Mexiko. *Natur und Museum* 106: 289–298, Frankfurt.

Heine, K. (1976): Schneegrenzdepressionen, Klimaentwicklung, Bodenerosion und Mensch im zentralmexikanischen Hochland im jüngeren Pleistozän und Holozän. *Z. Geomorph. Suppl.- Bd.* 24: 160–176, Berlin-Stuttgart.

Heine, K. (1976): Blockgletscher- und Blockzungen-Generationen am Navado de Toluca, Mexiko. *Die Erde* 107: 330–352, Berlin.

Heine, K. (1978): Höhenstufung von Zentral-Otago (Neuseeland) – geoökologischer Überblick. *Erdwissenschaftliche Forschung* 11: 537–560, Steiner, Wiesbaden.

Heine, K. (1978): Ökologische Katastrophe in Mexiko? Bodenerosion seit über 2.500 Jahren. *Umschau in Wissenschaft und Technik* 78 (16): 491–496, Frankfurt.

Heine, K. (1978): Jungquartäre Pluviale und Interpluviale in der Kalahari (südliches Afrika). *Palaeoecology of Africa* 10: 31–39, Rotterdam.

Heine, K. (1978): Radiocarbon chronology of late Quaternary Lakes in the Kalahari, Southern Africa. *Catena* 5: 145–149, Braunschweig.

Heine, K. (1980): Quartäre Pluvialzeiten und klimamorphologischer Formenwandel in den Randtropen (Mexiko, Kalahari). *Arb. Geogr. Inst. Univ. Saarland* 29: 135–157, Saarbrücken.

Heine, K. (1980): Wann regnete es in der Kalahari? *Umschau in Wissenschaft und Technik* 80: 250–251, Frankfurt.

Heine, K. (1980): *Studien zur jungpleistozänen Klima- und Landschaftsentwicklung der Kalahari, südliches Afrika. 42.* Deutscher Geographentag 1979, Göttingen, Tagungsberichte und wissenschaftliche Abhandlungen: 281–283, Steiner, Wiesbaden.

Heine, K. and Siebertz, H. (1980): Abriss der paläogeographischen Entwicklung des unteren Niederrheingebietes. *Arb. Rhein. Landeskunde* 46: 1–13, Bonn.

Golte, W. and Heine, K. (1980): Fossile Rieseneiskeilnetze als periglaziale Klimazeugen am Niederrhein. *Arb. Rhein. Landeskunde* 46: 15–26, Bonn.

Heine, K. (1981): Aride und pluviale Bedingungen während der letzten Kaltzeit in der Südwest-Kalahari (südliches Afrika). Ein Beitrag zur klimagenetischen Geomorphologie der Dünen, Pfannen und Täler. *Z. Geomorph. N.F., Suppl.-Bd.* 38: 1–37, Berlin-Stuttgart.

Heine, K. (1981): Sintflutartige Niederschläge in Mexiko vor 12.000 Jahren. *Ibero-Amerikanisches Archiv, NF.* 7: 69–76, Berlin.

Heine, K. (1982): The Main Stages of the Late Quaternary Evolution of the Kalahari Region, Southern Africa. *Palaeoecology of Africa* 15: 53–76, Rotterdam.

Heine, K. (1982): Das Mündungsgebiet der Ahr im Spät-Würm und Holozän. *Erdkunde* XXXVI: 1–11, Bonn.

Heine, K. (1982): The main stages of the late Quaternary evolution of the Kalahari region, southern Africa. *Palaeoecology of Africa*, 15, 53–76.

Heine, K. (1983): Spät- und postglaziale Gletscherschwankungen in Mexiko: Befunde und paläoklimatische Deutung. In Schroeder-Lanz, H. (ed.): *Late- and postglacial oscillations of glaciers.* Glacial and Periglacial forms, 291–304, Balkema, Rotterdam.

Heine, K. (1983): Führt die Quartärforschung zu nicht-aktualistischen Modellvorstellungen in der Geomorphologie? *Colloquium Geographicum* 16: 93–121, Bonn.

Heine, K. (1983): Das Verhältnis von Relief- und Bodenentwicklungsphasen im Jungquartär in Zentralmexiko und in der Kalahari. *Z. Geomorph. N.F., Supp.-Bd.* 48: 145–153, Berlin-Stuttgart.

Heine, K. (1983): Der Vorstoß des nordeuropäischen Inlandeises am Niederrhein (Raum Kleve – Kalkar – Goch). *Arb. Rhein. Landeskunde* 51: 39–49, Bonn.

Heine, K. (1983): Ein außergewöhnlicher Gletschervorstoß in Mexiko vor 12.000 Jahren. Ein Beitrag zum Problem der spätglazialen Klimaschwankungen. *Catena* 10: 1–25, Braunschweig.

Heine, K. (1983): Outline of man's impact on the natural environment in Central Mexico. *Jahrbuch für Geschichte von Staat, Wirtschaft und Gesellschaft Lateinamerikas* 20: 121–131, Köln.

Heine, K. (1983): Mesoformen der Periglazialstufe der semihumiden Randtropen, dargestellt an Beispielen der Cordillera Neovolcánica, Mexiko. Poser, H. and Schunke, E. (Hrsg.): *Mesoformen des Reliefs im heutigen Periglazialraum.* Abh. Akad. Wiss. Göttingen, Math.-Phys. Kl. 3, Folge Nr. 35, 1983: 403–424, Vandenhoeck & Ruprecht, Göttingen.

Heine, K. (1983): Preliminary reconstruction of the Late Quaternary climatic history of the central Namib desert, SW Africa, based on new ^{14}C dates. Horie, S. (Hrsg.): *Paleolimnology of Lake Biwa and the Japanese Pleistocene* 11: 41–54, Tokyo.

Heine, K. and Geyh, M.A. (1984): Radiocarbon Dating of Speleothems from the Rössing Cave (Namib Desert) and Palaeoclimatic Implications. Vogel, J.C. (Hrsg.): *Late Cainozoic Palaeoclimates of the Southern Hemisphere*: 465–470, Balkema, Rotterdam-Boston.

Heine, K. (1984): The classical late Weichselian climatic fluctuations in Mexico. Mörner, N.-A. and Karlen, W. (Hrsg.): *Climatic changes on a yearly to millenial basis*: 95–115, Reidel, Dordrecht.

Heine, K. (1985): Jungquartäre Klimaschwankungen auf der Südhalbkugel. Zbl. *Geol. Paläont. Teil* I (11/12): 1751–1768, Stuttgart.

Heine, K. (1985): Late Quaternary Development of the Kuiseb River Valley and Adjacent Areas, Central Namib Desert, South West Africa/Namibia, and Palaeoclimatic Implications. *Z. Gletscherkde. u. Glazialgeol.* 21: 151–157, Innsbruck.

Heine, K. (1985): Holozäne Meerestransgression an der Namibküste bei Lüderitz/SWA. *Kieler Geogr. Schr.* 62: 127–143, Kiel.

Heine, K. (1987): Anthropogenic Sedimentological Changes during the Holocene in Mexico and Central America. *Striae* 26: 51–63, Uppsala.

Heine, K. (1987): Zum Alter jungquartärer Seespiegelschwankungen in der mittleren Kalahari, südliches Africa. *Palaeoecology of Africa* 18: 73–101, Rotterdam.

Heine, K. (1987): Anthropogenic Sedimentological Changes during the Holocene in Southern Africa. *Striae* 26: 41–50, Uppsala.

Heine, K. (1987): Jungquartäre fluviale Geomorphodynamik in der Namib, Süd-westafrika/Namibia. *Z. Geomorph. N. F., Suppl.-Bd.* 66: 113–134, Berlin-Stuttgart.

Heine, K. (1988): Klimagang, Geomorphodynamik und Kulturentwicklung in Zentralmexiko. Jahrbuch der Geogr. *Gesellschaft zu Hannover, Jahrbuch* 1988: 189–211, Hannover.

Heine, K. (1988): Klimavariabilität und Bodenerosion im südlichen Afrika. *Geogr. Rundschau* 40 (12): 6–14, Braunschweig.

Buch, M.W. and Heine, K. (1988): Klima- oder Prozeßgeomorphologie. *Geogr. Rundschau* 40 (5): 16–26, Braunschweig.

Heine, K. (1988): Late Quaternary Glacial Chronology of the Mexican Volcanoes. *Die Geowissenschaften* 6: 197-205, Weinheim.

Heine, K. (1989): Die letzteiszeitliche Vergletscherung mexikanischer Vulkane als Zeugnis hochglazialer Aridität in Mittelamerika. *Acta Albertina Ratisbonensia* 46: 93–106, Regensburg.

Heine, K. (1990): Some observations concerning the age of the dunes in the western Kalahari and palaeoclimatic implications. *Palaeoecology of Africa* 21: 161–178, Rotterdam.

Heine, K. (1990): Klimaschwankungen und klimagenetische Geomorphologie am Beispiel der Namib. *Berliner Geograph. Studien* 30: 221–234, Berlin.

Heine, K. (1991): Critical comments on Ice-Age Amazon history of climate and vegetation based on new field evidence. *Bamberger Geogr. Schr.* 11: 337–349, Bamberg.

Heine, K. (1992): On the ages of humid Late Quaternary phases in southern African arid areas (Namibia, Botswana). *Palaeoecology of Africa* 23: 149–164, Rotterdam.

Heine, K. (1993): Warmzeitliche Bodenbildung im Bölling/Alleröd im Mittelrheingebiet. *Decheniana* 146: 315–324, Bonn.

Heine, K. (1993): Zum Alter jungquartärer Feuchtphasen im ariden und semiariden südwestlichen Afrika. *Würzburger Geogr. Arb.* 87: 149–162, Würzburg.

Heine, K. (1994): The Mera site revisited. *Quaternary International* 21: 113–119, Tarrytown, N.Y.

Heine, K. (1994): Bodenzerstörung – ein globales Umweltproblem. *Akad. Wiss. Lit. Mainz, Abh. math.-naturwiss. Kl., Jg.* 1994 (2): 65–91, Stuttgart.

Heine, K. (1994): Eis am Äquator – Wasser in der Wüste. Forschungsmagazin der Universität Regensburg, 3. *Jahrg.* 1994, H. 4: 46–61, Regensburg.

Heine, K. (1994): Present and past geocryogenic processes in Mexico: A synopsis. *Permafrost and Periglacial Processes* 5: 1–12, Chichester U.K.

Heine, K. (1994): The late-glacial moraine sequences in Mexico: is there evidence for the Younger Dryas event? - *Palaeogeography, Palaeoclimatology, Palaeoecology* 112: 113–123, Amsterdam.

Heine, K. (1995): Bedded slope deposits with respect to the late glacial sequence in the high Andes of Ecuador and Bolivia. Slaymaker, O. (Hrsg.): *Steepland Geomorphology:* 257–278, Chichester.

Buch, M.W. and Heine, K. (1995): Fluvial geomorphodynamics in the Danube River valley and tributary river systems near Regensburg during the Upper Quaternary – thesis, questions and conclusions. *Z. Geomorph. N.F. Suppl.-Bd.* 100: 53–64, Berlin – Stuttgart.

Heine, K. (1996): The extent of the last glaciation in the Bolivian Andes (Cordillera Real) and paleoclimatic implications. *Z. Geomorph. N.F., Suppl.-Bd.* 104: 187–202, Berlin – Stuttgart.

Heine, K. and Walter, R. (1996): Die Gipskrustenböden der zentralen Namib (Namibia) und ihr paläoklimatischer Aussagewert. *Petermanns Geogr. Mitteilungen* 140: 237–253, Gotha.

Heine, K. and Walter, R. (1996): Gypcretes of the central Namib desert (Namibia). *Palaeoecology of Africa* 24: 173–201, Rotterdam.

Heine, K. and Heine, J.T. (1996): Late-glacial climatic fluctuations in Ecuador: Glacier retreat during Younger Dryas time. *Arctic and Alpine Research* 28: 496–501, Boulder.

Heine, K. (1998): Late Quaternary climatic changes in the central Namib Desert, Namibia. Alsharhan, A.S., Glennie, K.W., Whittle, G.L. and Kendall, C.G.ST.C. (Hrsg.): *Quaternary Deserts and Climatic Change*: 293–304, Rotterdam.

Heine, K. (1998): Climate Change over the Past 135,000 Years in the Namib Desert (Namibia) Derived from Proxy Data. *Palaeoecology of Africa* 25: 171–198, Rotterdam.

Heine, K. (1998): Klimawandel und Desertifikation im südlichen Afrika – ein Blick in die Zukunft. *Geograph. Rdsch.* 50: 245–250, Braunschweig.

Heine, K. (1998): Geomorphologische Beobachtungen zur Besiedlungsgeschichte der Neuen Welt: Das Problem des "eisfreien Korridors" – *Regensburger Beiträge zur Prähistorischen Archäologie* 5: 1–22, Bonn.

Heine, K. and Nuber, T. (1998): Der Bogenberg. Breuer, T. and Jürgens, C. (Hrsg.): *Regensburg und das östliche Bayern*: 80–85, München.

Heine, K. (1998): Neue Ergebnisse der Paläoklimaforschung in den Tropen und Subtropen (Lateinamerika, südliches Afrika). Karrasch, H., Gamerith, W., Schwan, T., Sachs, K. and Krause, U. (Hrsg.): *Globaler Wandel – Welterbe. HGG-Journal* 13: 19–44, Heidelberg.

Heine, K. (1999): Der Kleine Süden Chiles – eine "klassische" Glaziallandschaft. Neue Feldforschungen und Ergebnisse zum Problem der interhemisphärischen Korrelation jungpleistozäner glazialer Ereignisse. *Bamberger Geogr. Schriften* 19: 77–105, Bamberg.

Heine, K. (1999): Fluvial response of the Danube river to climate change in Bavaria During the Weichselian and the Holocene. *Boletim Goiano de Geografia* 19, no. 1, Special Issue (Araguaia 99, Intern. Symp. on Geomorphology and Paleohydrology of Large Rivers): 82–93, Goiânia (Brasil).

Heine, K., Heine, C. and Kühn, T. (2000): Slackwater Deposits der Namib-Wüste (Namibia) und ihr paläoklimatischer Aussagewert. *Zbl. Geol. Paläont. Teil I* (1999), H. 5/6: 587–613, Stuttgart.

Heine, K. (2000): Fließgewässer und Flußauen. Eine Einführung zum Symposium "Fließgewässer und Flußauen – schützenswürdige Naturlandschaften und/oder moderne Wasserstraßen?" – *Z. Geomorph. Suppl.-Bd.* 124: 1–24, Berlin – Stuttgart.

Heine, K. (2000): Tropical South America during the Last Glacial Maximum: evidence from glacial, periglacial and fluvial records. *Quaternary International* 72: 7–21, Oxford.

Bao, H., Thiemens, M.H. and Heine, K. (2001): Oxygen-17 excesses of the Central Namib gypcretes: spatial distribution. *Earth and Planetary Science Letters* 192: 125–135, Amsterdam.

Heine, K. (2001): Und trocken sind seine Riviere. Wasser als Mangelfaktor. Hüser, K., Besler, H., Blümel, W.D., Heine, K., Leser, H. and Rust, U. (Hrsg.): *Namibia – Eine Landschaftskunde in Bildern.* Göttingen/Windhuk, 153–163.

Heine, K. and Pfeffer, K.-H. (Hg.) (2001): *Angewandte Geomorphologie in verschiedenen Geoökosystemen. Beiträge der Symposien der Kommission für Geomorphologie der Bayerischen Akademie der Wissenschaften: "Fließgewässer und Flußauen" und "Tropische Geomorphologie"; 9 Tabellen. Bayerische Akademie der Wissenschaften.* Berlin, Stuttgart (Zeitschrift für Geomorphologie Supplementband, 124).

Heine, K. and Geyh, M.A. (2002): Neue ^{14}C-Daten zur Jüngeren Dryaszeit in den ecuadorianischen Anden. *Eiszeitalter und Gegenwart* 51: 45–61, Hannover.

Heine, K. (2002): Sahara and Namib/Kalahari during the late Quaternary – inter-hemispheric contrasts and comparisons. *Z. Geomorph. N.F., Suppl.-Bd.* 126: 1–29, Berlin – Stuttgart.

Heine, K. and J.T. Heine (2002): A paleohydrologic reinterpretation of the Homeb Silts, Kuiseb River, central Namib Desert (Namibia) and paleoclimatic implications. *Catena* 48:107–130, Amsterdam.

Heine, K. (2003): Paleopedological evidence of human-induced environmental change in the Puebla-Tlaxcala area (Mexico) during the last 3,500 years. *Revista Méxicana de Ciencias Geológicas* 20(3): 235–244, Mexico.

Heine, K. and Niller, H.-P. (2003): Human and climate impacts on the Holocene landscape development in southern Germany (South Bavarian loess rolling hills and adjacent areas). *Geographia Polonica* 76(2): 109–122, Warszawa.

Heine, K. (2004): Late Quaternary Glaciations of Ecuador. In: Ehlers, J. and P.L. Gibbard (Hrsg.): *Quaternary Glaciations – Extent and Chronology, Part III (Latin America, Asia, Africa, Australia, Antarctica):* 165–169, Amsterdam.

Heine, K. (2004): Late Quaternary Glaciations of Bolivia. In: Ehlers, J. and P.L. Gibbard (Hrsg.): *Quaternary Glaciations – Extent and Chronology, Part III (Latin America, Asia, Africa, Australia, Antarctica)*: 83–88, Amsterdam.

Vázquez-Selem, L. and Heine, K. (2004): Late Quaternary Glaciation of Mexico. In: Ehlers, J. and P.L. Gibbard (Hrsg.): *Quaternary Glaciations – Extent and Chronology, Part III (Latin America, Asia, Africa, Australia, Antarctica)*: 233–242, Amsterdam.

Heine, K. (2004): Flood Reconstructions in the Namib Desert, Namibia, and Little Ice Age Climatic Implications: Evidence from Slackwater Deposits and Desert Soil Sequences. Kale, V.S., Gregory, K.J. and Joshi, V.U. (eds.), *Special Issue of the Journal Geological Society of India* 64(4): 535–547.

Heine, K. and Niller, H.-P. (2004): Die Anden Südamerikas – Geoarchive für Umweltveränderungen und Klimawandel. *Geogr. Rdsch.* 56(3): 4–12, Braunschweig.

Reuther, A., Geiger, C., Urdea, P., Niller, H.-P. and Heine, K. (2004): Determining the glacial equilibrium line altitude (ELA) for the northern Retezat Mountains, southern Carpathians and resulting paleoclimatic implications for the last glacial cycle. *Geografie (Analele Universitatii de Vest din Timisoara)* XIV: 11–34.

Heine, K., Niller, H.-P., Nuber, T. and Scheibe, R. (2005): Slope and valley sediments as evidence of deforestation and land-use in prehistoric and historic eastern Bavaria. *Z. Geomorph. N.F., Suppl.-Bd.* 139: 147–171.

Reuther, A., Ivy-Ochs, S. and Heine, K. (2006): Application of surface exposure dating in glacial geomorphology and the interpretation of moraine ages. *Z. Geomorph. N.F., Suppl.-Bd.* 142: 335–359.

Leopold, M., Völkel, J. and Heine, K. (2006): A ground-penetrating radar survey of late Holocene fluvial sediments in NW Namibian river valleys: characterization and comparison. *Journal of the Geological Society, London*, 163: 923–936.

Reuther, A.U., Herget, J., Ivy-Ochs, S., Borodavko, P., Kubik, P.W. and Heine, K. (2006): Constraining the timing of the most recent cataclysmic flood event from ice-dammed lakes in the Russian Altai mountains, Siberia, using cosmogenic in-situ ^{10}Be. *Geology* 34(11): 913–916.

Reuther, A.U., Urdea, P., Geiger, C., Ivy-Ochs, S., Niller, H.-P., Kubik, P.W. and Heine, K. (2007): Late Pleistocene glacial chronology of the Pietrele Valley, Retezat Mountains, Southern Carpathians, constrained by 10Be exposure ages and pedological investigations. *Quaternary International* 164–165: 151–169.

Heine, K. and Völkel, J. (2009): Desert flash flood series. Slackwater deposits and floodouts in Namibia: their significance for palaeoclimatic reconstructions. *Zbl. Geol. Paläont. Teil I. Jg.* 2007, H. 3–4: 287–308.

Heine, K., Reuther, A.U., Thieke, H.U., Schulz, R., Schlaak, N. and Kubik, P.W. (2009): Timing of Weichselian ice marginal positions in Brandenburg (northeastern Germany) using cosmogenic in situ ^{10}Be. *Z. Geomorphologie, N.F.* 53(4): 433–454.

Heine, K. and Völkel, J. (2010): Soil clay minerals in Namibia and their significance for the terrestrial and marine past global change research. *African Study Monographs (Kyoto)*, Suppl. 40: 15–34.

Heine, K. (2010): How it all began – Eduard M. van Zinderen Bakker and Palaeoecology of Africa. *Palaeoecology of Africa* 30: 1–7.

Heine, K. (2010): Climate reconstructions based on fluvial deposits in hyper–arid desert environments: the Namib case. *Palaeoecology of Africa* 30: 27–52.

Heine, K. (2010): Dürren, Bevölkerungskollaps und Bodenerosion: Epidemien und Umweltveränderungen in Zentralmexiko vor 400 bis 500 Jahren. *Geoöko* 31: 162–197.

Heine, K. (2011): Late Quaternary Glaciations of Ecuador. Ehlers, J., Gibbard, P.L. and Hughes, P.D. (eds.): *Quaternary Glaciations – Extent and Chronology, Part IV – A Closer Look.* 803–813. Developments in Quaternary Science 15. Amsterdam (Elsevier).

Heine, K. (2011): Late Quaternary glaciations in Bolivia: Comments on some new approaches to dating morainic sequences. Ehlers, J., Gibbard, P.L. and Hughes, P.D. (eds.): *Quaternary Glaciations – Extent and Chronology, Part IV – A Closer Look*: 757–772. Developments in Quaternary Science 15. Amsterdam (Elsevier).

Vázquez-Selem, L. and Heine, K. (2011): Late Quaternary Glaciation of Mexico. Ehlers, J., Gibbard, P.L. and Hughes, P.D. (eds.): *Quaternary Glaciations – Extent and Chronology, Part IV – A Closer Look.* 849–861. Developments in Quaternary Science 15. Amsterdam (Elsevier).

Reuther, A.U., Fiebig, M., Jerz, H., Ivy-Ochs, S., Kubik, P.W., Reitner, J. and Heine, K. (2011): Deglaciation of a large piedmont lobe glacier in comparison with a small mountain glacier – new insight from surface exposure dating. Two studies from SE Germany. *Eiszeitalter und Gegenwart* 60 (2–3): 248–269.

Heine, K. and Völkel, J. (2011): Extreme floods around AD 1700 in the northern Namib Desert, Namibia, and in the Orange River catchment, South Africa, forced by a decrease of solar irradiance during the Little Ice Age? – *Geographica Polonica* 84 (Special Issue Part 1): 61–80.

Heine, K., Rust, U. and Hilgers, A. (2014): Why 'Younger Dryas', why not 'Antarctic Cold Reversal'? Eksteenfontein revisited. *Palaeoecology of Africa* 32: 163–183.

Geyh, M.A. and Heine, K. (2014): Several distinct wet periods since 420 ka in the Namib Desert inferred from U-series dates of speleothems. *Quaternary Research* 81: 381–391.

Heine, K. (2019): *Das Quartär in den Tropen – Eine Rekonstruktion des Paläoklimas.* Springer Spektrum. Berlin, 1–754.

BIBLIOGRAPHY

German National Library. (n.d.). *Klaus Heine (GND 122138066).* GND Explorer. https://explore.gnd.network/gnd/122138066

Namibiana Buchdepot. (n.d.). *Klaus Heine.* https://www.namibiana.de/namibia-information/who-is-who/autoren/infos-zur-person/klaus-heine.html

ResearchGate. (n.d.). *Klaus Heine.* https://www.researchgate.net/profile/Klaus-Heine-3

Universität Regensburg. (2023). *Nachruf auf Professor Dr. Klaus Heine.* https://www.uni-regensburg.de/universitaet/nachrufe/2023/prof-dr/index.html

Rock Engravings at Taar Doo (Tibesti)

People, Animals, and their Migrations

2

Tilman Musch

INTRODUCTION

Taar Doo is an isolated sandstone ridge whose southwestern end lies about 60 km NNE of Oozu (Aouzou) in the Republic of Chad, and 40 km distant from the Libyan border (see Figures 2.1, 2.2). It belongs to the northern Tibesti and extends for about 20 km from SSW to NNE. Its base is about 750 m above sea level. The lowest step of the sandstone ridge, on which most of the engravings to be described are located, reaches a height of about 12–15 m, sometimes up to 20 m (estimated). The Taar Doo lies on the eastern bank of the dry delta fan of the Tanoa (cf. Figure 2.1), which is formed by the confluence of the Oozu and Yebbige rivers. These are two main valleys (called *enneri* in the Tibesti) that drain the north-northeastern part of the mountain range (cf. Musch 2021).

In March 2021, I was able to visit Taar Doo, accompanied by Teda from Aouzou. During this visit, we walked and drove along the southern part of the eastern wall of the mountain ridge for about 1,500 m, looking for engravings that were already known to my companions. I tried to photograph all recognizable engravings and took some GPS data. It is of major importance to publish information about these engravings because the site is almost unknown in the scientific literature and because it is difficult, if not impossible, for "Western" researchers to visit it, due to the current situation in Libya.

The probably only "Western" researcher who has previously visited Taar Doo in detail is Christoph Staewen (cf. Staewen and Schönberg, 1969). Staewen came to Taar Doo in spring 1964 as part of a study of African wildlife. His aim was to find places in the northern Tibesti where early hunters depicted their game that could have formed a link between other places in the central Tibesti (Gonoa and Yira-Yira) and in the Fezzan (In Habeter and Tel Issaghen). Staewen did not find such places – and thus also not the one described in the following – but he did find unique abstract engravings (Staewen and Schönberg, 1969).

DOI: 10.1201/9781003727781-2

FIGURE 2.1 Landscape in the dry delta fan of the Tanoa with the sandstone ridge of Taar Doo in the background. In the foreground, one of the numerous stalk remnants of a withered tamarisk (*Tamarix aphylla)* and in the center of the picture, pastureland that still looked very green in March 2021 (all pictures by the author)

Ironically, I, who was looking for such abstract engravings, did not find them, but instead found what Staewen had searched for in vain. That the site described here was not discovered by Staewen is surprising, as it is clearly visible on the outside of the mountain massif. In fact, as supposed by Staewen and as I will show in the following, the engravings of Taar Doo, in particular those of humans, can be directly related to engravings of the central Tibesti (Bardai), which, together with clues about the behavior of the wild animals depicted, will allow us to draw conclusions about migrations of humans and animals in an early Holocene environment. Such migrations can be seen as a strategy to react to the increasingly arid environmental conditions of the Central Sahara from 5,000 BCE onwards (cf. Kröpelin *et al.*, 2008).

The engravings of Taar Doo include 52 depictions of people with round heads, 70 depictions of cattle, 40 depictions of wild animals and around 70 depictions of animals that cannot be precisely identified, as well as 41 abstract signs. In rock paintings from sites further east, roundheads are characterized as "a small, featureless, round head, long neck, and short, stubby arms" (Riemer *et al.*, 2017: 12). Examples are representations from Jebel Arkenu (cf. Menardi Noguera and Zboray, 2012) or Jebel Uweinat, where some of the figures wear some kind of plait, lock of hair or feather, and are sometimes depicted with tails, tassels, or oversized penises (cf. Zboray, 2018). According to a scheme by Le Quellec (2013), the epoch of the roundheads extends roughly from 5,500 BCE to 4,500 BCE, i.e., over a thousand years. Although the concept of roundheads was introduced to characterize rock paintings from the Tassili and defined in more detail on the basis of these (cf. Muzzolini 1986: 120–123), it is also repeatedly used to describe representations from other areas. Riemer *et al.*, (2017: 16) distinguish three different styles of roundheads in rock paintings at Jebel Uweinat, which could have been created in the period between 6,500 BCE and 4,500 BCE (Riemer *et al.*, 2017: 16).

According to a scheme given by Le Quellec (2013), the period of domesticated cattle in the Central Sahara begins slightly before the 5th millennium BCE and ends slightly after the 2nd millennium BCE. According to Kuper and Kröpelin (2006), evidence for cattle husbandry varies widely locally. Most evidence originates from between 6,000 and 3,500 BCE. Evidence for cattle in the Tibesti was found by Gabriel (1972: 161), who gives the C^{14} age of a bone excavated in the dry delta fan of the Bardage as 7455 ± 180 years bp.

Scientists assume that African megafauna were particularly abundant in a humid Sahara, i.e., in the period before 5,000 BCE (cf. Kröpelin *et al.*, 2008). According to Le Quellec (2013), African megafauna were present in diminishing numbers up to the 2nd millennium BCE. In the search for clues to date Saharan rock art, attempts have been made to establish a chronological relationship between African

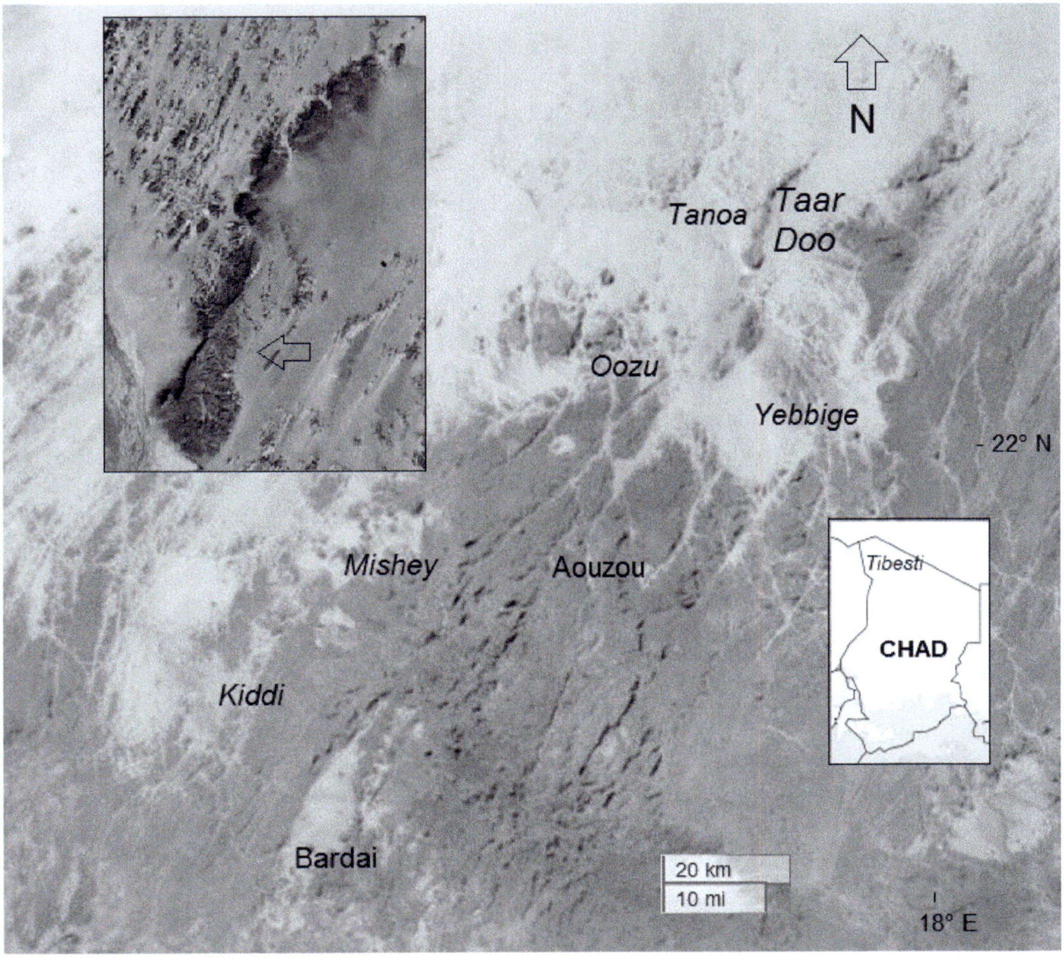

FIGURE 2.2 Map of the region, showing the villages of Bardai and Aouzou, and the valleys of Oozu, Mishey and Kiddi. In the north is the sandstone ridge of Taar Doo, west of it the dry delta fan of the Tanoa. The left inset shows Taar Doo with an arrow pointing to the site of the engravings; the right inset provides a general map of the Republic of Chad. Map base after www.geoplaner.de and Google Maps, modified by the author.

megafauna and their environmental requirements and cattle pastoralism. Zboray (2018: 709–721) reproduces a scientific debate concerning the chronological relationship between the giraffe representations from Jebel Uweinat and those of pastoral cattle husbandry. That is, the question is asked whether the giraffe hunters appeared before, during, or after the cattle herders. The relatively large number of cattle representations at Djebel Uweinat leads Zboray to conclude that at the time of pastoral cattle husbandry, the population may have reached a maximum, and that then, due to climate conditions becoming drier, there was a decline in cattle husbandry and a "gradual shift from pastoralism to hunting". Giraffes could still have been hunted by the inhabitants of Jebel Uweinat until at least 2,000 BCE, when the climate was already too dry for cattle husbandry (Zboray 2018: 709–721).

In this context, Zboray (2018: 725) also highlights the unnaturalness of the elephant representations from Jebel Uweinat and concludes that this animal species was no longer present in the area at all when the rock paintings were created, but that they were made from memory or oral descriptions. On the contrary, elephant bones have been found by Gabriel (1972) in the Tibesti – the youngest dates from shortly before our era. The age of this femur from the Enneri Direnao was determined to be 2690 ± 435 years bp by C^{14} dating. In the case of an elephant skeleton from the Enneri Bardague, C^{14} dating revealed 6,435 ± 225 years bp (Gabriel 1972: 161).

The aim of this paper is twofold: firstly, to describe the engravings of Taar Doo (presumably for the first time in "Western science"). Secondly, in the final discussion, to place the engravings in a scientific context raises the question of the chronological relationship between the occurrence of the African megafauna and pastoral cattle husbandry. I will argue that a roughly contemporaneous occurrence of the megafauna and domesticated cattle, or hunting and pastoralism, at Taar Doo, is quite conceivable. This assumption is based on patterns of contemporary or recent pastoral migrations in the region and on observations of the behavior of West African populations of elephant, giraffe, and ostrich in arid environments.

MATERIAL AND METHODS

Based on my photographic material, I found 276 figures and signs engraved in the stone, of which I would like to describe a representative selection. With such a large number of representations and the shortness of my stay, engravings may well have been overlooked. For this reason, the quantities given are not intended to be completely accurate, but to represent orders of magnitude. When converting the color photos into grayscale images, I increased the contrast and changed the brightness in some of the photos to make the engravings taken in the light of the midday sun more visible.

Humans

All 52 representations of humans have round heads with no facial features. One head shape resembles a mushroom (cf. Muzzolini, 1986: figs. 5 + 6). At least 30 of the humans carry a tail or tassel; in some of the others, tails or tassels may be present in hints but can no longer be accurately identified as such. Some humans hold a spear, a bow, or other objects. In some of the engravings, the head is adorned with a kind of upward-pointing plait, which can also be interpreted as a lock of hair or a feather (cf., for example, Figures 2.3, 2.8, 2.9, 2.15).

Of particular interest is a scene (cf. Figure 2.3), which bears an extreme resemblance to an engraved composition of the "Four Rocks" at Bardai (cf. Figure 2.4). The two scenes are so similar that one gets the impression that they are depictions of the same situations and possibly even of the same or an identical group of people. In these scenes, people with round heads, tails, and plaits are standing in a row by cattle with individually different coat patterns. The people hold their arms open upwards and hold a bow-like object in their left hand. It is possible that the engraving of Taar Doo once extended further down, so that it had approximately the same circumference as that of Bardai; however, the sandstone below the engravings of Taar Doo is heavily flaked, so that it is no longer possible to reconstruct what was once engraved here.

Cattle

In most of the 70 engravings of cattle, bovines are depicted without a hump (presumably *Bos taurus*). Thirty-three of these animals have forward-curved horns, and 26 have lyre-shaped horns (Figures 2.3, 2.5, 2.12). Of nine animals, one horn points forward and the other one backward (Figure 2.6). Two of the cattle engravings are probably depicted with riders (Figures 2.5, 2.11). Horns with one horn pointing forward and the other one backward may be a deformation deliberately caused by pastoralists, which was still practiced in the 20th century in the Nile Valley basin, for example, among the Dinka (Schwabe, 1984). In the case of animals with both horns curved forward, this may also be a deformation deliberately brought about by pastoralists, or it may be a racial trait brought about by selection (Muzzolini, 1991: 1553). In contrast to the lyre-shaped horns, the forward-curved horns are considered to be a frequent "Mediterranean" element, which does not seem to occur in sub-Saharan Africa today (Muzzolini, 1991).

FIGURE 2.3 This scene with the typical Taar Doo depictions of people and cattle with individually different coat patterns is very similar to a scene at Bardai (all rock art pictures by the author).

FIGURE 2.4 The similar scene with people and cattle from Bardai for comparison.

In some cases, cattle are displayed with possibly ritual attributes. Two of the Taar Doo animals bear a kind of disc or oval between their horns, others a double circle or spirals on their rumps (Figures 2.7 and 2.8). In this context, the symbolism of such representations of cattle with an oval (cf. Frobenius, 1978: 38–41) and of so-called "sun rams" with round or oval objects on their heads is brought into connection with the cult of the Egyptian deity Amon-Ra (cf. Camps, 1991).

Wildlife

Respectively, one and two of the depicted fourteen giraffes (*Giraffa camelopardalis*) are part of a hunting scene, together with a human carrying a bow (Figure 2.9). At least one of the five clearly identifiable elephants (*Loxodonta africana*) is engraved in perspective and naturalistically in the rock (Figure 2.14).

FIGURE 2.5 Representations of cattle with forward-curved and lyre-shaped horns. The biggest bovine may have a slight hump. The human figure on the cattle at the bottom right could represent a rider.

FIGURE 2.6 Depiction of a bovine with possible intentionally deformed horns.

FIGURE 2.7 One of the bovines with a possible cultic attribute on the rump, which is sometimes interpreted as a sun.

FIGURE 2.8 Bovine with an oval between the horns and depiction of a human.

FIGURE 2.9 Depiction of a giraffe hunt.

FIGURE 2.10 Possibly depiction of an aardvark.

FIGURE 2.11 Triedents with a dark patina.

FIGURE 2.12 Scene with cattle, ratites, and giraffe.

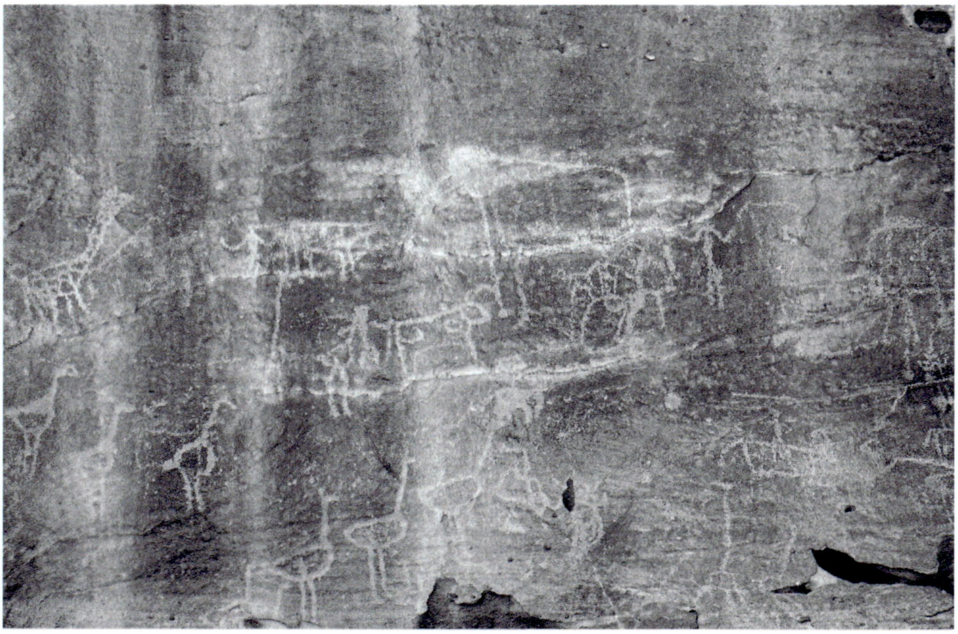

FIGURE 2.13 Scene with people, cattle, ratites, giraffe, and elephant. At the bottom right, a riding human.

FIGURE 2.14 Naturalistic depiction of an elephant.

Most (if not all) 20 engravings of a ratite probably depict an ostrich (*Struthio camelus*; Figures 2.12, 2.13, 2.15), but some of them could also resemble a bustard (e.g. *Neotis nuba*, *Ardeotis arabs*).

Of particular interest is an engraving that could represent a poorly drawn dog (Figure 2.10). However, the animal most closely resembles an aardvark (*Orycteropus afer*), especially because of its prominently displayed claws of the forefeet, which allow it to search for ants and termites in the hard ground (cf. Taylor,

2013: 292). Borda, describing the possible representation of an aardvark at Djebel Uweinat, suggests that its symbolism from the Western Desert may have influenced the religion and mythology of Predynastic Egypt (Borda, 2011: 132). Thus, there seems to be "strong ethnographic and iconographic evidence" that allows us to relate Seth, the god of the underworld associated with the desert, to the aardvark (cf. de Maret, 2005, cited by Borda, 2011).

Many of the approximately 70 engravings which cannot be clearly identified are presumably cattle. Other animals depicted whose identities can be presumed are an antelope, a donkey, two hyenas (or dogs?), a baboon, two big cats, and a camel with a rider. The latter, however, must date from a much later period, as camels only came to Africa around the beginning of the CE (cf. Le Quellec 2013).

Abstract signs

Finally, many abstract signs are also found at Taar Doo, although not in the variety encountered by Staewen at his site (cf. Staewen and Schönberg 1969). Forty-one times, a kind of trident is depicted, in most cases pointing downwards. In most places, this sign appears in clusters. Three times a four-pointed circle is depicted, and four times a circle enclosing a cross. Many of the marks are, as in the case of the ones Staewen examined (cf. Staewen and Schönberg, 1969), very heavily patinated and clearly differ from the neighboring (and thus probably younger) representations of living beings (cf. Figure 2.11).

DISCUSSION

Two hundred and seventy-six engravings and signs could be identified at Taar Doo. Numerically, the most important ones are engravings of cattle, followed by representations of humans and abstract signs. Far less frequent but still relatively numerous are representations of ostriches and giraffes. If one assumes that roundheads at Taar Doo were created at approximately the same time as in other areas, one could say, according to the references mentioned in Chapter 1, that the roundheads at Taar Doo originate from the period between 6,500 and 4,500 BCE. However, the time span at which the engravings of roundheads were made at Taar Doo is certainly a much shorter one. In contrast to the paintings at Jebel Uweinat, where different styles can be distinguished, the human depictions from Taar Doo are characterized by a very strong homogeneity, suggesting that they were produced in a relatively short period of time. As I will show in the following discussion, this period could have been around 5,000 BCE or later, i.e., when the formerly humid Sahara dried out, and migrations became part of survival strategies.

The above underlined similarity between two human scenes, one of Taar Doo and one of Bardai, is of major interest. The site at Bardai lies within the Tibesti Mountains and not, like Taar Doo, on its edge. In a straight line, Bardai is about 130 km away from Taar Doo and is located at an altitude of about 1,000 m above sea level, i.e., about 250 m higher than the site of Taar Doo. Direct routes lead over mountain passes; an easier route leads over wide desert valleys a little further west (*enneri* Aozou, Arab Sõ, Michey, Ohuni, Kiddi, etc.; cf. Figure 2.2). Nowadays, this route is even traveled by motorized vehicles. The vegetation in these desert valleys is still far more abundant than that of the surrounding desert areas. There may therefore have been migrations (and possibly even regular seasonal nomadism) from Taar Doo through vegetation-rich valleys to the central Tibesti as a refuge area in the desiccating Sahara. As mentioned above, such migrations can be seen as a strategy to react to the increasingly arid environmental conditions of the Central Sahara from 5,000 BCE onwards (cf. Kröpelin *et al.*, 2008).

D'Arbaumont reports (1979: 278) that the inhabitants of Tibesti reacted to the arid environmental conditions of the preceding centuries and to droughts that lasted for years by migrating to neighboring regions further south. If, in some years, particularly favorable climatic conditions had led to an abundance of natural resources, there have been remigrations to the Tibesti. Thus, one can assume that the people who left the engravings at Taar Doo could have migrated to the Tibesti, which would have offered a seasonal or permanent refuge for humans and animals. In years with more rainfall, people could then have migrated with their cattle through the river valleys to Taar Doo. The dry delta of the Tanoa River, which runs along the west of Taar Doo, still provides grazing grounds for the camels of the Teda of the Tibesti (cf. Figure 2.1). In a similar context, Riemer *et al.* (2017: 18) describe, for the period between 5,300 and 4,400 BCE, a progressive abandonment of the lowlands of the northern and central Libyan desert and the appearance of a different cultural unit in the highlands. This is probably what allowed cattle pastoralism to persist in the Gilf Kebir until 3,500 BC, despite increasing desiccation (Riemer *et al.*, 2017: 12).

Migrations could also have been a strategy for the animals at that time to cope with the increasing drought. Although African elephant populations normally require less arid areas with sufficient green mass as a food source, there are several populations that have adapted to highly arid environments. In north-west Namibia, an elephant population, adapted to scarce forage and water resources, copes with its arid environment by moving seasonally over ranges of about 650 km with home ranges up to 12,600 km^2 (Poole, Kahumbu and White, 2013: 184; Lindeque and Lindeque, 1991). The home range of female elephants from the Malian Gourma is even larger than that of the NW-Namibian population, with a minimum convex polygon of 21,801 km^2 (Blake *et al.*, 2003: 37–38). It seems that the home ranges of female African elephants appear to increase with decreasing rainfall (Thouless, 1996: 289–90).

The presence of elephants at the time of a pronounced cattle pastoralism in the Tibesti and its surrounding areas is thus quite conceivable. The same is conceivable for the giraffe. According to Le Pendu and Ciofolo (1999), West African giraffes (*Giraffa camelopardalis peralta*) and cattle still occur under the same environmental conditions in western Niger. These giraffes, which migrated from a national park in northern Benin, settled mainly in the tiger-bush savannah of Kouré, and now seem to migrate seasonally as far as the departments of Baleyara and Filingué in the northern Sahel (personal observations and local sources). Here, Tuareg herders, and further south, mainly Fulbe, engage in pastoral cattle nomadism. In the 1990s, a survey found that the home range of Niger's giraffes was about twice as large (90.7 km^2) in the dry season as in the rainy season (46.6 km^2). One-third of the giraffe population embarked on migrations of between 80 km and 200 km in three different directions during the same decade, and two individuals from an isolated population in Mali migrated as far as 300 km southward along the Niger River (Le Pendu and Ciofolo, 1999: 345–349).

As in the case of elephants, giraffes could also have grazed alongside cattle at the same time in the Tibesti and at Taar Doo. In this context, a coexistence of hunting and pastoralism could be a realistic scenario: The main activity of the people of Taar Doo may have been cattle herding, accompanied by the sporadic hunting of giraffes. Hunting and pastoralism can certainly complement each other, as the example of the ostrich among the Tuareg shows. Until the recent past, the ostrich was an appreciated game animal among the latter, and most parts of it were used: meat and eggs were a source of protein, fat was used for traditional medicine, tendons and skin were made into cultural artifacts, and feathers were a commodity in the caravan trade (Lhote, 1951: 140–158; for giraffe hunting: 69–75). Ostriches may also have been hunted at Taar Doo. At least one engraving scene resembles an ostrich hunt (Figure 2.15).

Of the three wild animal species described in detail here, ostriches are certainly the ones that can persist the longest under arid environmental conditions. In the 1990s, the red-necked ostrich (*Struthio camelus camelus*) was still present in the Air Mountains (Central Sahara, Niger) with about 1,600 individuals (cf. Giazzi *et al.*, 2006: 17). Ostriches can survive in partially or completely waterless areas (Cooper *et. al.*, 2009: 1675). In the case of the ostriches of the Air Mountains (Central Sahara/Niger), Giazzi *et al.* (2006: 20) describe the migrations of these animals, which react to different climatic conditions. It was observed that in difficult, dry years, the birds retreated to "inter-montaneous landscapes", i.e., mountain wadis with sometimes heavy runoff and relatively dense vegetation. In more favorable environmental conditions, the birds came out of the mountains, onto the lower plateaus and hills, and in good years, the ostriches were even observed in the desert-like landscapes of the regs.

How do the Taar Doo engravings fit into the discussion summarized in the introduction about the simultaneous occurrence of cattle pastoralism and the African megafauna used for hunting? In line with Zboray (2018), one can assume that giraffes, ostriches, and possibly even elephants occurred at the same

FIGURE 2.15 Probably a depiction of an ostrich hunt (all pictures by the author).

time as cattle pastoralism. At Taar Doo, there are at least two compositions in which wild animals and cattle seem to exist side by side. In one of these scenes, there are at least four cattle together with a giraffe and at least seven ratites (Figure 2.12). The second scene shows two bovines (each with a human), a giraffe, six ostriches, and an elephant (Figure 2.13). In the case of the latter, however, it is not clear whether it was engraved in the rock at an earlier time. In a second scene, there is at least one naturalistic elephant representation at Taar Doo, already mentioned above (Figure 2.14). This naturalistic depiction provides evidence for the occurrence of elephants at the moment when the engravings were made, in contrast to the case of Zboray's (2018: 725) unnatural elephant representations of Djebel Uweinat.

To summarize, one can state that the homogeneity of the humans depicted at Taar Doo provides evidence for the assumption that the engravings were made during a relatively short span of time. During this moment, wild animals and domesticated cattle, or hunting and pastoral cattle husbandry, occurred in parallel, which can be read from the engravings. However, this was a transitional moment, during which environments were becoming increasingly dry, with a decrease in the number of African megafauna and a majority of cattle representations in rock art. In the ecologies of that moment, humans and animals responded to increasing desiccation, as they still do today, with strategies of migration.

REFERENCES

Arbaumont, J. (de), 1979, Le Tibesti et le domaine Teda-Daza, *Bulletin de l'Institut français d'Afrique noire*, **16**(1–2), pp. 255–306.

Blake, S., Bouché, Ph., Rasmussen, H., Orlando, A. and Douglas-Hamilton, I., 2003, *The Last Sahelian Elephants. Ranging Behaviour, Population Status and Recent History of the Desert Elephants of Mali.* file:///C:/Users/bt302427/AppData/Local/Temp/Ranging_Behavior_Population_Status_and_Recent_Hist.pdf (accessed 27/07/2021).

Borda, M., 2011, Rock art finds at Garet Shezzu and an aardvark? *Sahara*, **22**, pp. 130–133.

Camps, G., 1991, Bélier à sphéroïde (Gravure rupestre de l'Afrique du Nord). *Encyclopédie berbère*, **9**, https://journals.openedition.org/encyclopedieberbere/1511 (accessed 27/07/2021).

Cooper, R.G., Mahrose, Kh.M.A., Horbanńczuk, J.O., Villegas-Vizcaíno, R., Kennou Sebei, S. and Faki Mohammed A.E., 2009, The wild ostrich (*Struthio camelus*): A review. *Tropical Animal Health and Production*, **41**, pp 1669–1678.

Frobenius, L., 1978. *Ekade Ektab. Die Felsbilder Fezzans* (Graz: Akademische Druck und Verlagsanstalt).

Gabriel, B., 1972, Zur Vorzeitfauna des Tibestigebirges. In *Palaeoecology of Africa & of the Surrounding Islands & Antarctica*, vol. VII, edited by van Zinderen Bakker, E.M., (Cape Town: A.A. Balkema), pp. 161–162.

Giazzi, F., Ciofolo, I. and Alvès, K., 2006, In favor of reintroducing the Niger Ostrich (*Struthio camelus camelus*) into the Aïr-TénéréReserve (Aïr-Niger massif). *Journal of Alpine research*, **4**, pp. 17–25.

Kröpelin, S., Verschuren, D., Lézine, A.-M., Eggermont, H., Cocquyt, C., Francus, P., Cazet, J.-P., Fagot, M., Rumes B., Russell, J.M., Darius, F., Conley, D.J., Schuster, M., Suchodoletz, H. von and Engstrom, D.R. 2008. Climate driven ecosystem succession in the Sahara: The past 6000 years. *Science*, **320**, pp. 765–768.

Kuper, R. and Kröpelin, S., 2006, Climate-controlled Holocene occupation in the Sahara: Motor of Africa's evolution. *Science*, **313**, pp. 803–807.

Le Pendu, Y. and Ciofolo, I., 1999, Seasonal movements of giraffes in Niger. *Journal of Tropical Ecology*, **15(3)**, pp 341–353.

Le Quellec, J.-L., 2013, Périodisation et chronologie des images rupestres du Sahara Central. *Préhistoires Méditerranéennes*, **4**, https://journals.openedition.org/pm/715 (accessed 27/07/2021).

Lhote, H., 1951, *La chasse chez les Touaregs*, (Paris: Amio-Dumont).

Lindeque, M. and Lindeque, P.M., 1991, Sattelite tracking of elephants in north-western Namibia. *African Journal of Ecology*, **29**, pp. 196–206.

Maret, P. (de), 2005, L'oryctérope, un animal "bon à penser" pour les Africains, est-il à l'origine du dieu égyptien Seth ? *Bulletin de l'Institut français d'archéologie du Caire*, **105**, pp. 107–128.

Menardi Noguera, A. and Zboray, A., 2012, Elongated human figures, large cows and tethered wild animals from the Northern Jebel Arkenu (Libya). *Sahara*, **23**, pp. 133–146.

Musch, T., 2021. Exploring environments through water: An ethno-hydrography of the Tibesti Mountains (Central Sahara). *Ethnobiology Letters*, **12(1)**. https://doi.org/10.14237/ebl.12.1.2021.1709

Muzzolini, R., 1986, *L'art rupestre préhistorique des massifs centraux sahariens*, (Oxford : BAR).

Muzzolini, R., 1991, Le bœuf à cornes en avant. In *Encyclopédie berbère*, **X**, Aix-en-Provence, edited by Union Internationale des Sciences Pré- et Protohistoriques, (Aix-en-Provence: Édisud), pp. 1553–1554.

Poole, J., Kahumbu, P. and White, I., 2013, *Loxodonta africana* Savanna Elephant (African Bush Elephant). In *Mammals of Africa*, Vol. I, edited by Kingdon, J., Happold, D. Hoffmann, M., Butynski, T., Happold, M. and Kalina, J. (London, New Dehli, New York, Sydney: Bloomsbury), pp. 181–194.

Riemer, H., Kröpelin, S., and Zboray, A., 2017, Climate, styles and archaeology: an integral approach towards an absolute chronology of the rock art in the Libyan Desert (Eastern Sahara). *Antiquity*, **91**, pp. 7–23.

Schwabe, C.R., 1984, A unique surgical operation on the horns of African bulls in ancient and modern times. *Agricultural History*, **58(2)**, pp. 138–156.

Staewen, C. and Schönberg, F., 1969, Schematische Felsgravierungen am Taar Doi in Nord-Tibesti (Rep. Du Tchad). *IPEG. Jahrbuch für prähistorische und ethnographische Kunst*, **22**, pp. 93–97.

Taylor, A., 2013, *Orycteropus afer.* Aardvark (Antbear). In *Mammals of Africa*, Vol. I, edited by Kingdon, J., Happold, D. Hoffmann, M., Butynski, T., Happold, M. and Kalina, J. (London, New Dehli, New York, Sydney: Bloomsbury), pp. 290–295.

Thouless, C.R., 1996, Home ranges and social organization of female elephants in Northern Kenya. *African Journal of Ecology*, **34**, pp. 284–297.

Zboray, A., 2018, The petroglyphs of Jebel Uweinat. Many questions and a few answers… In *Desert and the Nile. Prehistory of the Nile Basin and the Sahara. Papers in Honour of Fred Wendorf. Studies in African Archaeology 15*, edited by Poznán Archaeological Museum, (Poznán: Poznán Archaeological Museum), pp. 707–733.

Rock Engravings of the Tibesti

An Inventory of Sites (Re-)visited from an Environmental Perspective

3

Tilman Musch

INTRODUCTION

The systematic exploration of the rock engravings of the Tibesti (Central Sahara, northwestern Republic of Chad) already began in French colonial times (for a review, cf. Gabriel, 1973: 80–86). Often, officers of the colonial army were involved, such as Jean d'Alverny (e.g., 1950), or explorers and geographers, such as Marius Dalloni (1935), Ardito Desio (1941), or Théodore Monod (1945, 1947). Rock engravings of the Tibesti and other places in the Sahara were extensively recorded and annotated by Paul Huard (Huard 1953a, 1953b, 1957, 1959, 1966, Huard and Léonardi 1966, Huard and Massip 1966). Other scientists followed, such as the ethnologist Peter Fuchs (1957) or Ottar Rønneseth (1986), who carried out research for the research station of the Free University of Berlin in Bardai. The doctor Christoph Staewen (Staewen and Striedter 1987), who was stationed in Bardai at the time, is known for his extensive documentation of the Gonoa engraving site and was also the first to discover a site with ancient abstract signs far to the north of the Tibesti (Staewen and Schönberg 1969).

Most researchers, however, were mainly interested in the discovery itself, and later also in the supposed symbolism of the engravings (cf. Huard, 1959, 1966); often, no reference was made to the relationship of the latter to the respective environment. The principal aim of this article is to provide a descriptive inventory of engraving sites discovered or revisited during his ethnographic research in the Tibesti (Table 3.1), to point out some particularities which have not been mentioned in already existing descriptions (see above), and, in particular, to draw attention to some environmental features. Through the latter, an

DOI: 10.1201/9781003727781-3

environmental perspective (cf. Musch, 2021, 2023) can be outlined (and further deepened by the reader), which allows putting engravings in relation to their respective environments.

INVENTORY OF SITES (RE-VISITED)

The following inventory gives a description of sites of the northern, north-western and south-western Tibesti (cf. Figure 3.1). The already extensively documented site of Gonoa (cf. Staewen and Striedter, 1987) and the much-visited rocks of Bardai (Staewen and Striedter, 1987: 282–297) are not treated here. A table at the end of the text provides GPS data and some keyword information on the sites. In particular, the GPS data will allow a reader interested in deepening the environmental perspective to have a closer look at the respective environments by means of remote sensing.

During my fieldwork, I always tried, with the help of locals, to find and visit the largest number possible of engraving sites in the vicinity of places where I carried out ethnographic research. However, a systematic search over large distances was simply not possible for budgetary reasons: it would be difficult to find funding that would allow the costly hiring of a vehicle for weeks in order to visit a large number of sites spread over a vast and difficult terrain. And riding camels are practically nowhere to be found today in the Tibesti and among its contemporary inhabitants, the Tubu, who, after a gold rush in the mountains, now prefer motorized vehicles. Nevertheless, over the years, my trials to document engravings have

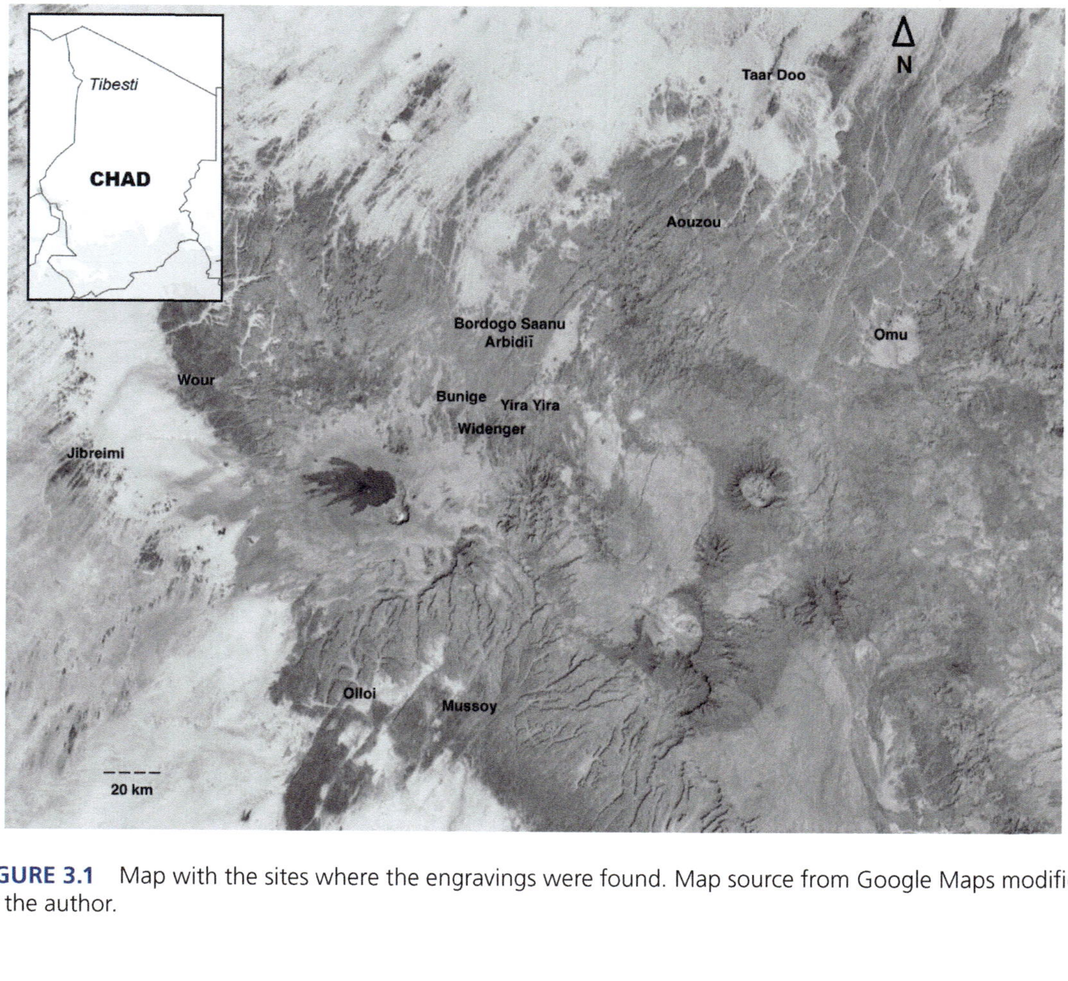

FIGURE 3.1 Map with the sites where the engravings were found. Map source from Google Maps modified by the author.

resulted in a certain number of (re-)visits of sites, which are definitely worth being described here – due to the difficult accessibility of the terrain, scientists may not reach them again soon.

When (re-)visiting the sites, I documented photographically all engravings that I could find. Based on the photographs, it was possible to calculate the numbers of the respective motifs shown below. These numbers are merely intended to illustrate the approximate numerical relationship between the different motifs. Images that seem important to illustrate an issue mentioned here will be reproduced. To make the engravings more visible, the contrast and illumination were increased in the photographs reproduced here in black and white.

I will occasionally refer to the patina of the engravings, which allows a relative chronological classification in comparison with other engravings. For a more general chronological classification, the reader may refer to Le Quellec (2013). According to the latter, the big African mammals, like elephants, appear before the 7th millennium BCE in the rock art, then become increasingly weaker and sometimes end around the 2nd millennium BCE. Their occurrence thus largely coincides with the Humid Optimum, which ended before the 3rd millennium BCE. Cattle husbandry began around the 5th millennium BCE and ended around the 2nd millennium BCE. Horses appeared after the 2nd millennium BCE, and camels a little before the year 0.

Aouzou

The valley south of the settlement of Aouzou was already explored by Monod. However, as Monod writes, he focused in particular on the depictions of humans (Monod, 1947). Staewen and Striedter reproduce the engravings of seven boulders from Aouzou Ellao (1987: 312–319), which do not seem to correspond to the site described here. When walking along both sides of the valley, it was possible to document the depictions of 105 cattle, 34 people, 14 camels (including one recent one and one with a rider), five giraffes (*Giraffa camelopardalis*), three elephants (*Loxodonta africana*), two antelopes, a pair of ostriches (*Struthio camelus*) or bustards, and 23 other animals that could not be recognized exactly. An engraved map and a Tifinagh inscription (Tuareg letters) could also be found, as well as clan signs of the Tubu.

Depictions of cattle clearly predominate in Aouzou. The plains around the mountains may have served as grazing grounds for the cattle herders, and the valleys as places of retreat. As is still the case today with the Tubu, the valleys could have been the territory of certain clans. The engraved map may have served to depict such territories (cf. Musch, 2021). Cattle are often depicted with individualized coat patterns (cf. Figures 3.2, 3.3). Figure 3.2 also shows a recent camel engraving in the lower left part of the image, bearing a Tubu clan mark. The camel is said to have been engraved into the rock by an inhabitant of Aouzou known to the author. At the bottom right of Figure 3.3, a human with the hairstyle typical of many engravings (plaits?) is depicted.

At the lower right edge of Figure 3.3, there are depictions of five giraffes, whose patina does not show any difference to that of the cattle – cattle husbandry and giraffe hunting could in fact have been practised at the same time (cf. Musch in the same volume). The animal in the upper right of the scene could represent an antelope – possibly a waterbuck (*Kobus ellipsiprymnus*) for which the forward-curved horns are characteristic (cf. Musch, 2023). The scene in Figure 3.4 could also be related to hunting. It shows a man with a lance and a dog-like animal next to a bovine, which could represent an aurochs (*Bos primigenius africanus*).

Monod wrote concerning Aouzou: 'I have never seen a single tifinagh inscription' (1947: 34). The tifinagh inscription (Figure 3.5) dates in fact from a later period, namely that which followed the occupation of the Aouzou Strip by the Libyan army in 1973. At that time, Tuareg fighters made up the majority of the Libyan troops in Aouzou. The inscription reads: *Aljamahiria Alarabia Al Libia,* which means 'Arab-Libyan People's Republic' (transliteration and translation by Dida Badi, Algeria).

FIGURE 3.2 Depictions of cattle from Aouzou. The contemporary engraving of a camel with a clan sign at the bottom left.

FIGURE 3.3 Depictions of cattle from Aouzou. Giraffes at the bottom left and right. Possibly a waterbuck on the top right. At the bottom right, there is a human with a typical hairstyle.

FIGURE 3.4 Possibly a depiction of a hunt for aurochs from Aouzou.

FIGURE 3.5 Tifinagh *inscription Aljamahiria Alarabia Al Libia* from Aouzou.

Taar Doo

Taar Doo is an isolated sandstone ridge, about 60 km NNE of Aouzou and 40 km from the Libyan border. Staewen (cf. Staewen and Schönberg, 1969) came here in spring 1964 in search of depictions of wild animals. He did not find any, but he did find unique abstract signs. I discovered here many engravings, including those of wild animals. The depictions of 70 cattle, 52 humans, 20 ratites (probably ostriches), 14 giraffes, and 70 other animals difficult to identify (including possibly an antelope, a donkey, two hyenas or dogs, a monkey, two large feline species, a camel with rider, and an aardvark), as well as 48 abstract signs, could be documented. The site and its engravings are presented in detail in a separate contribution (Musch in the same volume).

Omu

When driving along the Enneri ('valley') Omu in the south of Aouzou, the path leads through a narrow passage between rocks. Here, on the western rock face, in addition to numerous smaller depictions of humans, cattle, a feline, and other animals that are difficult to identify, are six large and heavily patinated engravings of cattle (Figure 3.6).

Yira Yira

Yira Yira lies to the left (south) of the route from Bardai to Zouar used by contemporary drivers. Some of its engravings were reproduced in drawings by Staewen and Striedter ('Gira Gira'; 1987: 272–281). From the site, the view reaches northwards to the Enneri Yira Yira, which joins the Enneri Widenger further to

FIGURE 3.6 Large depiction of cattle from Omu (all rock art photos by the author).

he northeast. A little northwest and east of Yira Yira, two smaller valleys flow into the former. Mountains are located to the south of the site. The people who settled in Yira Yira had chosen a strategically favourable position where they were protected by mountains behind them and could look over the valleys to the north. The mountains, the valley of Yira Yira, and the smaller ones form a kind of triangle that could have delimited their cattle pasture.

The depictions of 12 cattle and one human, as well as 12 other animals that are difficult to recognize could be counted. One depiction shows a large cattle with a carefully worked out coat pattern (Figure 3.7). There is also a large depiction of an elephant (cf. Staewen and Striedter, 1987: 273, Figure 3.3). Other elephants, ostriches and rhinoceroses (*Diceros bicornis*) were located not far away (cf. Staewen and Striedter, 1987). Tubu clan marks are carved into the rock in several places. The engravings of a camel (Figure 3.8) and a horse, both with riders, are contemporary. The rider wears the turban of nowadays Tubu. On the horse and camel, reins can be recognized. Strokes on the camel's cheek and neck could represent clan marks. Also recognizable is the fringed decoration of a travelling bag seemingly attached to the side facing away from the viewer. The front end of a whip extends behind the camel's hump. The latter converges in a point, as it is common in contemporary camel depictions.

Widenger

There are numerous curved lines on one of the boulders, but they do not reveal any geometric pattern. The abundance of engravings of wild animals here could be explained by the proximity of the site to the Enneri Widenger (Figure 3.9). Shortly after the site, a descent leads into the valley, which, according to the etymology of its name, is said to have been rich in gazelles (widen = Gazella dorcas). This site could have been an 'outpost' cattle herders used for hunting, as it may have been the case in the nearby village of Bunige too (cf. Musch, 2023).

FIGURE 3.7 Boulder with engravings of cattle from Yira Yira.

FIGURE 3.8 Contemporary depiction of a camel with rider from Yira Yira.

FIGURE 3.9 Engravings of antelopes and other animals near the descent to the Enneri Widenger.

Bunige

The site of Bunige is located about 20 kilometres south-southwest of Bardai and only about 10 kilometres from Gonoa. It is close to a rock formation of ignimbrite with a sandstone base, on a river terrace of the Enneri Bunige. The latter flows into the Enneri Gonoa. The valley, as well as seeps and springs in the rock formation may have been the cause of an abundance of water in the settlement in the days of the engravers. 'Buni-ge' means 'at the well' in the language of the contemporary Tubu, and in 2021, the inhabitants of the village were still getting their water from small, shallow wells in the riverbed. The site has already been described in detail (Musch, 2023).

Bunige can be found as 'Bénigué' in Huard and Léonardi (1966). It is surprising that the two authors pay little attention to the almost man-sized figure of an archer, although this engraving is sensational in such proximity to the 'Man of Gonoa' and in the Tibesti in general (where most human depictions are kept small). In addition to the large archer, the Bunige engravings include another small archer, 45 cattle or other bovids, four elephants (three of them in a row – Figure 3.10), three animals whose horns resemble those of mouflons (*Ammotragus lervia*), possibly a rhinoceros, and six other animals that are difficult to identify. In addition, there are engravings of 24 round objects that can be interpreted as traps (cf. Musch, 2023). Among the 45 depictions of bovids are probably those of the waterbuck with its characteristic forward-curved horns. The two depictions of an archer, the numerous depictions of traps and a relatively large number of wild animals could indicate the importance of hunting for the prehistoric inhabitants of Bunige. It is possible that the water-rich and remote site was an outpost of the settlers of Gonoa, where they hunted (cf. Musch, 2023).

FIGURE 3.10 Elephant engravings in Bunige.

Bordogo Saanu Arbidiĩ

The site of Bordogo Saanu Arbidiĩ (rough translation: 'Entrance gate to Bardai stone with clan signs') is located about 18 km northwest of Bardai in the Bardage valley. In the immediate vicinity, a boulder with marks of the Tubu clans Odoboya and Terendria can be found. This boulder demarcates the territories of the latter. At the site, there are, on a rock close to the ground (Figure 3.11), engravings of five bovids, a feline, and a camel. On walls at the upper edge of the rock face (Figure 3.12), depictions of ten cattle or other bovids, four humans, a ratite, a giraffe, and perhaps an elephant (the latter may have been engraved more recently) can be found. Dalloni (1935: 335, fig. 137) reproduces in drawings the engraving of a lion (*Panthera leo*) that he found, also in the north of Bardai, at a place called 'Bini Drousso' (*buni durusu* = the deep well). The site discovered by Dalloni could not be identified more precisely. Depictions of felines are rare in northwestern Tibesti; Staewen and Striedter reproduce only three engravings of felines for Widenger, two for Bardai (1987: 271, 287) and none for Gonoa. As it is the case for other sites too, the present one also allowed the prehistoric engravers to overview partly the valley. Perhaps the site was already important for territorial organization before the Tubu engraved here their clan marks on the nearby boulder.

Wour

Coming from the east, the Enneri Madige flows through Wour and then, outside the mountains, ends in a dry delta fan. Rock paintings in the enneri are documented by very vague information from Huard and Massip (1966: 75) who give the following information about the site: '1 hour's walk upstream from the post. Small paintings. Cap. Massip, 1963, unpublished'. In the valley's part eastwards of Wour, I could record engravings in three places over the first few kilometres: a total of 12 cattle (first site), two camels severely damaged by wind erosion (second site), and a probably recent camel (third site). Close to the cattle depictions of the first site (many of them again with individual fur patterns; Figure 3.13), recent Arabic

FIGURE 3.11 Depictions of bovids and possibly a feline (bottom centre) at Bordogo Saanu Arbidiĩ.

FIGURE 3.12 The engravings at the upper edge of the rock face site at Bordogo Saanu Arbidiĩ.

FIGURE 3.13 Cattle engravings in the Madige Valley near Wour.

FIGURE 3.14 Contemporary depiction of a camel with rider in the Madige Valley near Wour.

graffiti are engraved. The lack of patina, the pointed hump, and the hanging fringes of a travelling bag suggest that one of the camel depictions is recent (Figure 3.14). The protruding hair plaits (in contrast to the turban in Figure 3.8), appear to be inspired by prehistoric engravings.

Jibreimi

About 46 km to the southeast of Wour and only about 8 km from the Niger border, on the north side of a rock face, is the site of Jibreimi, which does not appear to have been documented until now. Here, a wide valley, draining the western flank of the Emi Tousside, narrows to flow through a small mountain massif into a dry delta fan on today's Niger border. A small wadi flows into this valley, which seems to originate at Jibreimi and which is still covered with acacia trees. Judging by the rock formation, there may once have been a waterfall with gueltas at the site. From the top of the waterfall, one overlooks a terrain almost completely enclosed by rocks in the shape of an oval. This location of the site, close to water and secured by rocks, was probably a strategically favourable choice for the settlers who made the engravings

The engravings at Jibreimi, which are spread over a relatively wide area, include depictions of 14 cattle (e.g., Figure 3.15), 14 camels (including one with a rider), five further humans, and two giraffes. There are also 16 other animals that are difficult to identify (possibly also cattle or camels). Clan marks were found in two places. The fact that the scene of a giraffe hunt is only slightly or not at all patinated (Figure 3.16) allows us to presume a recent age of the engraving. This supports the hypothesis, stating that giraffes were hunted in the Tibesti until relatively recent times (cf. Musch in the same volume).

Olloi

Access to the wide plain around Zouar is only possible via the mountains or through narrow valleys. The place was thus particularly well protected for prehistoric herders who probably may have feared raids.

FIGURE 3.15 Engravings of two cattle and a human on a stone lying on the ground at Jibreimi.

FIGURE 3.16 Depiction of a giraffe hunt at Jibreimi.

Further valleys leading into the mountains and gueltas on the mountains themselves offered additional places of refuge.

The valley of Olloi runs in the north of the plain of Zouar, draining the mountains located here. Before the rocks flanking the valley open up to the plain of Zouar, there is a site in the scree of the rock face with many human depictions. Here, engravings of 25 people, including warriors with shields and spears, 12 cattle, two giraffes, and two animals resembling monkeys and elephants, respectively (cf. figs. 17 + 18), could be documented. In addition, there are eight other unidentifiable animals and at least two abstract signs.

Close to the valley's westward extension, the Enneri Areun, other engravings can be found in various places. Here, the depictions of 67 unmounted camels (also foals), 23 mounted camels, and 36 people could be found, as well as engravings of 20 other animals difficult to identify. Four times, a hand is carved into the rock (see Figure 3.19). Horses appear to be depicted seven times, six of them with riders (e.g., Figures 3.19, 3.20). Huard, (1957) points out the special interest of horse engravings in the valley of Olloi (197 + plate XXX), because depictions of this animal are, in fact hardly to be found in the Tibesti.

A pentagram is engraved high in one of the rock faces flanking the valley – a sign that is rarely encountered among the engravings of the Tibesti (Figure 3.21). Huard (1957: 195–197) also reports a pentagram from the valley of Olloi, without describing the site in more detail. However, Huard's reproduction of this sign does not correspond to the one reproduced here. In Huard's case, one spike of the pentagram is pointing directly upwards, and the star itself is in the immediate vicinity of the depiction of a warrior. Seemingly, there is another side in the valley with another pentagram. Huard assumes that the pentagram was engraved into the rock before the introduction of Islam.

Another site is located in a smaller valley running parallel to the valley of Olloi. Here, one can recognize, in the mountains northwest of the valley, a 'goat-path' (*ñatere*) 'carved' by the hoofs of goats into the rocks over the centuries. The path leads far into the massif to a guelta. Until the recent past, small livestock were watered and brought to refuge in such high and hidden places. Along the path, the remains of dwellings are scattered, and on the rocks alongside it, there are heavily eroded engravings. These include

FIGURE 3.17 Engravings of warriors, cattle, and wild animals in the valley of Olloi.

FIGURE 3.18 Engravings of cattle and possibly monkeys in the valley of Olloi.

FIGURE 3.19 Engravings of humans, camels, hands, and a horse (centre) in the valley of Olloi.

FIGURE 3.20 Engraving of a horse with rider in the valley of Olloi.

FIGURE 3.21 Engraving of a pentagram in the valley of Olloi.

FIGURE 3.22 Engravings on the 'goat path' near Olloi. The depictions of cattle, camels, and humans show no difference in the intensity of their patina.

depictions of 25 camels without riders, nine camels with riders, nine further depictions of humans, seven engravings of cattle, and 19 depictions of animals that cannot be further identified, as well as abstract signs or drawings. In the case of the camel engraving in the centre of Figure 3.22, the object hanging from its body could represent a saddlebag, as still used by the Tubu today and of which one can see parts in the contemporary engraving of Yira Yira (cf. Figure 3.8). One should also note the apparently simultaneous appearance of camel and cattle, both of which are depicted similarly here and neither of which appears to have a patina.

When following the Enneri Areun further west and then southwest, one reaches another site with mostly very poorly preserved engravings, including those of 31 camels without riders and six more camels with riders. There are also depictions of nine cattle, four people, and five animals that cannot be precisely identified. The darker patina of some of the cattle engravings here clearly shows the difference in date to the camel engravings (Figure 3.23).

It is not possible to describe all the other sites that the Olloi valley is rich in. In particular, there are several rocks near a store hidden high in the rocks called *muši doã* ('ghost house'), on which engravings of 37 camels, four camels with riders, 35 people, 11 cattle, and eight other animals could be counted. More recent engravings in this valley show camels with pointed humps, another one soldiers with machine guns (Figure 3.24).

Mussoy

The site is located at an entrance to the Mussoy valley, near the confluence of the Mussoy, Kayoge, and Tegahan valleys. A site at Tegahan is mentioned by Huard (1953b: 58) as 'Tegaham'.

The site is exposed, and many engravings are heavily eroded. Depictions of 20 camels (including one camel with rider and a foal), 10 cattle, nine humans, five ratites, and 30 animals that are difficult to

FIGURE 3.23 Depictions in the valley of Areun. Here, the differences in the intensity of the patina of cattle and camel engravings can be clearly recognized.

FIGURE 3.24 Contemporary engravings of soldiers with machine guns at one of the sites close to the store of *muši doã*.

FIGURE 3.25 Detail of a scene with cattle and camels from Mussoy.

identify (probably cattle) could be counted. Of particular interest is a scene in which both camels and animals that strongly resemble cattle are depicted mixed together (Figure 3.25). Judging by the patina of these engravings, camels and cattle must have been depicted at the same time. These cattle engravings also differ from other cattle engravings at the same site – some of which are depicted above the scene – whose patina is much darker (Figure 3.26). It thus seems that cattle and camel husbandry were practised here at the same moment, and that camels were used to herd the cattle. It seems thus quite possible that cattle were still able to survive in the central Tibesti, while in more peripheral parts the environmental conditions had already become too dry. According to the reports of older Tubu, there were still cattle even in the valleys of Miski and Maru before the drought that began in the 1970s. Some of them were then moved to North Borku (Budo, Kirdimi), where cattle still exist today. The author was able to see these cattle in swamp areas in 2021 and 2022.

DISCUSSION AND CONCLUSION

If one tries to apply an environmental perspective to rock engravings or the places where they were found, three criteria become clear according to which the people at whose time the engravings were made could have chosen a place: (1) Overview of the terrain, for example from higher vantage points (e.g. Yira Yira, Mussoy, Jibreimi); (2) the presence of water, due to the proximity to river valleys, seeps or springs (e.g., Mussoy, Bunige, and Jibreimi), or the vicinity of other resources like pastures (Aouzou) or game (Widenger, Bunige); and (3) the protection from enemies.

FIGURE 3.26 The entire scene of Mussoy. In the upper part of the picture, one can recognize cattle engravings, which have a much darker patina than the cattle and camel engravings in the lower part of the picture.

The importance of sites for protection from enemies can be illustrated in particular well by the case of Olloi. Until the recent pre-colonial past, camel herds grazing in the plains to the west of the Tibesti were threatened by raids. The Zouar plain, for example, which is difficult to access, offered protection. But here, too, protection could be further strengthened, if necessary, by leading camels into the narrow mountain valleys and by watering and hiding goats at gueltas in the mountains.

One function the engravings may have had is this one of marking territories, or, in a wider sense, of appropriating environments or landscape (cf. Lenssen-Erz 2012). The case of the map of Aouzou (Musch, 2021) provides evidence for this function, as does the proximity of the Bordogo Saanu Arbidiĩ site to a boundary stone of recent clan territories. Even today, clan affiliation is still of great importance to the Tubu of the Tibesti, as the careful depiction of the clan mark on the contemporary camel engraving of Aouzou or the insinuated clan marks on the contemporary camel engraving of Yira Yira may show. Rock art, as a kind of marker, may thus still today serve to appropriate environments. Contemporary engravings, like the camel of Aouzou with its brand, the camel and the horse of Yira Yira engraved with much detail, the camel of Wour, the soldiers with machine guns of Olloi, and, not least, Arabic graffiti, may show this.

A future in-depth analysis of engravings and engraving sites from an environmental perspective could confirm the hypothesis outlined in the Mussoy chapter that small-scale environmental conditions were different from those depicted by large-scale climate models. For example, herding cattle in the more humid mountains could still have played an important role in the times of camel herders, and hunting big game such as giraffes, buffalo, or wild cattle could also have been very important still in the times of cattle husbandry.

TABLE 3.1 Table of Sites by the author

SITE	*COORDINATES*	*ALTITUDE*	*DATE OF VISIT AND NOTES*
Aouzou	N21° 48.731′ E17° 25.949′	3.024 m asl.	2019, 2021. Sites on both sides of the valley, from the waypoint approx. 700 m to S and 500 m to E.
Bordogo Saanu Arbidiĩ	N21° 29.451′ E16° 54.755′	3.114 m asl.	2021. Site on the rock face to the west of the waypoint
Bunige	N21° 19.357′ E16° 48.180′	3.614 m asl.	2021. Engravings at the rock wall and on boulders close to it.
Jibreimi	N21° 07.642′ E15° 35.718′	2.065 m asl.	2021. Sites in the rocks to the south of the waypoint.
Mussoy	N20° 22.321′ E16° 46.857′	2.449 m asl.	2020. Two sites at the waypoint.
Olloi	N20° 28.357′ E16° 26.898′	2.474 m asl.	2020. First site of Olloi described here.
Olloi 'Goat path'	N20° 28.815′ E16° 28.137′	2.508 m asl.	2020. The ascent to the 'goat path' is to the north-west of the waypoint.
Olloi Areun	N20° 24.619′ E16° 21.834′	2.378 m asl.	2020.
Omu	N21° 29.318′ E18° 03.923′	3.535 m asl.	2019. Site in the west of the waypoint.
Taar Doo	N22° 17.246′ E17° 43.898′	2.458 m asl.	2021. Sites on the east face of the rock massif.
Widenger	N21° 12.873′ E16° 49.824′	4.026 m asl.	2022.
Wour	N21° 21.084′ E16° 00.326′	2.519 m asl.	2021. Site of the cattle engravings. Camel engravings 2–3 km (as the crow flies) upstream.
Yira Yira	N21° 19.056′ E16° 54.165′	3.468 m asl.	2022.

ACKNOWLEDGEMENTS

I would like to thank my local Tubu guides who took me to the respective engraving sites: Adoum Gonei (Zouar), Adoum Adeli (Aouzou), Ali Moidi (Bardai), Brahim Malloumami (Wour), Galmai Issouf (Faya), Koki Sidi (Aouzou), and Souleymane Yaya (Zouar). Thanks also to many inhabitants of the Tibesti, not mentioned here, for their hospitality and their indications of engravings.

REFERENCES

Dalloni, M., 1935, Palethnologie. In *Mission au Tibesti. Mémoires de l'Académie de France* **2**, edited by Dalloni, M. (Paris: 62), pp. 93–403.

D'Alverny, F., 1950, Vestiges d'art rupestre au Tibesti oriental. *Journal de la Societé des Africanistes*, **20(2)**, pp. 239–272.

Desio, A., 1941, Sculture rupestri di nuove località del Tibesti settentrionale e del deserto libico. *Gli Annali dell'Africa italiana*, **IV(I)**, pp. 203–206.

Fuchs, P., 1957, Felsmalereien und Felsgravuren in Tibesti, Borku und Ennedi. *Archiv für Völkerkunde*, **XII**, pp. 110–135.

Gabriel, B., (1973): Von der Routenaufnahme zum Weltraumphoto. Die Erforschung des Tibesti Gebirges in der zentralen Sahara, Kartographische Miniaturen, 4, pp. 1–96.

Huard, P., 1953a, Gravures rupestres de la lisière nord-occidentale du Tibesti. *Travaux de l'Institut de Recherches Sahariennes*, **10**, pp. 75–106.

Huard, P., 1953b, Répertoire des stations rupestres du Sahara oriental français. *Journal de la Société des Africanistes*, **23**, pp. 43–76.

Huard, P., 1957, Nouvelles gravures rupestres du Djado, de l'Afafi et du Tibesti. *Bulletin de l'Institut Français d'Afrique Noire*, **19B(1–2)**, pp. 184–223.

Huard, P., 1959, Les cornes déformées sur les gravures du Sahara sud-oriental. *Travaux de l'Institut de Recherches sahariennes*, **18**, pp. 109–131.

Huard, P., 1966, Contribution à l'étude des spirales au Sahara central et nigéro-tchadien. *Bulletin de la société préhistorique française. Études et travaux*, **63(2)**, pp. 433–464.

Huard, P. and Léonardi, Lt., 1966, Nouvelles gravures rupestres des chasseurs du Fézzan méridional, du Djado et du Tibesti. *Rivista di scienze preistoriche*, **31**, pp. 135–156.

Huard, P. and Massip, J.-M., 1966, Nouveaux centres de peintures rupestres au Sahara nigéro-tchadien. *Bulletin de l'Institut Français de l'Afrique Noire*, **28B(1–2)**, pp. 44–81.

Lenssen-Erz, T., 2012, Pastoralist appropriation of landscape by means of rock art in Ennedi highlands, Chad. *Afrique : Archéologie & Arts* **8**, pp. 27–43.

Le Quellec, J.-L., 2013, Périodisation et chronologie des images rupestres du Sahara Central. *Préhistoires Méditerranéennes*, **4**, https://journals.openedition.org/pm/715 (accessed 10/05/2024).

Monod, T, 1945, Note pré́liminaire sur quelques gravures rupestres de la ré́gion d'Aozou (Tibesti). *Comptes rendus*, **II**, pp. 443–446.

Monod, T., 1947, Sur quelques gravures rupestres de la région d'Aouzou (Tibesti). *Rivista di Scienze Preistoriche*, **2**, pp. 30–47.

Musch, T., 2021, Exploring environments through water: An ethno-hydrography of the Tibesti Mountains (Central Sahara). *Ethnobiology Letters*, **12(1)**, https://doi.org/10.14237/ebl.12.1.2021.1709 (accessed 10/05/2024).

Musch, T., 2023, A neighbor of Gonoa: 'The Man of Bunige 'and his game.' *Journal of Global Archaeology*, **2023(04)**, pp. 2–15. https://doi:10.34780/lc92-3ecc

Rønneseth, O., 1986, Felszeichnungen in Tibesti (Tschad) und Klimaänderungen in der Ostsahara. *Praehistorische Zeitschrift*, **61(1)**, pp. 64–83.

Staewen, C. and Schönberg, F., 1969, Schematische Felsgravierungen am Taar Doi in Nord-Tibesti (Rep. du Tchad). *IN IPEG. Jahrbuch für prähistorische und ethnographische Kunst*, **22**, pp. 93–97.

Staewen, C. and Striedter, K. H., 1987, *Gonoa. Felsbilder aus Nord-Tibesti (Tschad).* (Stuttgart: Franz Steiner Verlag).

A Fisherman and His Wife? *Rock-Paintings from Širiê/Ennedi*

4

Tilman Musch

INTRODUCTION

Fish are rare in the Sahara. This sentence, which could be considered a commonplace, is a remarkable fact in relation to rock art: While the millennia-old paintings and engravings of the Sahara depict humans as well as domestic and larger wild animals, fish are extremely rarely represented. According to a publication from 2018, of the approximately 10,000 known rock art sites in the Sahara (excluding those of Pharaonic Egypt), only about 40, including four in Chad, show one or more depictions of fish (Gauthier and Gauthier 2018: 287).

In this chapter, I will present the assumed depiction of a catfish (*Clarias sp.* or *Arius gigas*) that I found on a rock painting in the Ennedi in November 2022 (Figures 4.1, 4.2). At the same time, I will consider whether another composition from the same location could be interpreted as a scene of collective fishing (Figure 4.3). Both paintings are drawn on the walls of a rock shelter at a place called Širiê[1] which is located in the immediate vicinity of Nehi, about 55 kilometres southeast of Fada (Figures 4.4, 4.5). In the same rock shelter, there is a painting of granaries that were probably used to store wild cereals (Figure 4.6), which I will also present in the following and a painting of a hut (Figure 4.7). The juxtaposition of fish-related scenes and granaries may seem arbitrary. However, this is by no means the case, because, according to a recent study (Manning *et al.*, 2023), fish represented a protein-rich diet for the cattle pastoralists who probably left the rock paintings. The carbohydrates of the wild cereals would be a perfect complement to this. Wild cereals (e.g., *Panicum turgidum* or *Panicum laetum*) are still gathered by women in the valleys of the Ennedi; it is a practice that is widespread throughout the Sahara (cf. e.g., Gast, 2000: 37–67; Tubiana, 1969). The aim of the present contribution is thus to document, for the first time, the presumed catfish from Širiê and to relate the paintings from Širiê to a wider context of environment and nutrition in the formerly water-rich Sahara.

There is no doubt that fish and fishing existed in a formerly green Sahara with its numerous lakes before it began to dry out 5,000 to 6,000 years B.P. (cf. Kröpelin *et al.*, 2008). Harpoon points or hooks

DOI: 10.1201/9781003727781-4

FIGURE 4.1 Possible depiction of a fisherman with his wife from Širiê I in original colours and treated with DStretch_crgb. November 2023 (all figures by the author).

FIGURE 4.2 Possible depiction of a fisherman with his wife from Širiê I in original colours and treated with DStretch_crgb. (November 2023).

FIGURE 4.3 Possible depiction of fishing from Širiê I treated with DStrech_yre and converted into a black and white photograph. (November 2023).

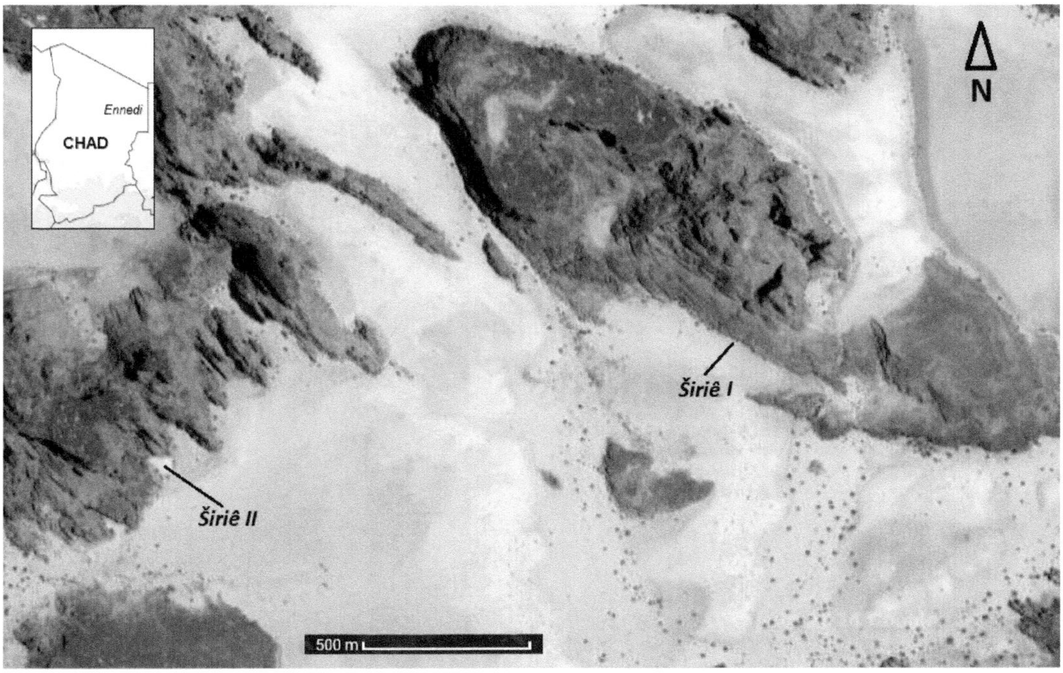

FIGURE 4.4 Aerial photograph of Širiê. Map from Google Maps modified by the author.

FIGURE 4.5 The rock shelter of Širiê I. (November 2022).

FIGURE 4.6 Composition of granaries with a woman in a grinding posture, processing perhaps wild cereals from Širiê I, treated with DStretch_lds and converted into a black and white photograph. (November 2022).

FIGURE 4.7 Depiction of a hut with a women grinding possibly wild cereals and cattle from Širiê I, treated with DStretch_lds and converted into a black and white photograph.

made of bone found in northern regions of the Republic of Chad, for example, indicate fishing (Courtin, 1965, 1967; Mauny, 1952).[2] Researchers have also come across fish skeletons in the beds of paleolakes that have since dried up (cf. Jousse and Van Neer 2009, Musch, 2017). Fish still live in the water bodies of the Sahara today. Recently, the presence of 20 species has been documented (Trape 2013, 2016, 2018). These include also catfish like *Clarias* cf. *gariepinus*, the occurrence of which has been discovered by me in a high mountain guelta of the Tibesti in 2020 (Musch, 2023b) and recently also in a pond in the valley of Tebi/Ennedi (Figure 4.8).

Ethnographic reports on collective fishing are available from the south of Chad (Blache and Miton, 1962), the inland delta of the Niger (Daget, 1949) and the Algerian Sahara (Cauvet, 1913), among others. It is reported from the south of Chad that "whole villages go into the water" during collective fishing (Blache and Miton, 1962: 28). Cauvet (1913: 705–706) describes methods of collective fishing for the Algerian Sahara. At Tatahouine the fishing of *Chromis* was preceded by an annual ceremony in which a donkey was sacrificed to the Bahr Tatahouine. "Then all the men entered the pond and splashed about in a way that completely muddied the water. Panic-stricken chromis came to the surface, where they were easily captured in every possible container" (Cauvet, 1913: 706).

Boats, which have been present in the wider region for thousands of years, may also have been involved in such collective fishing.[3] Rock carvings of boats, some of them with boatmen, have been documented for Borku and Ennedi (Arcangioli 1996, Huard and Massip, 1964, 1966, Jacquet, 1999). Many of these boats are simplified and only outlined, in most cases as a kind of semicircle lying on its back, without any details being reproduced (cf. Gauthier and Gauthier, 2018: 287–288). One exception is a petroglyph from Soboro (Ennedi), in which a man in a boat is holding a paddle (Simonis *et al.*, 2017: 226). Zigzag lines can be seen on the hull of the boat, which could be decorations, but could also be the binding cords of a reed boat.

The fishing scenes presented here have apparently never been described by "Western" scientists, and the rock shelter in which they are located may also still be undiscovered. In fact, most authors who have

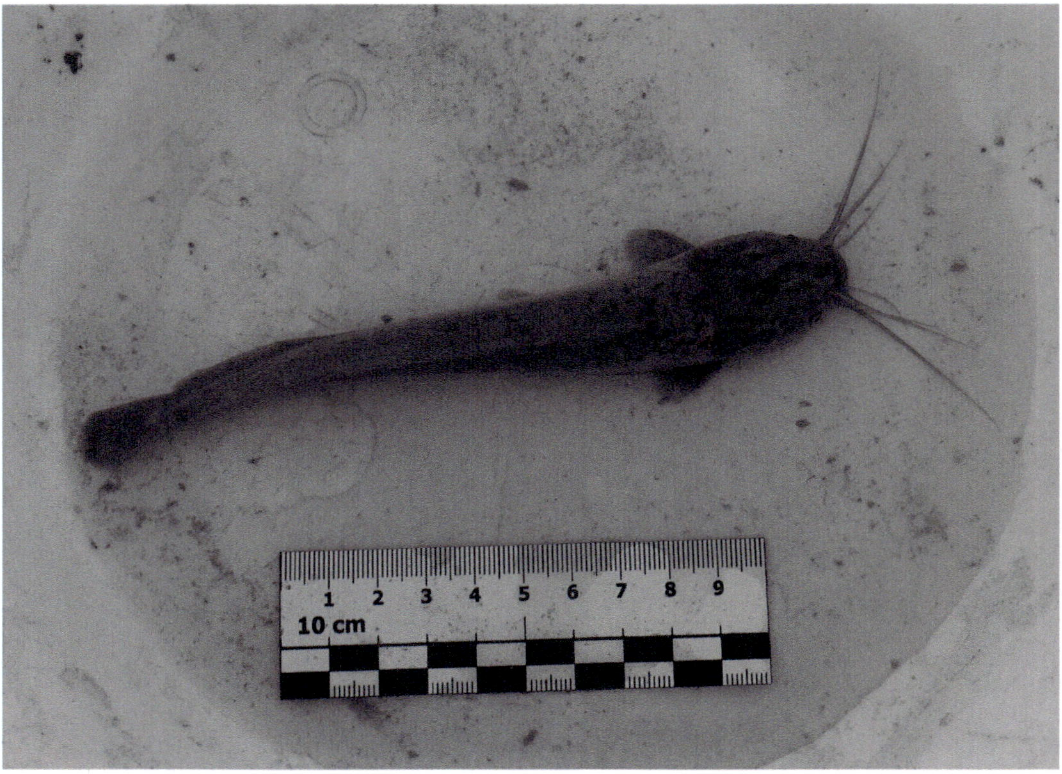

FIGURE 4.8 Juvenile individual of a catfish (Clarias gariepinus or C. anguillaris) from a pond in the valley o
Tebi, approx. 90 km north-east of Fada. (November 2023).

searched for and described rock paintings in the Ennedi (Bailloud, 1960, 1997, Huard, 1963, Jacquet, 2000 do not mention Širiê. On an exhaustive map of places prospected by Bailloud (Bailloud, 1997: 20), there i no entry for a site with rock art in the immediate vicinity of Nehi (in Bailloud: "Nohi"), and the wadi o Nehi itself is only described as a geographical feature, but not as a place of discovery. Simonis *et al.* (2017 51–53) mention six sites near Nehi (NH01-NH06[4]; also called "Nohi" here). Širiê II can be clearly identifie as NH03 due to the image of a cow, of which only the white horns and the white coat pattern are still visibl (2017: 53). Based on the brief descriptions given by Simonis *et al.*, (2017: 51–53) of the other sites at Nehi an on the pictures they published, it appears that they did not visit Širiê I. This would however be surprising because the site is in the immediate vicinity of NH03. But even if Širiê I was among the sites described b Simonis *et al.*, (2017) neither of the two possibly fish-related scenes caught the authors' attention.[5]

METHODS AND MATERIAL

The rock-shelter of Širiê is facing south-west; less than 2 kilometres west-southwest of the shelter, whic I will call Širiê I here, there is another rock shelter which faces south-east.[6] This site will be referred t here as Širiê II. On my first visit in 2022, I had taken incomplete photographs of the paintings presumabl related to fishing, which only became clear afterwards when I processed them with DStretch software an image enhancement technique which makes features that are no longer recognizable partially visibl again. I therefore returned to Širiê II in November 2023 and was able to photograph the composition entirety.

Methods

During my visits at Širiê I and II, I tried to photograph the rock paintings as completely as possible with a digital camera. I also did this in places where the presence of the drawings could only be assumed due to their poor state of preservation. In Širiê II in particular, there was hardly anything left to recognize on the rock face. I later processed the photographs with DStretch.

As I did also in former articles (cf. Musch, 2021, 2023a), I attempt to examine rock art in an "environmental perspective", taking also into account the input of several scientific disciplines, as, for example, ethnology, geology or zoology. Through triangulation of such disciplinarily different inputs, assumptions made about the rock paintings can become more tenable. Of course, there can hardly be complete certainty in such assumptions about the millennia-old rock art, as almost nothing is known about those who created it. Thus, hypotheses will always have to remain hypotheses.

Description of the Rock Art at Širiê

I would like to briefly describe the rock art of Širiê I and II here, insofar as this is still possible due to the difficulty of recognizing many of the depictions. Such an inventory is intended to give an approximate overview of how often which motif appears; however, it is very likely that the respective motifs appear more frequently than mentioned here but are no longer recognizable as such today. The relationships between the number of motifs may provide an insight into the thematic priorities of those who created the paintings at Širiê.

The paintings of Širiê I include 42 cattle, 17 further quadrupeds (possibly also cattle), three camels, the object identified here as a catfish and numerous other unidentifiable items. There are also 43 depictions of humans. A "mushroom-head hairstyle"[7] is clearly recognizable in ten of the human depictions. In two cases, a couple is depicted with one figure wearing a skirt, the other one not. Some of these depictions of people belong to compositions: a couple with the presumed catfish, four schematic people together with two presumed granaries, a hut with a woman in its inside and cattle outside, two people with a presumed net or fish trap and a person in a presumed boat. There is also a composition of five people, four of whom are each holding a lance.

Among the depictions of Širiê II, 24 cattle, four other quadrupeds (possibly also cattle), clearly one and possibly up to three giraffes, as well as ten camels (a few of them probably with riders) can be recognized. There are also 15 depictions of humans, some of them with mushroom-head hairstyles (Figure 4.9). In addition, there are many other depictions that are no longer recognizable even after processing with DStretch.

The first scene I will discuss in the following shows two people with mushroom hairstyles holding hands (Figures 4.1, 4.2). The presumed woman is standing to the right of the presumed man; she is wearing a kind of collar and a kind of belt. The man is also wearing some kind of belt; his legs are decorated with zigzag lines (tattoos?); he appears to be wearing a loincloth. Feathers or other headdresses are recognizable on the hairstyles of both. The man is holding an object in his left hand. Attached to this is a deformed circle, the surface of which is filled with colour.

The second scene of Širiê I which I will relate to fishing shows two slightly bent figures on the left who appear to be holding something in their hands. Below them is an elongated oval whose surface is divided into rectangles by a vertical line and further horizontal lines (Figures 4.3, 4.10). Further to the viewer's right is a circular structure with a dot at its centre. This is followed by a group of people with their legs spread apart[8], each holding an elongated object in their right hand. Above these people are two further depictions of humans. One of them appears to be sitting or standing in a semi-circular structure. This one also holds an elongated object in his right hand.

In another scene of Širiê (Figure 4.6), one can recognize two oval objects, one of which is standing on a kind of stilts. In the following, I will presume that these objects are granaries. One oval has seven

FIGURE 4.9 Very poorly preserved composition showing people with mushroom-head hairstyles from Širiê II, treated with DStretch_lds and converted into a black and white photograph. (November 2022).

FIGURE 4.10 Two people pulling a fish trap ashore or hauling in a net. Detail of Figure 4.3.

ircular structures representing either its contents or openings. The other has only one such opening. It is ivided by a diagonal. There are three human figures next to the ovals. In one oval there appears to be a uman figure in a bent posture wearing a skirt. In the depiction of a hut by Širiê I (Figure 4.7), one can also ecognize a human figure with a skirt in such a bent posture, as well as possibly two children and utensils.

DISCUSSION

rom the above outlined inventories of Širiê I and II, it becomes clear that humans and cattle make up he largest number of depictions, from which it can be concluded that most of the images were probably reated during the "cattle period". They could therefore have been painted between 7,000 and 4,000 B.P. cf. Le Quellec 2013). This is also the time when there were numerous paleolakes in the Central Sahara. Their number is said to have been at its highest around 6,500 B.P. (cf. Le Quellec 2013). The mushroom airstyles could provide a further chronological clue: Depictions of people with expansive hairstyles, ome of them wearing skirts, are described by Bailloud (1997: 75) as belonging to the "style of Tamada" named after a site in the Ennedi, located 19 kilometres WSW of Fada) and attributed to the later cattle eriod ("bovidien recent"), where depictions of couples can also be found. The drawings of camels were reated at a later date, namely after 4,000 B.P. (cf. Le Quellec 2013). As far as the depictions of giraffes re concerned, dating them is difficult, as giraffes were still to be found in the Ennedi in colonial times, ccording to locals.

Several details of the object in the hand of the man from the couple of Širiê I (Figures 4.1, 4.2) are uggestive of a fish. Firstly, there is the plump shape and the broad head (pointing downwards here). Both haracteristics could indicate a catfish, although its barbels are not depicted.[9] The animal's body has "outgrowths", most of which point towards the right of the scene. The two "outgrowths" directly following the ead could represent pectoral fins, followed by one of the ventral fins and then the anal fin. Only three of he caudal fin rays would be shown then. Finally, either the dorsal or the adipose fin is shown on the back f the animal. In the depiction of three fish from the northern Tibesti (Boccazzi and Boccazzi 1999), one f them has a similarly plump shape that tapers towards the tail. In this fish, the two ventral and pectoral ins (or the two protruding gill covers?) are apparently represented graphically; the caudal fin is depicted s a unit in which the rays are clearly recognizable; the barbels are also recognizable.

In the second scene of Širiê I, presumably related to fishing (Figures 4.3, 4.10), an elongated oval vhose surface is divided into rectangles by a vertical line and further horizontal lines is represented. This bject could be a fish trap or another device for catching fish, and the two people aside could be pulling his trap or device ashore on a rope or hauling in a net. The semi-circular structure with a person in the ame scene could represent a boat, and the other people depicted could take part in a collective fishing, luring which they are driving fish into a net.

The oval structures in another scene mentioned above (Figure 4.6) could show probably granaries. n the so-called "Grande Riparo", at the entrance to the Guelta of Archei, there is also a depiction of an val figure on stilts, which Simonis *et al.* (2017: 20) describe as "granary" with "foodstuff". At Hohou Dahon, granaries are as well depicted on stilts; figures with skirts and broad hairstyles are also depicted here (Simonis *et al.* 2017: 168). Still today, granaries are used in Borku and Ennedi in order to store dates Figure 4.11), some of them stand on a kind of stilts too (Figure 4.12). In Širiê II, granaries for the storing f wild cereals are built directly under the rock overhang where the paintings are located (Figure 4.13); rinding stones lie on the ground next to them. The stilts of some granaries may have had the function of rotecting the stored food from soil moisture.

The granaries at Širiê II, or the grinding stones lying around there, are used until today by women vho still collect, as above mentioned, wild cereals. When women grind wild cereals on a somewhat elevated grinding stone, they adopt the same bent posture and hold their hands in the same way as the figures lepicted in the granary and in the hut of Širiê I.[10] One can thus assume that the scenes described here are

FIGURE 4.11 Granaries for dates in Yarda / Northern Borku. (March 2023, see cover of the book).

FIGURE 4.12 Granaries for dates in Deymi / Ennedi. The granaries are located near a date grove and a salt lake. The stilts are possibly against the soil moisture. November 2023.

FIGURE 4.13 One of several granaries at Širiê II which is still in use for the storage of wild cereals. Close to the granaries several grinding stones could be found. (November 2022).

related to the processing of wild cereals. If this is the case, then Dupuy (2014: 46) is wrong when he writes that there is "no scene" in the rock art of the early cattle herders between the 6th and 2nd millennium BC "that can be linked to harvesting practices". Dupuy puts forward several hypotheses for the lack of such scenes that he assumes. According to his first hypothesis, plant-based food could have been neglected by the cattle herders of the time, whose diet was based on milk and meat, and only in times of scarcity could supplementary plant-based food have played a role. Therefore, according to Dupuy, the cattle herders may have avoided depicting a diet of "misery" in their "idealized images". Or, according to Dupuy's second hypothesis, wild cereals were such a widespread food that the rock art painters saw no interest in depicting them alongside their highly esteemed cattle.

According to Dupuy's third hypothesis, the harvesting and processing of wild cereals was possibly a task for women, while the men moved around in the bush with their cattle and depicted their "concerns directed outwards from the camps". Finally, in a fourth hypothesis, Dupuy suggests that cattle-herding and the gathering of wild cereals were practised by different groups, and that only the group of cattle-herders documented their activities on rock paintings. However, neither this last hypothesis nor the third one seems to apply to the context depicted here, in which scenes of cattle herding and those of the presumed processing of wild cereals can be found side by side.

CONCLUSION

Why might the depictions of fish be so rare among the rock art of the Sahara? In this context, Paugy *et al.* (2015: 40) pose the question of whether there may have been some kind of taboo ("interdit") that forbade the depiction of fish. Perhaps the opposite was the case: fish was such a "banal" everyday food in the late Cattle Age when the number of paleolakes had reached its peak, that it did not seem interesting to treat it artistically. This would be a hypothesis similar to that of Dupuy (2014), who assumes that scenes of wild cereals were not depicted because they would have been too banal to be artistically represented.

The innumerable depictions of cattle, on the other hand, may have had a cultic significance. It is generally known from the contemporary Ful'Be, especially from the nomadic group of the Wodaabe (cf. e.g. Schareika 2003), that cattle are of great cultural importance in the traditional human-animal relationship; they are not simply considered as a source of meat and are only reluctantly and very rarely sold. It has already been hypothesised that today's Ful'Be could have originated from the Neolithic cattle herders who lived in the Sahara (c.f. Hampâté-Bâ and Dieterlen, 1966). It could therefore be assumed that meat was not the main source of proteins in the diet of the prehistoric cattle herders, but rather other foods such as fish (cf. Manning *et al.*, 2023). Fish, as an everyday food and source of protein, could have been supplemented by the carbohydrates of wild cereals. In this way, a link could be established between the presumed fishing scenes and the scene with the depictions of the granaries. Of course, for the reasons already mentioned in the introduction, this link must remain highly hypothetical.

ACKNOWLEDGEMENTS

I would like to thank my guide Kodi Togoi from Fada, who took me to Širiê for the first time in November 2022, Galmai Issouf from Faya, with whom I was able to visit the site a second time in November 2023, and all other locals who received me well. I would also thank the following colleagues for their expertise on topics relevant for my article: Barbara Sponholz (Würzburg University) commented the aerial photographs of Širiê; Roland Baumhauer (Würzburg University) analysed the substrate samples; Erhard Schulz (Würzburg University) identified wild cereals I documented at Nehi; and Timo Moritz (Deutsches Meeresmuseum Stralsund) identified, based on my photos from the field, several fish species occurring in the Sahara. I also thank the Max Planck Institute for the History of Science (Berlin), which financed my research in and on Širiê.

NOTES

1. Širiê I: N16° 50.279' E21° 56.542', Širiê II: N16° 50.096' E21° 55.545'.
2. Arčana in North Borku with its 18 rock engravings of harpoons or barbed spearheads (which may, however, have been used for hippopotamus hunting), is said to be the only place where depictions of such objects can be found (Gauthier and Gauthier 2018: 286).

3. In Dufuna, a village on the Komadugu Gana River (Nigeria), a pirogue made of *Khaya* wood was found in a seasonally flooded basin and was determined to be about 8,000 years old (Breunig *et al.* 1996: 116).
4. Abbreviations according to Simonis *et al.* (2017).
5. NH01 to NH06 lie on a 25-kilometer-long zigzag line running from west to east (Simonis *et al.* 2017: 51–53). Širiê I could be identical to NH04 or NH05 in Simonis *et al.* (2017). However, none of the sparse descriptions given by Simonis *et al.* (2017) of the rock art from Nehi seem to apply to Širiê I. These descriptions are: "a few faded white images of cattle and a vanishing giraffe" (NH01), a "large frieze (...) animated with a number of bovines and some roughly drawn camels", "a galloping mounted horse" (NH02), "a few petroglyphs of cattle", "a herder and five more similar bovines with forward-pointing horns" (NH04), "mainly (...) mounted camels" (NH05), "a hunter elaborately equipped with weapons", "a striking set of circular and elongated vertical cupules" (NH06).
6. The aerial photograph of Širiê shows bleached sands in the immediate vicinity of the two sites. One could therefore assume the presence of a paleolake. For this reason, two substrate samples (one taken slightly south of the rock shelter of Širiê I and the other one in the small tree-lined valley that runs around the rocks of Širiê I to their east and north) were analysed by Roland Baumhauer (University of Würzburg) under the light microscope in January 2024. This analysis revealed however the presence of dune sands that have hardly been fluviatilised or limnified and that may have just been redeposited over short distances, and therefore does not allow us to assume the existence of a paleolake right on the rock artists' "doorstep".
7. This term may have been coined by Huard; however, his chronological classification does not seem appropriate in the present case: "The dominant 'mushroom' hairstyles (…) are cultural features of the Chadian Sahara in the 1st millennium BC" (Huard and Massip 1966: 58).
8. Figures whose legs are depicted partly with bent knees, partly as in the splits, are also found at Etakou and are interpreted by Simonis *et al.* (2017: 46) as "running human figures". During my stay in the Ennedi in November 2023, I found more such figures in other places, without however identifying any connection with fishing in these cases.
9. Besides *Tilapia* sp. and *Lates niloticus*, *Clarias* sp. can be found among fish depictions among the "roundheads" of the Tassili-n-Ajjer and the Tadrat of the Algerian Sahara (Gauthier and Gauthier 2011: 127–129).
10. If the grinding stone is on the ground, the woman could be kneeling in front of it. A rock painting of Egmé Chima shows such a person kneeling in a hut (cf. Simonis *et al.* 2017: 132).

REFERENCES

Arcangioli, S., 1996, Tohil. In *Arte rupestre nel Ciad: Borku, Ennedi, Tibesti*, (Milano: Centro Luigi Nero), pp. 63–65.

Bailloud, G., 1960, Les peintures rupestres archaïques de l'Ennedi (Tchad). *L'anthropologie*, **64(3/4)**, pp. 211–234.

Bailloud, G., 1997, *Art rupestre en Ennedi*, (Saint-Maur: Éditions Sépia).

Blache, J. and Miton, F., 1962, *Première contribution à la connaissance de la pêche dans le bassin hydrographique Logone – Chari – Lac Tchad. Aspect général des activités de la pêche et de la commercialisation des produits. Description des engins de pêche et de leur employ*, (Paris: ORSTOM).

Boccazzi, A. and Boccazzi, D., 1999, La fauna selvaggia nelle pitture della regione di Ouri (Tibesti orientale). *Sahara*, **11**, p. 130.

Breunig, P., Neumann, K. and van Neer, W., 1996, New research on the Holocene settlement and environment of the Chad basin in Nigeria. *African Archaeological Review*, **13(2)**, pp. 111–145.

Cauvet, G., 1913, Les poissons du Sahara algérien. *Bulletin de la Société géographique d'Alger et de l'Afrique du Nord*, **18**, pp. 696–717.

Courtin, J., 1965, Découverte de harpons en os au Borkou, République du Tchad. *Bulletin de la Société préhistorique française*, **62(2)**, pp. 70–75.

Courtin, J., 1967, Engins de pêche du Nord Tchad. *Travaux de l'Institut de Recherches Sahariennes*, **26**, pp. 103–111.

Daget, J., 1949, La pêche dans le delta central du Niger. *Journal de la Société des Africanistes*, **19(1)**, 1–79.

Dupuy, C., 2014, Des céréales et du lait au Sahara et au Sahel, de l'épipaléolithique à l'âge des métaux. *Afriques* [online], **5**, consulted 13/01/2024 2024. http://journals.openedition.org/afriques/1376; https://doi.org/10.4000/afriques.1376.

Gast, M., 2000, *Moissons du désert. Utilisation des ressources naturelles au Sahara central*, (Paris: Ibis Press).

Gauthier, Y. and Gauthier, C., 2011, Des poissons chez les *Têtes Rondes*? *Cahiers de l'AARS*, **15**, pp. 119–132.

Gauthier, Y. and Gauthier, C., 2018, Nouveaux documents sur les milieux aquatiques au Sahara tchadien : le site d'Artchana II revisité. In *What Ever Happened to the People? Humans and Anthropomorphs in the Rock Ar of Northern Africa*, edited by Huyge, D. and van Noten, F. (Bruessels: Royal Academy for Overseas Scieces), pp. 265–292.

Hampâté-Bâ, A. and Dieterlen, G., 1966, Les fresques d'époque bovidienne du Tassili n'Ajjer et les traditions de Peuls : hypotheses d'interprétation. *Journal de la Société des Africanistes*, **XXXVI(1)**, pp. 7–27.

Huard, P., 1963, Gravures rupestres de l'Ennedi et des Erdis. *Bulletin de l'Institut de Recherches Scientifiques au Congo*, **2**, pp. 25–39.

Huard, P. and Massip, J.-M., 1964, Les peintures rupestres de la falaise d'Artiena (Borkou). *Bulletin de la Société préhistorique française*, **61(1)**, p. 10.

Huard, P. and Massip, J.-M., 1966, Nouveaux centres de peintures rupestres au Sahara nigéro-tchadien. *Bulletin de l'IFAN*, **XXVIII(B, 1–2)**, pp. 45–81.

Jacquet, G., 1999, Voyage d'exploration : Haut Ennedi (Tchad) : nouvelles peintures rupestres. *Le Saharien*, **151**, pp 44–52.

Jacquet, G., 2000, Piste oubliée en Haut-Ennedi. *Sahara*, **12**, 141–149.

Jousse, H., and van Neer, W., 2009, Palaeoecology of the Giant Catfish (*Arius gigas*, Ariidae) in Holocene Saharan and tropical West African waters. *Holocene Palaeoenvironmental History of the Central Sahara, Palaeoecology of Africa*, **29**, pp. 159–177.

Kröpelin, S., Verschuren, D., Lézine, A.-M., Eggermont, H., Cocquyt, C., Francus, P., Cazet, J.-P., Fagot, M., Rumes B., Russel, J.M., Darius, F., Conley, D.J., Schuster, M., von Suchodoletz, H. and Engstrom, D.R., 2008, Climate driven ecosystem succession in the Sahara: The past 6000 years. *Science*, **320**, pp. 765–768.

Le Quellec, J.-L., 2013, Périodisation et chronologie des images rupestres du Sahara Central. *Préhistoires méditerranéennes*, **4**, pp. 1–47.

Manning, K., Breeze, P.S., Drake, N., Dunne, J., Casanova, E., and Evershed, R.P., 2023, Habitat fragmentation and the sporadic spread of pastoralism in the mid-Holocene Sahara, *Quaternary Science Reviews*, **309**, 8070 https://doi.org/10.1016/j.quascirev.2023.108070

Mauny, R., 1952, Harpon en os du territoire du Tchad. *Bulletin de la Société préhistorique francaise*, **XLIX**, pp 469–471.

Musch, T., 2017, Footprints in the mud of Agadem. Eastern Nigers way towards the Anthropocene. *Modern Africa* **5(2)**, pp. 105–126. https://doi.org/10.26806/modafr.v5i2.198

Musch, T., 2021, Exploring environments through water: An ethno-hydrography of the Tibesti mountains (Central Sahara). *Ethnobiology Letters*, **12(1)**. https://doi.org/10.14237/ebl.12.1.2021.1709

Musch, T., 2023a, A neighbor of Gonoa: The Man of Bunige'and his game. *Journal of Global Archaeology*, **2023(04)**, pp. 2–15. https://doi:10.34780/lc92-3ecc

Musch, T., 2023b, Three fish species from Horchi (Tibesti Mountains, Central Sahara): Rediscovery after decades of drought. *Bulletin of Fish Biology*, **20**, 23–27. https://www.ichthyologie.de/wp-content/uploads/2023/04/Musch-2023-Three-fish-species-from-Horchi-Bull-fish-Biol-20-23–027.pdf

Paugy, D., Lévêque, C. and Mouas, I., 2015, *Poissons d'Afrique et peuples de l'eau*, (Marseille: IRD).

Schareika, N. 2003, *Westlich der Kälberleine. Nomadische Tierhaltung und naturkundliches Wissen bei den Wodaabe Südostnigers*, (Münster: LIT).

Simonis, R., Ravenna, A. and Rossi, P.P., 2017, *Ennedi. Tales on Stone*, (Florence: All'Insegna del Giglio).

Trape, S., 2013, A study of the relict fish fauna of northern Chad, with the first records of a polypterid and a poeciliid in the Sahara Desert. *Comptes Rendus Biologiques*, **336(11–12)**, pp. 582–587.

Trape, S., 2016, A new cichlid fish in the Sahara: The Ounianga Serir lakes (Chad), a biodiversity hotspot in the desert. *Comptes Rendus Biologiques*, **339**, pp. 529–536. https://doi.org/10.1016/j.crvi.2016.08.003

Trape, S., 2018, *Epiplatys bifasciatus* (Steindachner, 1881) (Nothobranchiidae) and *Hemichromis fasciatus* Peters 1852 (Cichlidae), two relict fish species in the Sahara Desert. *Bonn Zoological Bulletin*, **67(1)**, pp. 37–40.

Tubiana, M.-J., 1969, La pratique actuelle de la cueillette chez les Zaghawa du Tchad. *Journal d'agriculture tropicale et de botanique appliquée*, **16(2–5)**, pp. 55–83.

Water Harvesting and Utilization in Iron Age, Northern Kordofan

5

Tim Karberg and Jana Eger-Karberg

INTRODUCTION

The semi-arid landscapes of Northern Kordofan, Sudan, unfold an interesting aspect of ancient water management highlighted by the presence of so-called hafirs. These installations were recently brought to light through the earth observation data analysis as well as ground survey campaigns of the Interregional Linkage Investigations in Northern Kordofan (InterLINK) project. While the Butana region has long been recognized as the epicenter of hafir use in ancient Sudan, these new findings in Northern Kordofan challenge previous geographical assumptions and shed light on an intricate network of water management that sustained past civilizations in this marginal environment.

Positioned strategically within the terrain alongside (and therefore making use of) natural surface drainage (so called "khors"), these hafirs offer a glimpse into climatic conditions that deviated significantly from those of the present day. Numerous reservoirs are located further north and along erstwhile drainage channels that have, over time, ceased to operate as active watercourses. This geographic distribution suggests a period when the region experienced climatic conditions amenable to supporting more extensive human settlement than is currently possible under today's arid climate. Such insights provoke a reevaluation of Northern Kordofan's historical role in accommodating human activities and socio-economic networks.

This study undertakes first steps towards an analysis of the use and significance of hafirs in Northern Kordofan, probing their role within a broader context of palaeoclimatic evidence and diachronic developments in human settlement and economic patterns. The chapter aims to explore how past societies adapted their resource management strategies to align with shifting environmental parameters, thus providing a framework for understanding the adaptability and resilience of these cultures.

By integrating a multi-faceted approach, this examination draws on the interplay between environmental change and human agency. It seeks to illuminate the socio-economic and ecological dynamics that allowed ancient communities to thrive in a challenging environment by strategically leveraging available

DOI: 10.1201/9781003727781-5

water resources. This chapter underscores the importance of hafirs as critical components in the interconnected systems of ancient water management and socio-economic innovation in Northern Kordofan.

Northern Kordofan as a Cultural Landscape and its Exploration

Kordofan is a vast and geographically diverse region in Sudan, positioned between the White Nile's course to the east and the rugged highlands of Darfur to the west. This region encompasses a rich tapestry of landscapes, each with its distinctive ecological and cultural features. To the south, the subtropical Nuba Mountains rise, characterized by their lush vegetation and varied topography, which support a biodiverse habitat and a mosaic of agricultural communities. This area transitions northward into extensive savannah rangelands, which form the heart of Kordofan's agricultural and pastoral economy. These fertile plains, dotted with acacia trees and grasslands, have long sustained local livelihoods through mixed farming and pastoralist practices.

As one moves further north towards the vicinity of the Jebel al-Ain, the landscape shifts dramatically into the arid semi-desert of Northern Kordofan. Here, the environmental conditions become harsher, with sparse vegetation and limited water resources defining the semi-desert terrain. This part of Kordofan presents unique challenges and opportunities for human habitation and resource management, prompting the development of innovative strategies to adapt to the environment.

The region's hydrography is dominated by the Wadi Melek, the principal wadi system, which originates near the town of Nahud. This wadi snakes its way northeastward, eventually draining into the great southern bend of the Nile. Wadi Melek and its tributaries play a crucial role in the area's hydrology, providing essential water resources that have historically supported both settlement and seasonal agricultural activities.

Northern Kordofan's soil landscape is diverse, with Arenosols and fluvisols comprising the primary soil types. Arenosols, known for their sandy texture and relatively low organic matter content, dominate the upland areas and are well-drained but poor in nutrients. In contrast, fluvisols, associated with alluvial plains, are finer-textured and more fertile, often found in low-lying areas along watercourses, but also in a large plain between the Wadi Mahbaz and the Jebel Awadun.

Archaeological research in and about Kordofan was rare. Apart from individual research projects such as the prospections of the Section française de la Direction des antiquités du Soudan (SFDAS) in Zankor (Gratien *et al.*, 2013), and the surveys of Khartoum University (Howeida and Khabir, 2011) and the University of Poznań together with the Sudanese National Corporation of Antiquities and Museums (Elmirghani and Drzewiecki, 2011) in the Sodiri region, knowledge of the northernmost parts of Kordofan, in particular, was limited for a long time to random individual observations by travelers passing through.

The InterLINK Project

The research project "InterLINK – Interregional Linkage Investigations in Northern Kordofan", conducted by the University of Münster's Research Centre Ancient Sudan, is an integral part of the wider framework of the Priority Programme (SPP) "Entangled Africa" funded by the German Research Foundation (DFG).

The project geographically focuses on the semi-desert northern part of Kordofan, specifically between the hills of Jebel al-Ain and Jebel Haraza (Figure 5.1). This area presents a compelling focus due to its distinct geological formations and historical significance as a crossroad of cultural and economic interactions. The semi-desert conditions offer a unique context for studying past and present adaptations to environmental stressors, particularly concerning water management and resource utilization.

The InterLINK project seeks to unravel the complex web of connections between the ancient and medieval cultures of the Nile Valley and their western neighbors. Its primary aim is to explore how and to what extent influential states such as the Napatan and Meroitic empires, along with the medieval states of Makuria and Alwa, extended their reach into the western peripheries, interacting with the early cultures

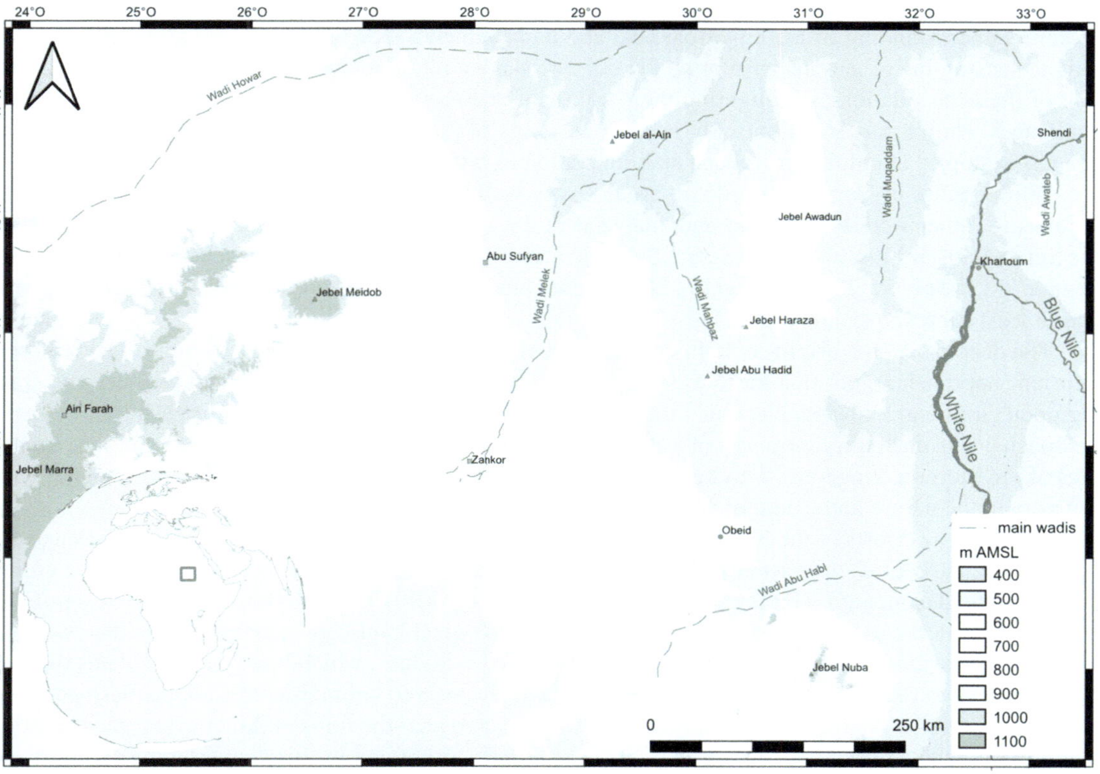

FIGURE 5.1 Working area of the InterLINK project in Northern Kordofan. Elevation derived from ALOS DTM (© JAXA).

of Kordofan, Darfur, and the Chad Basin. In this context, the Kushite and Nubian cultures are examined not merely as entities aligned along a north-south axis within the Nile Valley, but also as "Sudanic" Sahel cultures actively engaged in east-west cultural and economic exchanges. Through this investigation, InterLINK contributes to a nuanced understanding of the dynamic and interconnected nature of African civilizations, emphasizing the broader continental networks that shaped historical trajectories (Eger-Karberg and Karberg, 2024a). In its first project phase from 2017 till 2022, comprehensive earth observation data analysis (mainly of high-resolution panchromatic satellite images) was combined with two ground truth field campaigns (Eger-Karberg and Karberg, 2022), while during the ongoing second project phase the focus was shifted towards the analysis of multispectral and radar-based earth observation data.

Hafirs as Main Water Harvesting Installations in Ancient Sudan

Hafirs, traditional water storage structures, represent a cornerstone of ancient hydraulic engineering in Sudan. Distinguished by their characteristic round basins, hafirs are a unique and endemic feature of Sudanese landscapes, embodying the ingenuity of ancient societies in addressing water scarcity. These water harvesting installations were meticulously engineered to capture and store seasonal rainfall, a vital innovation in regions marked by arid and semi-arid climates. Unlike other water infrastructure more typical for arid cultural landscapes, such as dams and larger wells, which played only a minor role in Sudan at least during the Iron Age, hafirs were the predominant solution for water management and a central component of socio-economic life in ancient Sudan. The circular design of the basins optimized water retention and collection, and their positioning within natural surface water drainage systems reflected a profound understanding of local hydrological and environmental conditions. Recent archaeological and remote sensing investigations have provided deeper insights into the construction methods and distribution of these installations, highlighting their significance as a uniquely Sudanese adaptation to climatic challenges. Hafirs thus stand not only as vital elements of past water management systems but also as general examples of cultural resilience and innovation in Sudan's historical landscape.

While Hafirs in Northern Kordofan are not very well researched yet, these water harvesting installations are rather well documented and analyzed in the Butana region. Some Hafirs, like the Great and the Small Hafir at Musawwarat es Sufra (Scheibner, 2017) as well as the Hafir close to the so-called "Sun Temple" M250 (Wolf, 2015; Wolf *et al.*, 2015) were comprehensively excavated, while the great majority was at least surveyed (Hinkel, 2015).

The debate over the origins of hafirs primarily revolves around two schools of thought. The "communitarian" approach argues that these water storage units were the result of local communal efforts, drawing upon empirical knowledge accrued through generations (Scheibner, 2014). In contrast, the "etatistic" approach posits that hafirs were part of a strategically planned infrastructure initiative by the state authorities of the Meroitic kingdom (Weschenfelder, 2014). These differing interpretations reflect broader questions about the nature and extent of centralized governance and local agency in ancient Sudan.

Through the efforts of the InterLINK project, we have successfully documented 66 previously undiscovered hafir structures in Northern Kordofan (Table 5.1). Twelve of them were investigated on the ground during the field campaigns (Figure 5.2; cf. Eger and Karberg, 2019) These findings represent a significant advancement in the understanding of the region's archaeological landscape, particularly as they were not included in Marion Hinkel's influential work "Die Hafire im Sudan", which largely encapsulates the state of research from the 1980s (Hinkel, 2015). These newly discovered hafir structures are primarily concentrated at the middle Wadi Melek, adjacent to Jebel al-Ain, and near the ruins of Zankor. The geo-statistical analysis we conducted in the project, and which is in parts presented in this chapter, reveals significant clustering in these areas. The hafirs in Northern Kordofan closely resemble those found in the Butana and Keraba with regard to their functional and structural layout, raising the question of possible cultural and technological transfer during the Meroitic period. Despite the lack of diagnostic chronological artifacts, the structural similarity, severe erosion, and sedimentation indicate at least their considerable age, further supported by their locations in regions devoid of recent persistent water flow.

RESEARCH QUESTIONS

In the study of hafirs in Northern Kordofan, several research questions arise that aim to enhance our understanding of these ancient water management systems, their historical context, and their functional roles within the landscape. Addressing these questions provides insights into the socio-economic adaptations of past societies and the environmental conditions they faced. The following key research questions guide this inquiry:

- Temporal Context and Climatic Correlation: During ground truth campaigns, there were only a few indications of the chronological context of the hafirs of Northern Kordofan (Eger and Karberg, 2019). Given the challenges in directly dating hafirs through Earth observation satellite data, how can we leverage correlations between hafir volumes and precipitation estimates from climatic models to identify the likely time periods during which these structures were most effectively utilized?
- Functional Role and Soil Suitability: What is the primary function of hafirs in Northern Kordofan, especially concerning their distribution on arenosol soils versus more fertile

TABLE 5.1 Hafirs Discovered in Northern Kordofan by the InterLINK Project (from South to North).

Ø/M	*VOL. (EST.)/M³*	*INTAKES*	*INFLOW*	*SOIL*	*LAT. N*	*LONG. E*
120	30812	1	NW	CMvr	13,76119°	27,95317°
116	27858	1	SSW	CMvr	13,79431°	27,91404°
116	27858	2	NNW; S	CMvr	13,79780°	27,91157°
118	29310	1	SSE	CMvr	13,81001°	27,91850°
96	15880	1	E	CMvr	13,81075°	27,93285°
58	3577	1	SE	CMvr	13,81095°	27,91863°
166	80968	3	NW; E; SSW	CMvr	13,84794°	27,94381°
180	103076	2	SSW; S	CMvr	13,85046°	27,94385°
88	12268	1	SSW	CMvr	13,85337°	27,94518°
125	34790	1	E	CLha	14,95516°	30,34070°
82	9952	1	NE	LPli	15,34895°	27,91284°
62	4355	1	SW	LPli	15,35274°	27,91671°
28	425	1	SSE	LPli	15,51320°	27,84817°
72	6773	?	?	ARha	15,61330°	28,10428°
160	72554	1	SW	ARha	15,61947°	28,11237°
78	8582	1	SW	CLha	15,66076°	28,70530°
102	19011	1	SE	ARha	15,79362°	28,98433°
76	7947	?	?	FLeu	15,79581°	28,70285°
96	15880	?	?	FLeu	15,79594°	28,70448°
82	9952	2	SE; W	ARha	15,79896°	28,99130°
75	7642	?	?	CLha	15,79976°	28,69625°
76	7947	1	NE	ARha	15,81691°	28,04260°
73	7055	1	E	ARha	15,83560°	28,98921°
78	8582	1	E	ARha	15,83767°	29,01008°
96	15880	1	ESE	CLha	15,85560°	28,67416°
108	22528	1	SE	CLha	15,87011°	28,67556°
92	13997	1?	S?	ARha	15,92315°	28,76667°
108	22528	1	SSW	ARha	15,94761°	29,56135°
103	19570	1	SSW	ARha	15,94863°	29,56251°
95	15394	1	E	ARha	15,95792°	28,73839°
60	3953	1	SW	ARha	15,95993°	28,73744°
68	5720	?	?	ARha	15,98054°	28,74567°
110	23790	?	?	ARha	16,00832°	28,91233°
135	43743	1?	SW?	ARha	16,01416°	28,75201°
90	13113	1	SE	ARha	16,07404°	28,81864°
90	13113	1	ENE	ARha	16,09029°	28,84679°
88	12268	1	NE	ARha	16,09664°	28,84792°
16	86	?	?	FLeu	16,10025°	29,79103°
102	19011	1	ENE	ARha	16,10155°	28,85073°
112	25099	1	W	ARha	16,17663°	28,81594°
102	19011	1	NE	ARha	16,19202°	28,99909°
100	17925	1	NW	ARha	16,22691°	28,86690°
88	12268	?	?	ARha	16,25722°	28,90197°
98	16882	2	NW; SW	ARha	16,26226°	28,92370°

(Continued)

TABLE 5.1 (Continued)

Ø/M	VOL. (EST.)/M³	INTAKES	INFLOW	SOIL	LAT. N	LONG. E
98	16882	2	WNW; ESE	ARha	16,27522°	28,95421°
98	16882	1	S	ARha	16,29556°	29,06718°
80	9250	?	?	ARha	16,29850°	29,00626°
83	10316	1	E	ARha	16,49858°	29,38240°
39	1116	2	ENE; WSW	LPli	16,53358°	29,16981°
78	8582	2	NNE; ESE	ARha	16,55632°	29,09008°
70	6232	?	?	ARha	16,59132°	29,01460°
49	2178	?	?	ARha	16,59496°	29,02853°
50	2311	1	N	ARha	16,60196°	29,46542°
39	1116	1	S	ARha	16,60358°	29,09134°
55	3059	1?	W?	ARha	16,60382°	29,09607°
65	5006	1	WNW	ARha	16,60428°	29,09121°
50	2311	1	S	ARha	16,60428°	29,09593°
50	2311	2?	SW; SSE?	ARha	16,60536°	29,09405°
55	3059	?	?	ARha	16,60566°	29,02056°
49	2178	2	SW; SSE	ARha	16,60615°	29,02028°
68	5720	2	NE; SW	ARha	16,60654°	29,09536°
49	2178	2	NE; SE	ARha	16,60657°	29,02063°
52	2594	1?	N?	ARha	16,60669°	29,09586°
65	5006	2?	N?; SSE	ARha	16,62902°	29,41442°
53	2743	1?	NNE?	ARha	16,72104°	29,88495°
98	16882	2	NW; SE	ARha	16,97788°	29,05303°

FIGURE 5.2 Hafir close to the western pediment of the Jebel al-Ain (photo: Jana Eger-Karberg, 2017).

fluvisols? Are they predominantly used for supporting pastoral economies and extensive rain-fed agriculture, or do they play a crucial role in irrigation practices?

- Economic Strategies and Environmental Adaptations: How do the spatial distribution and architectural variations of hafirs reflect diverse economic strategies and environmental adaptations, considering the localized climate and soil conditions in Northern Kordofan? What can these patterns reveal about historical land use practices and resource management strategies in the region?

DATASETS AND METHODOLOGY

Earth Observation Datasets

Digital terrain modeling in this study is based on TerraSAR-X and TanDEM-X radar satellite datasets provided by the German Aerospace Center (DLR). They offer synthetic aperture radar (SAR) imagery in a raw data resolution of 3 × 3 m in StripMap mode, and a digital elevation model with a relative elevation resolution of 2 × 2 m combined with an absolute 3D spatial resolution of 12 × 12 m (Reinke and Werninghaus, 2009). This exceeds by far the spatial resolutions of other digital elevation models based on spaceborne radar imagery, like the widely used Shuttle Radar Topographic Mission SRTM (Albertz, 2007). This, therefore, comparably high-resolution radar data enables the detailed analysis of topographical and geological characteristics in diverse environments (Reinke and Werninghaus, 2009). Specifically, their ability to penetrate cloud cover and provide consistent, precise elevation data makes them particularly useful in semi-arid regions like Northern Kordofan. The DLR's TerraSAR-X mission, coupled with the TanDEM-X satellite, creates a comprehensive digital elevation model (DEM), allowing the modeling of terrain features with exceptional accuracy and also some limited ground penetrating capacity under very dry conditions on loose, sandy soils like Aeolic sand dunes. Additionally, for overview purposes, the ALOS DTM, which is provided free of charge by the Japanese space agency JAXA, was used.

The reconstruction of ancient drainage systems was complemented by multispectral satellite imagery from the Sentinel-2 mission of the European Copernicus Earth observation program. Sentinel-2, with its up to 10 × 10 m spatially resolved spectral capabilities (Fletcher, 2012), enables the analysis of current vegetation patterns through indices like the Normalized Difference Vegetation Index NDVI (Tucker, 1979). These patterns often reveal underlying hydrological features, as vegetation tends to grow more luxuriantly along historic watercourses where moisture is retained in the soil. By identifying these vegetative anomalies, we were able to infer the presence of paleo-drainage channels that may have been obscured by sedimentation over time, especially in conjunction with the TerraSAR-X datasets with their limited ground penetrating capacity.

Palaeoclimatic Modeling

The palaeoclimatic model used for this research follows the data provided by the TRaCE21ka experiment (Otto-Bliesner *et al.*, 2014; Liu *et al.*, 2014) using the Community Climate System Model (CCSM) version 3 (Collins *et al.*, 2006), and visualized by the PaleoView 1.5.1 software (Fordham *et al.*, 2017). The TRaCE21ka experiment generates palaeoclimate reconstructions spanning from 22,000 bp (before present) until 1990 CE. This global coupled atmosphere-ocean-sea ice-land general circulation model replicates major climatic changes, enabling the creation of continuous climate projections at regional and global scales. The TRaCE21ka experiment's simulation outputs, summarized as monthly means, provide comprehensive data on temperature, precipitation, humidity, and mean sea-level pressure. By employing outputs regridded to a spatial resolution of 2.5° × 2.5°, the TRaCE21ka experiment offers a temporal and spatial resolution resource sufficient for our research questions in an area where only a few direct palaeoclimatic proxies for the time frame in question were investigated so far.

Methodology

The methodology utilized in this study includes geo-statistical analysis of the spatial distribution patterns of the hafirs and their storage capacities, their spatial and functional contextualization with the natural precipitation and drainage regime under different palaeoclimatic circumstances, as well as different soil conditions. The integration of geographical information systems (GIS), earth observation data, and statistical analysis tools allows for a comprehensive understanding of the environmental and cultural factors influencing hafir placement and functionality.

In order to analyze the amount of rain and surface water necessary to feed the hafirs, we employed the deterministic 8 (D8) drainage model (O'Callaghan and Mark, 1984), a key method for simulating hydrological flow across the landscape. Derived from a three-dimensional digital terrain model (DTM, cf. previous paragraph), the D8 model facilitates the estimation of rainfall accumulation and flow direction towards each hafir. This model assumes that water from any given cell in the DTM flows to one of the eight surrounding cells, depending on which direction has the steepest descent. By applying the D8 model, the authors can predict the concentration of surface runoff and assess the potential water catchment efficiency, critical for understanding their historical water management roles. The modeling was calculated using the SAGA tools "Channel Network and Drainage Basins" and "Flow Accumulation (Top-Down)" from the "Terrain Analysis" toolbox of SAGA 7.6.2 (Conrad *et al.*, 2022). Alongside drainage modeling, paleo-hydrological features, now obscured by sedimentation, are reconstructed using plant markers derived from the Normalized Difference Vegetation Index (NDVI). This index, calculated from multispectral satellite imagery (Tucker, 1979; Tucker *et al.*, 2005), identifies vegetation differences that hint at ancient, sediment-covered watercourses, known as palaeo-khors (Eger-Karberg and Karberg, 2021). By detecting these vegetative patterns, we can infer the historical flow paths that may have influenced hafir placement and effectiveness.

The spatial distribution and volumes of hafirs are analyzed using geostatistical tools, specifically Moran's I (Conolly and Lake, 2010) and Getis-Ord G statistics (Conolly and Lake, 2010), accessed via the ArcGIS Geostatistics Toolbox. Moran's I assesses spatial autocorrelation, indicating the degree of clustering or dispersion within our dataset. A high Moran's I value suggests significant clustering of hafirs, influencing interpretations of regional settlement and water management strategies. Analysis was carried out with ArcGIS 10.2.1. "Spatial Statistics – Spatial Autocorrelation (Moran's I)" tool. The Getis-Ord G statistic, on the other hand, identifies hotspots or cold spots by evaluating the concentration of high or low values, providing further insights into localized environmental or cultural factors affecting hafir distribution. Analysis was carried out with the ArcGIS 10.2.1. "Spatial Statistics – High/Low Clustering (Getis-Ord G)" tool.

To analyze categorical data, such as the soil substrate present at different hafir locations, we employ the χ^2 (chi-squared) test (Drennan, 2009, pp. 182–188). For this analysis, the chisq.test package (Carlson, 2017, pp. 190–198) from the R software environment was used (R Core Team, 2020). This test evaluates whether the observed distribution of categorical variables differs from what might be expected under an even distribution. By applying this statistical method, we determine the significance of soil types in influencing hafir placement, a crucial factor in understanding agricultural potential and historical utilization patterns.

RESULTS

A crucial aspect of the archaeological record in our area of investigation is the role of hafirs as key indicators of historical water management practices. These newly identified hafirs, some of which were directly examined during our two field campaigns in 2017 and 2018 (Eger and Karberg, 2019), provide fresh insights into the distribution and function of these vital features.

To the right, a heat map illustrates the results of a kernel density analysis, highlighting the spatial concentration of these hafirs. It is important to note that this visualization represents not the number of hafirs, but the aggregated area covered by each structure. Noteworthy concentrations are found in the middle stretch of the Wadi Melek, the vicinity of Jebel al-Ain, and near the ancient site of Zankor, situated in the headwaters of the Wadi Melek. The latter location was previously explored by a French mission under the leadership of Brigitte Gratien in the 2000s (Gratien *et al.*, 2013).

The construction techniques of these hafirs mirror those observed in regions such as the Butana and Keraba, suggesting a potential cultural or technological exchange. However, accurate dating of the hafirs identified in the InterLINK project remains challenging. Pottery artifacts recovered during fieldwork have thus far been inadequate for providing definitive chronological markers. Nonetheless, the pronounced erosion visible at many of these sites, alongside the placement of hafirs in northern areas where persistent surface water no longer flows, suggests a considerable antiquity.

Despite these indicators of age, it remains uncertain whether these hafirs are contemporaneous with or related to those in the Meroitic heartland east of the Nile. Further research, including stratigraphic analysis and more precise dating methods, is needed to establish a clearer chronological and cultural relationship. This ongoing inquiry promises to enhance our understanding of the historical dynamics and environmental adaptations within Kordofan.

Hafir Volumes

Estimating the volumes of hafirs is a difficult task, involving a complex interplay of archaeological, geographical, and geometrical analyses. Generally, the volume of a hafir is dependent on several factors, including its diameter, depth, and the shape of its basin. While accurate assessments, especially of the depth and shape, require detailed archaeological excavations that provide specific data on the angles and other morphological characteristics of the basins, only a few hafirs in Sudan have been subject to such detailed studies. In the newly discovered hafirs in Northern Kordofan, we have so far not been able to carry out invasive archaeological excavations. Therefore, the volumes of these hafirs can be estimated using assumptions about their typical geometrical configuration acquired by comparison with similar, but more comprehensively studied examples, especially from the Butana and Keraba, like the hafirs at Musawwarat es Sufra, which were intensively studied already by Fritz Hintze in the 1960s (Hintze, 1963) and especially by Thomas Scheibner in the 2000s and 2010s (Scheibner, 2017).

A common estimation method involves modeling the hafir as a truncated combined with stepped wall segments (Scheibner, 2017, p. 110), which reflects the likely tapering of the basin from a wider circumference at the top to a narrower base according to the geological circumstances. This approach requires estimates of the upper and lower mean diameters, as well as the overall depth, to calculate an approximate volume. For instance, hafirs that are part of well-documented sites, such as those in the Butana region, offer a comparative basis for estimating volumes in less explored areas by assuming similar construction techniques and environmental adaptations. Very detailed investigations on the volume of the Great Hafir at Musawwarat es Sufra were carried out by Thomas Scheibner (Scheibner, 2017), which are used as a role model in this study where limited data, based on surface surveys and earth observation, make comparative analyses with similar structures whose dimensions have been reliably measured necessary.

The variability in hafir volumes is also dependent on local environmental conditions and community needs. Estimations might also consider the erosion patterns and sedimentation levels visible at the sites, as these can provide indirect evidence of the original dimensions and thus the potential capacity of the hafir, although these estimates carry a considerable degree of uncertainty.

Thomas Scheibner estimates a diameter of 213 m of the usable hafir basin of the Great Hafir of Musawwarat es Sufra, compared to the diameter of 250 m of the surrounding wall structures, which were only deposits of the debris of the excavation of the hafir basin without water-storing function (Scheibner, 2017). Using this estimation as a role model, a ratio between the diameter of the surrounding walls (which are the only observable parts of a hafir in satellite imagery) and the surface diameter of the hafir basin can be estimated as roughly 1:0.85. Therefore, Thomas Scheibner estimates an overall volume of the basin of the Great Hafir of Musawwarat es Sufra of roughly 300,000 m^3. When a mean slope angle of 14°–15° and a ratio between surface diameter and bottom diameter of 1:2.3, as well as a ratio between surface diameter and depth of 1:0.065, is estimated for Musawwarat es Sufra (Scheibner, 2017), these results can also be applied to the volume estimations for the newly discovered hafirs in Northern Kordofan (cf. Table 5.1).

Spatial Distribution of the Hafirs and their Volumes

In the analysis of the distribution of hafirs across Northern Kordofan (Figures 5.3, 5.4), particularly in the vicinity of Jebel al-Ain, spatial patterns can be contextualized with different geo- and ecofactors. The natural environment of this region is shaped by geofactors, including soil formation and underlying

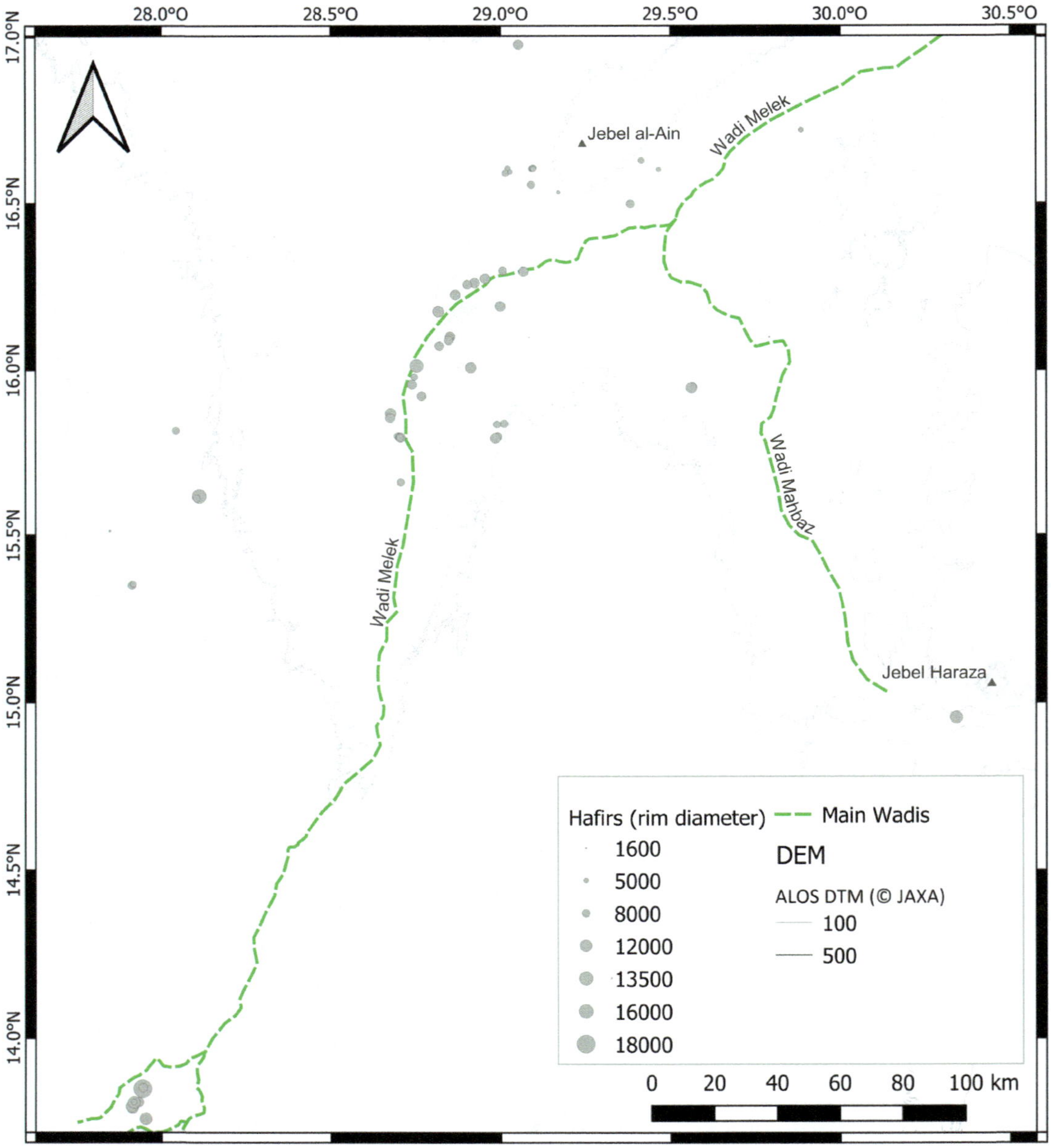

FIGURE 5.3 Map of the hafirs discovered in the InterLINK project in Northern Kordofan.

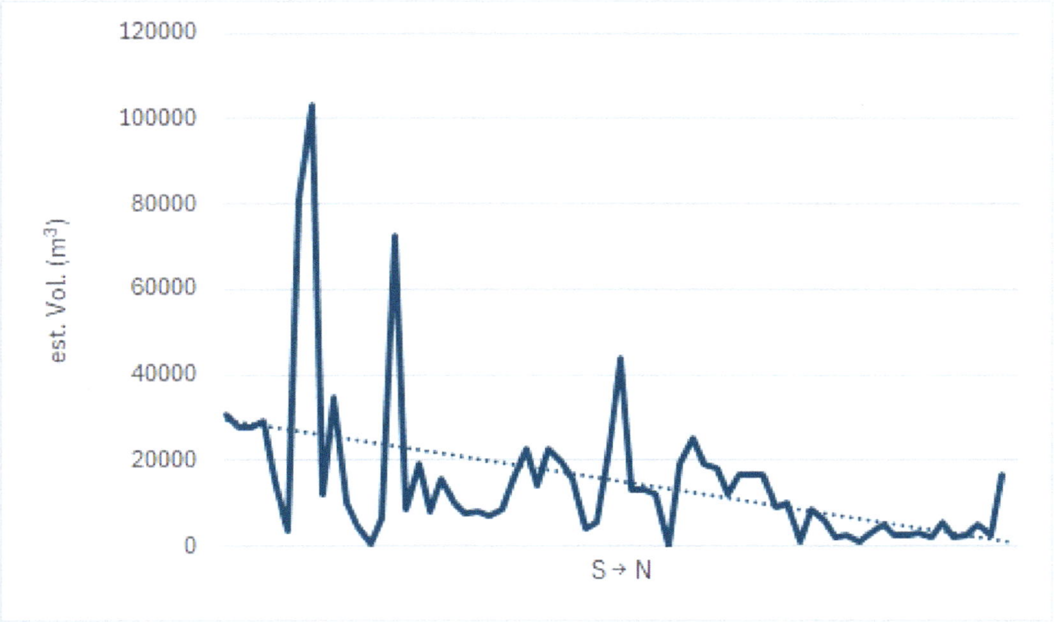

FIGURE 5.4 Spatial distribution (South → North) of estimated hafir.

geological bedrock, as well as ecofactors like climatic elements such as total annual precipitation and rainfall patterns. Among these, soil formation stands out as a crucial geofactor when considering the geographical distribution of hafirs. Leveraging data from the Africa Soil Mapping Project conducted by the European Soil Data Center (Jones *et al.*, 2013) and the Soil Spectroscopy for Global Good project based at the Woodwell Climate Research Center and the University of Florida (Sanderman *et al.*, 2021), we gain insights into the relationship between soil characteristics and hafir locations.

The area surrounding Jebel al-Ain is predominantly characterized by Arenosol soils, known for their sandy composition. These soils typically exhibit low nutrient concentrations but offer high water permeability. While generally deemed unsuitable for intensified agriculture making use of irrigation, Arenosols can support pastoral activities as well as extensive, rain-fed agriculture of undemanding crops like sorghum if adequate water supply is ensured (Born, 1965). South of Jebel al-Ain, sections of Fluvisol soils are also present. These soils result from sedimentation processes influenced by standing or flowing surface water, some of which are associated with the course of the Wadi Melek, although other formation causes like the presence of a number of palaeo-lakes west of the Jebel Awadun (Eger and Karberg, 2020, pp. 89–92) may also be at play.

Another distinct soil type found in the region is the lithic Leptosol, a stony soil that impedes root penetration. Its formation is closely linked to local topography, frequently appearing near rock ridges, escarpments, and their erosional zones. Given the steep and rugged terrain associated with Leptosol soils, these areas are generally unsuitable for cultivation or settlement, rendering them a negligible factor in terms of agricultural viability.

Other soil types dominating the more southern parts of the research area around the upper Wadi Melek are clay-rich Cambisols (similar to the local Fluvisols) and Calcisols (calcified Arenosols). Simplified, Cambisols and Fluvisols in the region can be regarded as soils rich in clay, Arenosols and Calcisols as sandy, and Leptosols as rocky soil structures.

When contextualising the distribution of hafirs with different soils, it becomes rather clear that the high-density distribution kernels of hafirs spatially correlate significantly with sandy soils, mostly Arenosols. Deploying a χ^2 test of the density of the hafir distribution at different soil types, it returns a χ^2 value of 97.333 against an assumed even distribution with 4 degrees of freedom (five soil types) and a p value of < 2.2 × 10^{-16}. When deploying the χ^2 test on the density of hafirs at the simplified soil supertype (sand, rock, and clay), it returns a χ^2 value of 70.182 against an assumed even distribution with 2 degrees c freedom (three soil supertypes) and a p value of 5.757 × 10^{-16}, proving the high significance of the results

The main ecofactor concerning the distribution of hafirs in Northern Kordofan is the mean annua precipitation, which directly results in the amount of flowing surface water which might contribute to th water to be stored in the hafirs. In general, the annual precipitation in the region concentrates in the mea rainy season between June and September, influenced by the annual circulation of the Atlantic as well a Indic monsoon systems. Diachronic shifts are mainly based on the increasing and declining influence c the Atlantic monsoon, which can be deduced from the northwestern shifting of the (ideally east-western course of the isohyetes during periods of precipitation optima (cf. fig. 5.7)

In examining the distribution of hafirs across Northern Kordofan, a significant pattern emerges con cerning their size and capacity. The InterLINK project has meticulously documented these water har vesting installations, categorizing them by diameter and storage volume. Notably, there is a pronounce concentration of smaller hafirs in the north, while larger ones are more prevalent in the south (cf. Eger Karberg and Karberg, 2024b, p. fig. 5.6). Although this distribution is not perfectly linear, statistica analysis provides a clearer understanding of its significance

The hafirs in our study area present a Moran's I index of 0.7525049, indicating a substantial degree o spatial clustering close to 1, and far from random (fig. 5.5). This clustering exhibits a p-value of 0.00007, significantly below the conventional threshold of 0.05 for statistical significance. A more nuanced exami nation of the question of whether small or large volumes are more relevant concerning these distributio patterns, the deployment of a Getis-Ord G analysis reveals that it is particularly the smaller diameters an capacities that contribute to this clustering, resulting in a Getis-Ord G z-score of roughly -2.3 (fig. 5.5) Although the specific p-value for clustering among smaller hafir metrics is 0.037 and therefore higher tha for the Moran's I index, it remains below the general threshold of significance, reinforcing our findings Practically, this implies that while larger hafirs can occur alongside smaller ones, the northern regio exhibits hafirs almost exclusively below a certain capacity threshold. This suggests a dependent variabl affecting this distribution pattern, likely linked to regional variations in annual rainfall and its seasona distribution. Such climatic factors could explain the observed clustering, underscoring the importance o integrating environmental variables when interpreting the archaeological distribution of hafirs. Thus, ou focus turns back to the climatic factors initially introduced as critical variables in understanding thes distribution patterns.

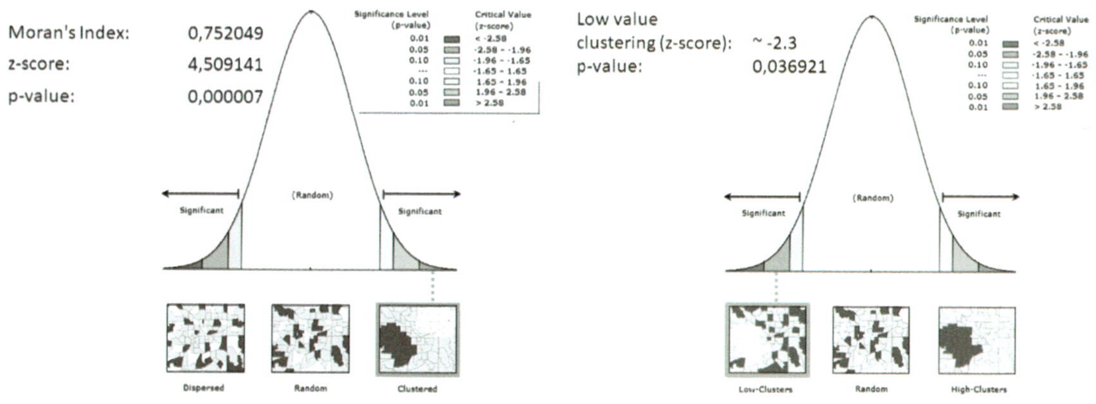

FIGURE 5.5 Visualisation of Moran's I (left) and Getis-Ord G (right) spatial autocorrelation coefficients c the spatial distribution of hafir volumes in the area of investigation (after ArcGIS 10.2.1 Spatial Statistics tool "Spatial Autocorrelation (Moran's I)" and "High/Low Clustering (Getis-Ord G)".

Palaeclimatic Contextualization

The palaeoclimatic circumstances surrounding the hafirs in Northern Kordofan, particularly in the cluster west of the pediment of the Jebel al-Ain escarpment, provide relevant insights into the diachronic development of water management strategies (Figure 5.6). The catchment area for this specific cluster spans approximately 19 square kilometers (calculated using SAGA 7.6.2 "Channel Network and Drainage Basin" tool), with a recorded maximum isohyete of 63 mm per rainy season during the 1800 bp time slice (Collins *et al.*, 2006). This results in a potential maximum precipitation input of 1,197,000 m^3 within the catchment area (calculated mainly for the part of the catchment area situated at the cuesta surface, since at the Arenosol plain at the Pediment, it can be assumed that infiltration, evaporation, and precipitation balance each other out, even during major precipitation events, resulting in a catchment area of 18.67397 km^2). The digital terrain model employed in this study facilitates an accurate estimation of the relationship between the volume of the hafirs and their associated local rain catchment. For comparative analysis, in the Butana region, it is estimated that large hafirs fill up with roughly 1/15 of the precipitation from their catchment area within a single rainy season (Scheibner, 2017).

Applying this ratio, at least roughly 460,500 m^3 of mean precipitation during one rainy season would be necessary for the practical use of these hafirs, which collectively hold approximately 47,000 m^3. This requirement translates to an average rainfall of about 45 to 50 mm per rainy season, or 0.5 mm/day (Figure 5.7). Based on these estimations, the construction of these hafirs can be dated to periods with adequate rainfall levels, notably between 2300 bp and earlier, 2100 bp to 1400 bp, and a smaller time window between 1200 bp and 1100 bp. A southward shift of the 0.5 mm/day isohyet marks a threshold of utility for these hafirs, as it significantly impacts their capacity to serve the water needs of the region.

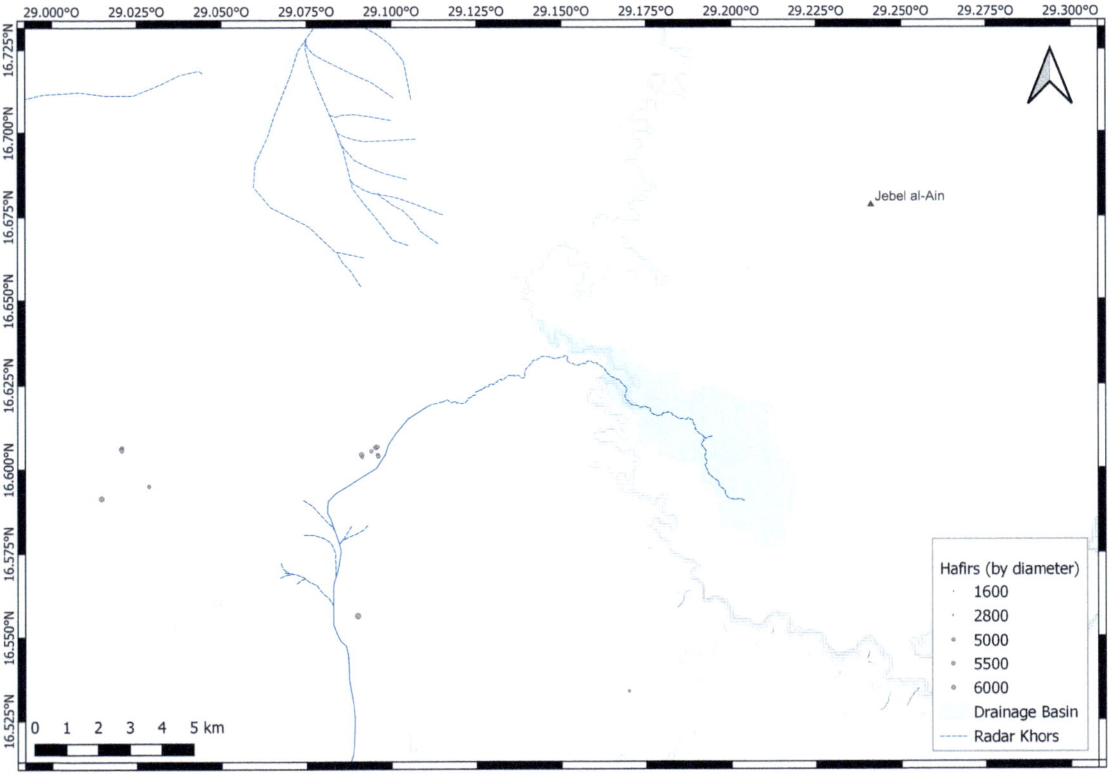

FIGURE 5.6 Map of the hafir cluster (indicating their different rim diameters) west of the pediment of the Jebel al-Ain cuelta, with palaeo-drainages ("radar khors") and catchment area on top of the cuelta.

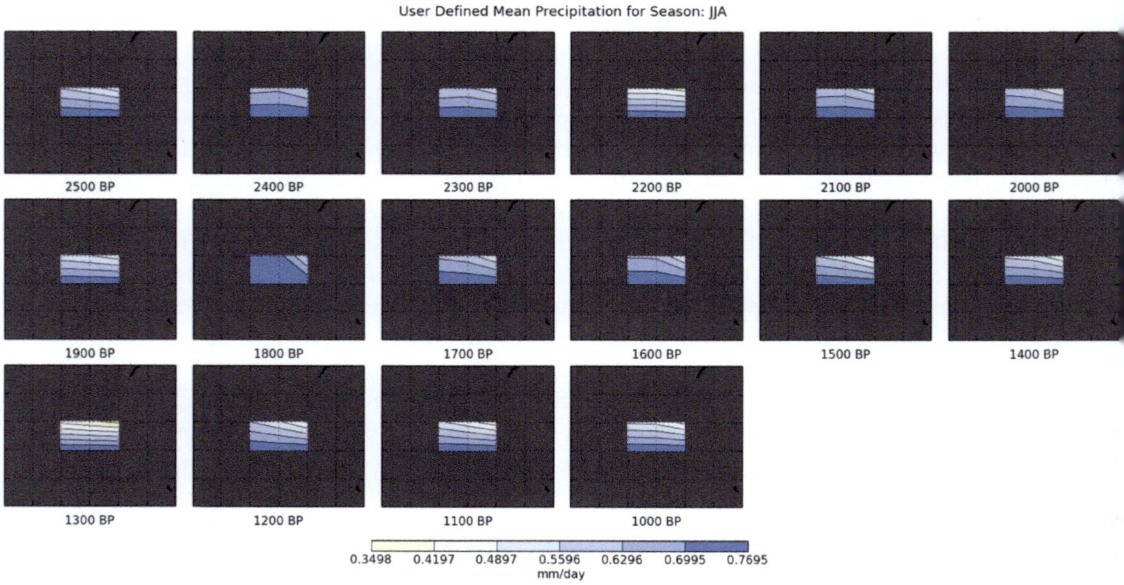

FIGURE 5.7 Diachronic visualization of absolute precipitation values (mm/day) at 15°-17.5°N 25°-30°E during the annual rainy season from 2500 bp till 1000 bp using CCSM3 (Collins *et al.*, 2006) TRaCE21 experiment (Liu *et al.*, 2014; Otto-Bliesner *et al.*, 2014), accessed and visualized via PaleoView 1.5.1 (Fordham *et al.*, 2017).

Even if this opens up a relatively wide time horizon for the use of the hafirs from a geo-hydrological point of view, there are some arguments in favor of a temporal setting during the precipitation optimum in the 100-year time slice around 1800 bp, because a generally favorable ecological situation (also for the carrying capacity of the surrounding pasture and cultivation areas) can also be assumed here.

In general, the carrying capacity of natural pasture in an entangled rangeland economy is a complex calculation, which (besides the availability of both pasture and water within walking distance of different kinds of livestock) also includes assumptions concerning herd management, which can at the current state of research only be assumed. Studies for the Butana (Karberg, 2017; Karberg, 2015) and the Bayuda region in Sudan provide some comparative material, but comprehensive studies for the circumstances in Northern Kordofan require fieldwork on the ground (including archaeo-zoological investigations).

CONCLUSIONS

The study of hafirs as water harvesting installations in Northern Kordofan reveals a complex interplay between environmental adaptation, cultural practice, and subsistence strategies in ancient Sudan. The analysis indicates that the primary function of these hafirs aligns more closely with a combined subsistence strategy focusing on rangeland pastoralism as the primary economic activity, supplemented by extensive, rain-fed seasonal agriculture, rather than intensified agriculture involving irrigation. This conclusion is supported by the spatial distribution of hafirs predominantly found on less fertile arenosol soils, which, despite their low nutrient concentration, benefit from high water permeability. This characteristic makes Arenosols more adaptable to small amounts of surface water, a critical factor in semi-arid regions.

Conversely, the more fertile fluvisol soils, which are conducive to intensified agriculture and the cultivation of more demanding crops, show a notable lack of hafir installations. This distribution suggests that such areas did not rely as heavily on hafirs, possibly due to their existing natural water retention and greater suitability for other forms of water management. This is a significant difference from Butana, where hafirs were also placed on fluvisol soils.

Furthermore, the spatial distribution patterns of hafirs hint at a reliance on localized precipitation rather than the conveyance of surface water from distant regions via major wadis like the Wadi Melek. This distinction underscores the importance of local rainfall in determining the operational viability of these installations, as opposed to simply channeling water over long distances from more humid southern areas.

Chronologically, the practical utility of the hafirs would have been contingent on specific time slices characterized by a minimum mean rainfall of approximately 0.5 mm per day. These conditions allow for dating the use of hafirs to periods around 2300 bp and earlier, 2100 bp to 1400 bp, and a smaller window between 1200 bp and 1100 bp. However, the period around 1800 bp emerges as the most likely time of significant hafir use, given the favorable climatic conditions prevalent at that time. This aligns with the hypothesis of technology transfer from the eastern regions, particularly the Butana, where many hafirs have been dated to the Meroitic period (approximately 2300 bp to 1700 bp). These insights contribute to our understanding of the resource management strategies that sustained ancient communities in Northern Kordofan, their adaptation to changing ecological circumstances, and possible transfers of knowledge and technology on an east-western communication axis.

DISCUSSION

The findings presented in this study open several avenues for further research, which could significantly enhance our understanding of the socio-economic dynamics in ancient Northern Kordofan. One potential area of exploration is the colocation of hafirs with other categories of archaeological records that can indicate mobile versus sedentary lifestyles and economic subsistence strategies, as already done for the Bayuda area (cf. Karberg, 2021 and Karberg and Lohwasser, 2018). By examining these associations, we can gain some understanding of the extent to which hafirs supported pastoral nomadism versus more settled agricultural communities.

Future fieldwork in Northern Kordofan, currently limited by political circumstances, could provide invaluable insights once conditions permit. On-the-ground surveys and excavations may uncover additional artifacts or features that could offer more precise dating and contextual information for the hafirs. This work could also clarify the relationships between water management systems and the broader economic and cultural practices of the region.

Further research into the specific grazing capacity of Northern Kordofan's rangelands is another promising direction. Insights from studies in the Butana have shown that the storage volumes of hafirs, such as those at Musawwarat es Sufra, often exceeded the grazing capacity of the surrounding area accessible from each water point (Karberg, 2015); (Karberg, 2017). This suggests a sophisticated integration of mobile pastoralism with substantial agricultural activities. Similar investigations in Northern Kordofan could yield a better understanding of how these water management structures facilitated economic strategies in the region. Such research might also illuminate how rangelands were utilized on the peripheries of the Napatan, Meroitic, and medieval Nubian states and cultures, offering a more nuanced picture of how these societies adapted to and thrived in their environments.

By expanding our research focus to include these areas, we can develop a more comprehensive understanding of the historical dynamics at play in Northern Kordofan and contribute to broader discussions about ancient water management and subsistence strategies across Sudan and other arid cultural landscapes.

REFERENCES

Albertz, J. (2007). *Einführung in die Fernerkundung – Grundlagen der Interpretation von Luft- und Satellitenbildern* (3rd ed.). Darmstadt: Wissenschaftliche Buchgesellschaft.

Born, M. (1965). *Zentralkordofan. Bauern und Nomaden in Savannengebieten des Sudan.* Marburg: Selbstverlag des Geographischen Institutes der Universität Marburg.

Carlson, D. L. (2017). *Quantitative Methods in Archaeology Using R.* Cambridge/New York: Cambridge University Press.

Collins, W. D., Bitz, C. M., Blackmon, M. L., Bonan, G. B., Bretherton, C. S., Carton, J. A., ... Smith, R. D. (2006). The Community Climate System Model Version 3 (CCSM3). *Journal of Climate, 19*, pp. 2122–2143.

Conolly, J., and Lake, M. (2010). *Geographical Information Systems in Archaeology* (4th ed.). Cambridge: Cambridge University Press.

Conrad, O., Bechtel, B., Bock, M., Dietrich, H., Fischer, E., Gerlitz, L., . . . Böhner, J. (2022). *SAGA GIS Tool Library Documentation v.7.6.2.* Accessed 10. 05 2022 at https://saga-gis.sourceforge.io/saga_tool_doc/7.6.2/index.html.

Drennan, R. D. (2009). *Statistics for Archaeologists. A Commonsense Approach* (2nd ed.). Berlin/Heidelberg: Springer.

Eger, J., and Karberg, T. (2019). Neue Forschungen in Nord-Kordofan. Vorbericht über die Feldkampagnen des InterLINK-Projektes der Jahre 2017 und 2018. *Der antike Sudan. MittSAG, 30*, pp. 131–146.

Eger, J., and Karberg, T. (2020). Nord-Kordofan im Satellitenbild. Vorbericht über die Forschungen des InterLINK-Projektes 2020. *Der antike Sudan. MittSAG, 31*, pp. 87–98.

Eger-Karberg, J., and Karberg, T. (2021). Wasser für Gala Abu Ahmed. Ein Paläo-Drainagesystem zwischen Jebel Nagaschusch und Wadi Howar (Sudan) im Radar-Satellitenbild. *Der antike Sudan. MittSAG, 32*, pp. 99–108.

Eger-Karberg, J., and Karberg, T. (2022). Interregional linkage investigations in Northern Kordofan (InterLINK). Report on the first project phase (2017–2022). *Sudan and Nubia, 26*, pp. 82–93.

Eger-Karberg, J., and Karberg, T. (2024a). The management of resources in remote desert areas: Bayuda and North Kordofan. In R. Lemos, and J. Budka (eds.), *Landscape and Resource Management in Bronze Age Nubia. Archaeological Perspectives on the Exploitation of Natural Resources and the Circulation of Commodities in the Middle Nile* (pp. 34–45). Wiesbaden: Harrassowitz.

Eger-Karberg, J., and Karberg, T. (2024b). Transfer von Waren und Ideen - Kordofan als Wegekreuz. *Antike Welt 3/24*, 24–29.

Elmirghani, A., and Drzewiecki, M. (2011). *Archaeological Survey in the Sodiri Region, 2011. Preliminary Report.* Sodiri: unpublished.

Fletcher, K. (ed.). (2012). *Sentinel-2: ESA's Optical High-Resolution Mission for GMES Operational Services.* Noordwijk: ESA Communications.

Fordham, D. A., Saltré, F., Haythorne, S., Wigley, T. M., Otto-Bliesner, B. L., Ka, C. C., and Brook, B. W. (2017). PaleoView: A tool for generating continuous climate projections spanning the last 21 000 years at regional and global scales. *Ecography, 40*(11), pp. 1348–1358.

Gratien, B., Dissaux, R.-P., Evrard, J., Marchi, S., Nogara, G., and Usai, D. (2013). *Abou Sofyan et Zankor. Prospections dans le Kordofan occidental (Soudan).* Villeneuve d'Ascq: Presses Universitaires du Septentrion.

Hinkel, M. (2015). *Die Hafire im Sudan. The archaeological map of the Sudan Supplement 2* (W. F. Reineke, ed.). Bonn: Habelt.

Hintze, F. (1963). Musawwarat es Sufra. Vorbericht über die Ausgrabungen des Instituts für Ägyptologie der Humboldt-Universität zu Berlin, 1961–1962 (Dritte Kampagne). *Wissenschaftliche Zeitschrift der Humboldt-Universität zu Berlin (WZHU), Gesellschafts- und Sprachwissenschaftliche Reihe, XII*, 63–77.

Howeida M.A., and Khabir A.M. (2011). Pottery from sites surveyed in Sodari District, Kordofan Province. An interim report 2008–2009. *Sudan and Nubia, 15*, pp. 18–22.

Jones, A., Breuning-Madsen, H., Brossard, M., Dampha, A., Deckers, J., Dewitte, O., . . . Zougmoré, R. (eds.). (2013). *Soil Atlas of Africa.* Luxemburg: Publications Office of the European Union.

Karberg, T. (2015). Culture, crops, and cattle. Aspects of environmental assessment, seen through an archaeologist's eye. In M. Zach (ed.), *The Kushite World. Proceedings of the 11th International Conference for Meroitic Studies* (pp. 69–76). Wien: Verein der Förderer der Sudanforschung.

Karberg, T. (2017). Cattle in the Meroitic Hinterland exemplified by archaeological record from Musawwarat es Sufra. In E. Fantusani, and M. Baldi (eds.), *The Fifth Day for Nubian Studies* (pp. 105–116). Rome: Scienze e Lettre.

Karberg, T. (2021). Landnutzungsentwicklung und -modelle der Oase El Rum - Analyse multispektraler und SAR-Satellitenaufnahmen im Kontext raumbezogener Surveydaten. *Der antike Sudan. MittSAG, 32*, pp. 69–82.

Karberg, T., and Lohwasser, A. (2018). Pastoralism in the Wadi Abu Dom. Non-sedentary economical strategies in the Central Bayuda. *Ādāb al-Neelain, 3*(2), pp. 1–15.

Liu, Z., Lu, Z., Wen, X., Otto-Bliesner, B. L., Timmerman, A., and Cobb, K. M. (2014). Evolution and forcing mechanisms of El Niño over the past 21,000 years. *Nature, 515*, 550–553.

O'Callaghan, J. F., and Mark, D. M. (1984). The extraction of drainage networks from digital elevation data. *Computer Vision, Graphics and Image Processing, 28*, pp. 323–344.

Otto-Bliesner, B. L., Russel, J. M., Clark, P. U., Liu, Z., Overpeck, J. T., Konecky, B., and Lu, Z. (2014). Coherent changes of southeastern equatorial and northern African rainfall during the last deglaciation. *Science, 346*(6214), 1223–1227.

R Core Team. (2020). *R: A Language and Environment for Statistical Computing 4.0.* Accessed 06. 01. 2025 at https://cran.r-project.org.

Reinke, N., and Werninghaus, R. (2009). *TerraSAR-X – Das deutsche Radar-Auge im All.* Köln: Deutsches Zentrum für Luft- und Raumfahrt.

Sanderman, J., Hengl, T., Todd-Brown, K., Parente, L. L., and de Sousa Mendes, W. (2021). *Open Soil Spectral Library Manual.* Accessed 29. 04. 2022 at https://soilspectroscopy.github.io/ossl-manual/

Scheibner, T. (2014). Entstehung, Ursprung und Nutzung – Die Hafire in Musawwarat es-Sufra und in der Keraba als Wirtschaftsbauten. In A. Lohwasser, and P. Wolf (eds.), *Ein Forscherleben zwischen den Welten. Zum 80. Geburtstag von Steffen Wenig* (S. 299–322). Berlin: Sudanarchäologische Gesellschaft zu Berlin e.V.

Scheibner, T. (2017). *Wasserbauliche Infrastruktur und Wassermanagement in Musawwarat es-Sufra in kuschitischer Zeit.* Wien: Universität Wien (Dissertation).

Tucker, C. J. (1979). Red and photographic infrared linear combinations for monitoring vegetation. *Remote Sensing of Environment, 8*(2), S. 127–150.

Tucker, C. J., Pinzon, J. E., Brown, M. E., Slayback, D. A., Pak, E. W., Mahoney, R., . . . El Saleous, N. (2005). An extended AVHRR 8-km NDVI dataset compatible with MODIS and SPOT vegetation NDVI data. *International Journal of Remote Sensing, 26*(20), S. 4485–4498.

Weschenfelder, P. (2014). Who gets the lion's share? Thoughts on Meroitic water management and its role in royal legitimization. In A. Lohwasser, and P. Wolf (eds.), *Ein Forscherleben zwischen den Welten. Zum 80. Geburtstag von Steffen Wenig* (S. 335–350). Berlin: Sudanarchäologische Gesellschaft zu Berlin e.V.

Wolf, P. (2015). The Qatar-Sudan Archaeological Project – The Meroitic Town of Hamadab and the palaeo-environment of the Meroe Region. *Sudan and Nubia, 19*, 115–131.

Wolf, P., Schomaker, R., Nowotnick, U., Eichmann, V., Scheibner, T., and Salamanek, N. (2015). The 'Island of Meroe' and the palaeo-environment of the Greater Meroe Region. In K. Finneiser, and J. Helmboldt-Doyé (eds.), *Der andere Blick. Forscherlust und Wissensdrang. Museumsgabe zum 80. Geburtstag von Karl-Heinz Priese* (S. 247–291). Berlin: EBV.

Air-Kawar
A Buffet for the Sahara: Landscape, Food, and Medicine in Late Pleistocene and Holocene

6

Erhard Schulz, Aboubacar Adamou, Sani Ibrahim, Roland Baumhauer, Barbara Sponholz, Karl Heinz Striedter, and Tilman Musch

INTRODUCTION

In recent years, the interest in traditional food and in traditional agricultural products in arid areas has significantly increased (FAO, 2011). The FAO called for greater consideration and promotion of traditional crops, especially in the aftermath of the Ukraine-war and the resulting wheat crisis, (FAO, 1988; FAO, 2023a.) These plants have long been of interest for science and industry as genetic resources (see EBI, 2023; FAO, 2023b). Parallel to this, there is an intensive discussion about the nutritional situation in the Holocene of arid regions, especially the Sahara (Bernus, 1967, Dunne *et al.*, 2016; Fuller *et al.*, 2021; Garcea, 2004; Gast, 1968; Mercuri *et al.*, 2018; Sadig, 2009; Schulz and Adamou, 1997; Tubiana, 1969; Wasylikowa and Dahlberg, 1999) – in which the questions of early agriculture and the subsequent development of cultural landscapes play a major role. Moreover, there is a controversial discussion about the role of as an accelerator or as a retarding force towards the end of the AHP (African Humid Period) at about 5500bp (see Brierley *et al.*, 2018; Li *et al.*, 2022, Wright, 2017). It mainly concerns the dimension of the albedo. Either it was restricted by the ability of the Holocene pastoralists to manage pastures and thus landscape properly or a possible neglecting or even ignorance towards overgrazing accelerated it.

DOI: 10.1201/9781003727781-6

Moreover, there is another background. The rising insecurity in most parts of the Sahara and the Sahel does not allow the continuation of cooperation programs on environmental history. Thus, it is time for a conclusion of these projects and an intermediate comparison of the results of the different disciplines.

These questions led to a reevaluation of investigations on vegetation and its traditional use in the Air Mts. and its surroundings in northern Niger as well as to some recounted pollen diagrams from the Kawar in NE-Niger.

The Bilma pollen diagram covers the Late Pleistocene and the Holocene, and the Kawar Lake records stand for the last 3000 years (Schultze, 1994, Schulz *et al.*, 1990). This allows a reconstruction of both landscape evolution and food resources for the Late Pleistocene and Holocene population and their possible impact on the landscape dynamics. Two models are to be discussed: The simple shift of vegetation-landscape zones following Neumann (1989) and an interfingering model of a prevailing Saharan savanna with Sudanian typed gallery forests as "no analogue types" (Watrin *et al.*, 2009).

Fire was an ever-present phenomenon in the sediment records and so a leading element in landscape dynamics. Therefore, questions arise whether they were for the most intended fires or mainly caused by lighting (Ballouche, 2019; Sponholz, 2009).

Archaeological investigations depict the differences of a pure extraction economy and of a production economy of pastoralists (cf. Di Lernia, 2002; Manning *et al.*, 2023; Tauveron and Striedter, 2009). Agriculture was not part of the Saharan Neolithic (Garcea, 2004) and millet cultivation starts only around 4000 bp (Fuller *et al.*, 2021; Manning, 2010; Neumann, 1989, 2003).

This paper will evaluate the types of alimentations and of livelihood during the Late Pleistocene and Holocene of the Central Sahara. Thus, mountains and plains between the Tassili n-Ajer, Ahhagar, Tibesi and Air Mountains and the vicinities of Lake Chad (for the regional setting see fig. 6.1).

Finally, it will discuss the role of human impact in shaping domestic or cultural landscapes.

FIGURE 6.1 The regional setting. The relevant localities in the Central Sahara in satellite images. After Google Earth and Terra Cognita database. (Planet observer).

Methodically, it will be based on an investigation into the present vegetation, as well as their traditional use for food and medicine. Results of palynology, sedimentology and archaeology will be exploited for a history of food and landscape management. Maps, diagrams, and graphics are essential tools to explain terms and facts, in order to avoid linguistic and conceptual misunderstandings, as "Sahara" is sometimes not even recognized as a territorial entity (Thomas and Nigaru, 2018). Moreover, they shall elucidate which of the present landscape elements may serve as models for the past.

In respect to its headings, this paper is presented as a menu: Starter, three main courses and a dessert.

The *Starter* will introduce the Saharan landscapes in order to explain the terms which will be used for the former vegetation and its comparison with the present situation. Questions of resilience and regeneration will be of special interest.

The *First Course* will consist of a Saharan package of food and a medicine package. Both are based on investigations on the traditional use of plants and vegetation in the Air Mts. (N-Niger) and of the respective literature from the central and southern Sahara.

The *Second Course* will present the pollen diagrams of Bilma and the Kawar lakes (NE-Niger) and their interpretation for the former vegetation between the poles of a zonal model or of an interfingering between Sudanian and Saharan vegetation types. These are lake sediment records in front of the Bilma escarpment in NE-Niger, which cover the time of the Late Pleistocene and Holocene.

The *Third Course* will deal with the Late Pleistocene and Holocene population and their different economies.

The three models of Hunters-Gatherers-fishermen, Pastoralist and early cultivators will be discussed for their food resources and their different livelihoods. For the part of collection grasses, we will give for the first time a quantitative model. An important point will be landscape management, especially the handling of fire.

The *Dessert* will be a summing up of the vegetation and economy history of the Central and South Sahara. It will concern the resilience and adaptation potential of both man and landscape, the principal changes caused by the end of the AHP and the question of cultural landscapes and their dynamics in the Sahara.

A *postscript* will deal with the consequences of the present political situation, the chances of preservation cooperation, and the resources and the knowledge about them and of further developments.

STARTER: THE REGIONAL BACKGROUND OF THE RESOURCES – VEGETATION OF THE SAHARA AND ITS SURROUNDINGS

Vegetation is the best way to characterize a region when it comes to current and past food resources. Therefore, we start with a vegetation map of North and West Africa. It was designed as a modern model to compare and test past vegetation units.

The map (figs. 6.2A, 6.2B) is based on direct observations and on published maps of the respective regions. An extensive bibliography is given in Schulz (2020) in which the differences to similar maps are also discussed. A black line marks the boundaries of the Sahara. The black square shows the main area of interest; the black dots indicate the locations of the Bilma and Kawar sediment records.

In the following different vegetation units are explained using drawings to avoid confusion in nomenclature, e.g. "steppe", "semi-desert", "desert", or "savannah". This is the best method beyond language or long texts. A short north-south transect will characterize the main vegetation/landscape units of the Sahara and its neighbouring regions.

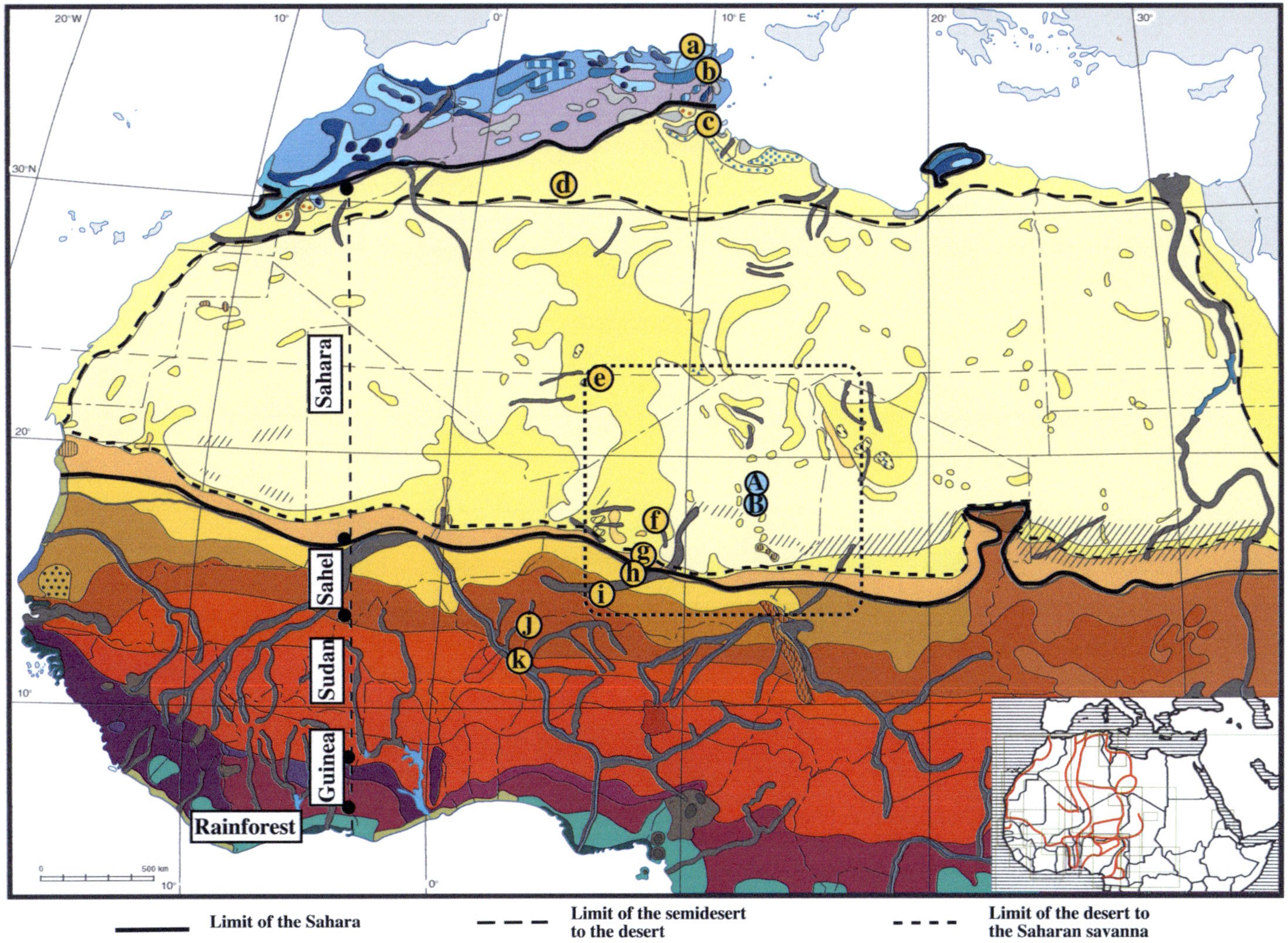

FIGURE 6.2A Vegetation map of northern and western Africa also showing the area of main interest of this study (Schulz, 2020, modified).

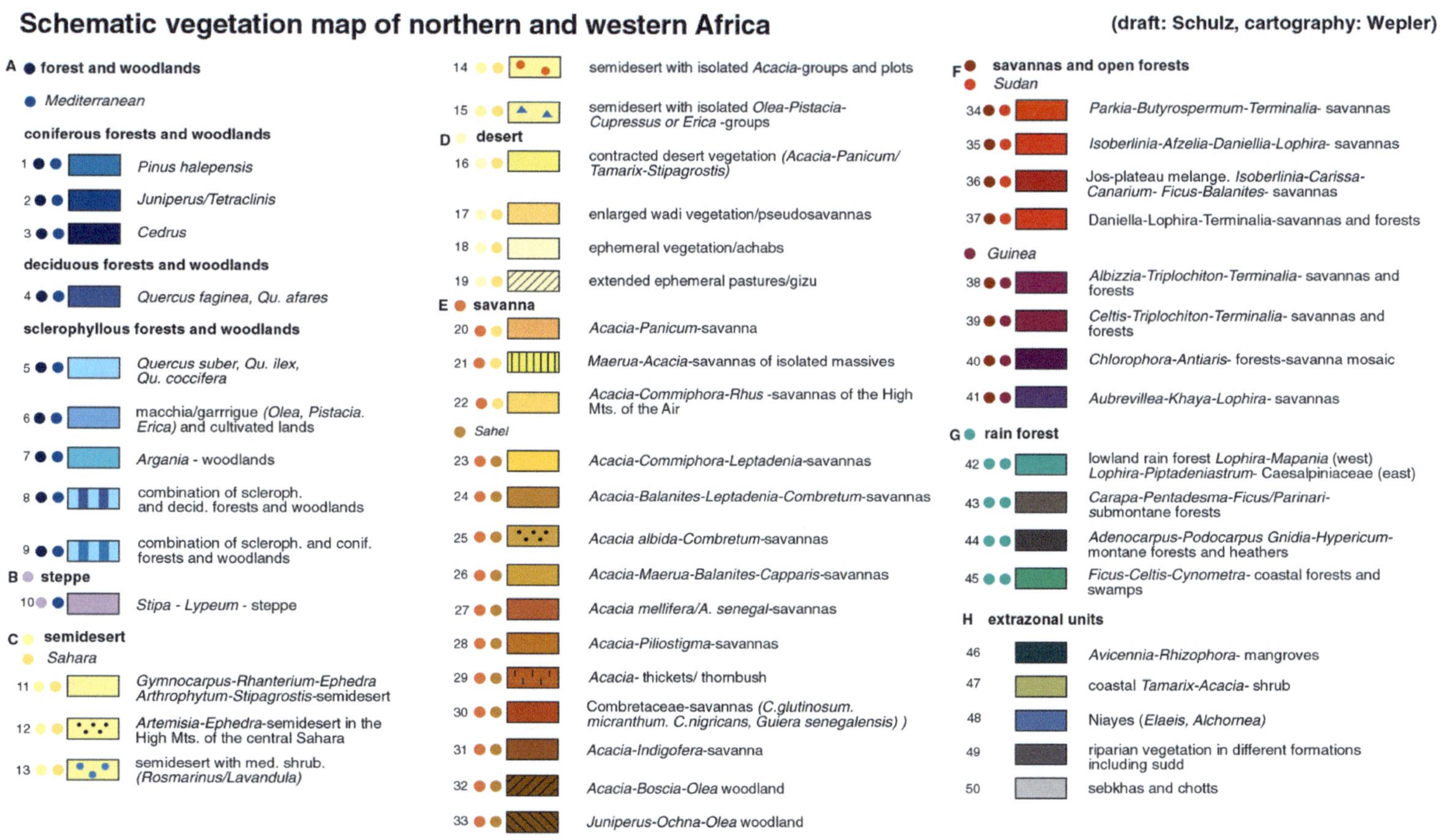

FIGURE 6.2B Legend to the vegetation map (Schulz, 2020, modified).

The Mediterranean Complex: from Forest to Steppe

The northern part of the map is characterized by Mediterranean formations ranging from forests, maquis shrubland (macchia), to steppe (including all stages of degradation). The term "steppe" is used for the tussock grass formation of *Stipa tenacissima* and *Lygeum spartum*. It refers to their degraded secondary formations (mostly based on *Artemisia*). The term is a borrowing from the Russian word *step* (the origin of which is unclear), describing winter cold continental grass formations mainly of the genus *Stipa* (Walter 1964) (see figs. 6.2A+B, 6.3).

The Sahara

The Sahara is a threefold landscape system that encloses the largest desert on Earth. Vegetation is the best tool to characterize and delimit its landscape types (see Monod, 1965, 1986; Quezel, 1965; Schulz, 2020). Comparisons can also be made with past states based on the vegetation in order to determine environmental change.

South of the Atlas Mts. (Morocco, Algeria, and Tunisia) and South of Djebel Aktar (Libya) the first relevant boundary is visible. It refers to the change between the Mediterranean and the Saharan domains. It follows the "Sahara-Line" (see Giessner, 1984), which combines the transition from mountains to plains with a transition to the Saharan floristic elements like *Gymnocarpus*, *Rhantherium,* or *Fagonia.*

The Semi-Desert

At the Sahara-Line (figs. 6.2A+6B), the landscape changes to diffuse small- and dwarf shrub-formations of the so-called "semi-desert" (Stokker, 1962; Stokker, 1976). Plants cover rarely 30 % of the soil, and

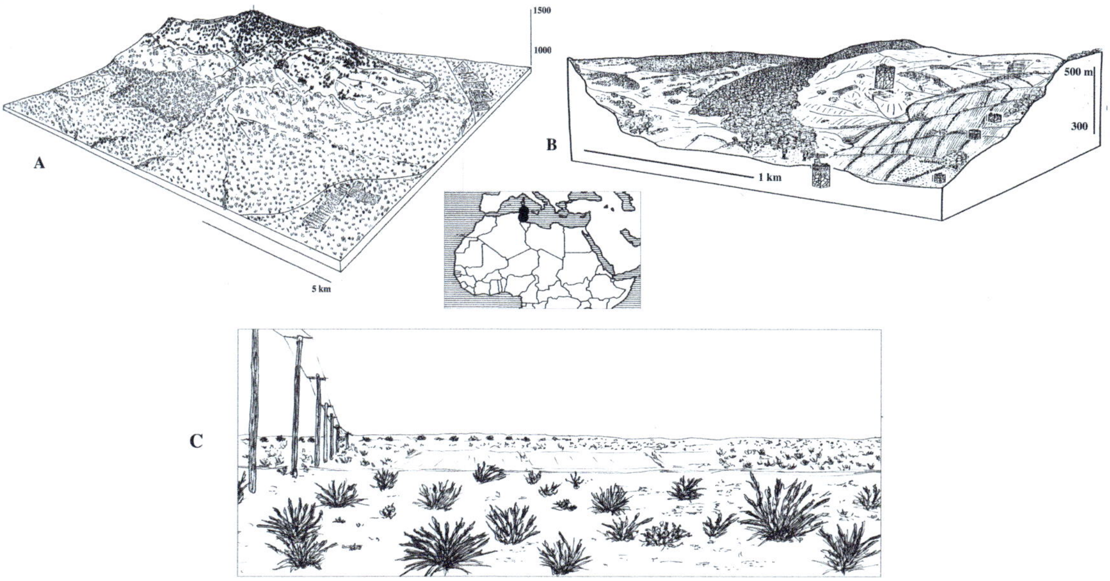

FIGURE 6.3 The Mediterranean complex in North Africa. A: Block diagram of Djebel Chaambi in Central Tunisia. In the upper part,the southernmost stands of Quercus forests, in the lower part, the Juniperus–Rosmarinus-shrubs, and on the plain, the Stipa–Lygeum–steppe or Artemisia-formation. B: Mediterranean oak (Quercus suber) forest with different stages of degradation like macchia or garrigue. The small block diagrams depict soil profiles of regosols or chambosols. C: The Stipa-Lygeum steppe near Kasserine, Central Tunisia (Schulz, 2020, modified).

most of the biomass is underground. There is a still ongoing discussion concerning the role of the *Acacia* trees (Jaquadi *et al.*, 2012).

The development of pellicular soil surface structures in sand (above) and silt/clay (middle). The different regeneration time of woody and herbacious vegetation on sand (a) and silt/clay (b) (Schulz, 2020 modified, Figure 6.4).

Here, in the semi-desert, the dual strategy of life in the Sahara becomes visible: A restricted number of permanent plants resilient to drought and frost is accompanied by "achabs". These ephemeral grass and herbaceous formations only need the amount of water from aleatory rainfall (see also below). This shows the importance of "chance" and "uncertainty" as "parameters" of Saharan vegetation (see Dubief 1959/1963). The semi-desert then reappears as the upper formation of the Sahara Mountains.

The Desert

Around 30°N, happens the definite change to the desert. Here it becomes visible throughout the continent by the transition from the diffuse permanent vegetation to the contracted or linear one, a model already presented by Monod (1965, 1986).

Thus, the desert has to be defined as the region where permanent life is restricted to places with the necessary amount of water. As also in the case of the semi-desert, the "achab"-strategy is important in the desert too.

The permanent part of desert vegetation is a tree-tussock grass formation, mainly characterized by *Acacia-*, *Maerua-*, *Balanites- Tamarix*-trees and *Salvadora-*, *Boscia-*, and *Leptadenia*-bushes. *Panicum turgidum* and *Stipagrostis pungens* are the typical grasses. These elements characterize the general aspect of the Saharan vegetation – as the *Acacia-Panicum*-realm.

Large wadis in the forelands of mountains – especially in the Southwest of Adrar des Iforas and Air Mts. might touch for a short distance but they separate afterward which is perfectly demonstrated for the western forelands of the Adrar des Iforas in northern Mali.

This dual strategy of life and repartition of vegetation (few permanent and aleatoric achabs) also determines the resources for animals and men and thus, the strategies to obtain food.

Soil, Resilience, and Regeneration

The lack of a diffuse and permanent plant cover and so a root carpet desert prevents the development of horizontal soils such as luvic or chambic soils. Regosols, in which ascending substances have dissolved and in which iron, gypsum, or other salts have accumulated near the surface, are therefore characteristic features (see Aguilar, 1987, Sys, 1987; Vogg and Meckelein, 1991). However, regeneration in first stages is quite regularly visible through the pellicular organization of soil surfaces (POS), a film-like system which covers the bare surface and closes it (see Figure 6.6). Sand or silt surfaces often favor the formation of algae/bacteria layers inducing biofilms (constituted by jelly substances) which lead to alternation with newly arrived sand/silt layers and so to biocrusts (Hahn and Kusserow, 1988; Pomel, 2008, Schulz *et al.* 2021, Williams *et al.*, 2012). As sandy layers, they support the infiltration of humidity and the germination of seeds (Figure 6.6) in contrast to the silt-clay layers (Figure 6.6 middle). These seal the surface for a long time. Sandy areas are therefore susceptible to the development of grass and herbaceous vegetation on them, regeneration of the plant cover only takes place in the form of achabs.

These different reactions described above have been documented by a long-term observation program, which took place in Mali (Hiernaux *et al.*, 2009). On sandy sediments (figs. 6.5, 6.6), the grass and herb vegetation recovered quite immediately, as did the woody plants after a short period of further decline. On silts or clays, on the contrary, the decline of woody plants continued; herbs recovered here only after six years and woody plants only after 15 years.

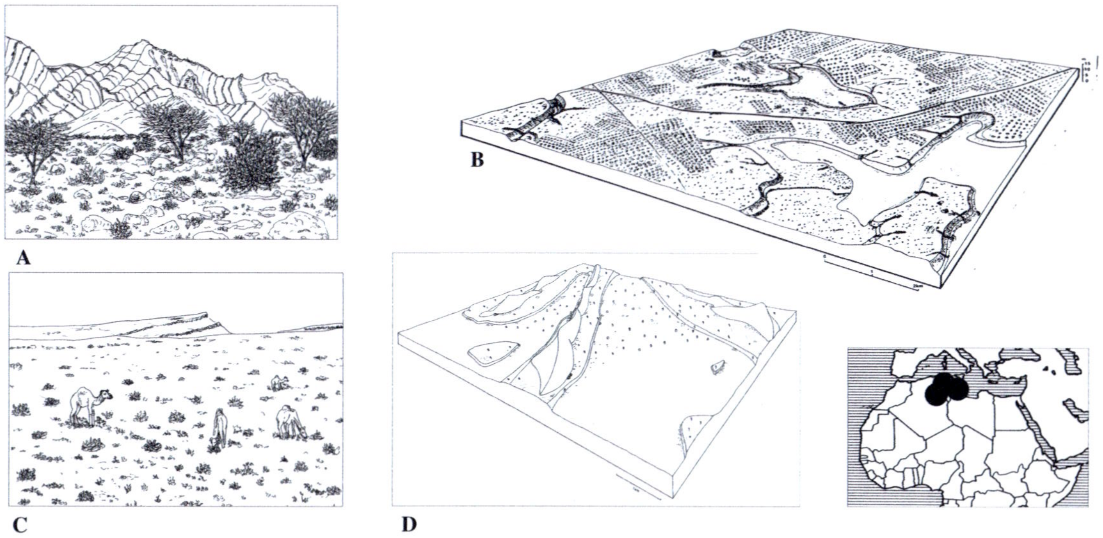

FIGURE 6.4 Aspects of the semi-desert. A: Acacia-Rhantherium stands in Bou Hedma, southern Tunisia. B: The Djeffara plain around Zarzis, southeastern Tunisia. Rhanatherium-Gymnocarpus shrubland and olive plantations around the sebkha Mehabeul. C: Rhanatherium semidesert south of Remada, southern Tunisia. C: The southern limit of the semi-desert with Calligonum-Ephedra at 30°N, south of El Golea, Algeria (Schulz, 2020, modified).

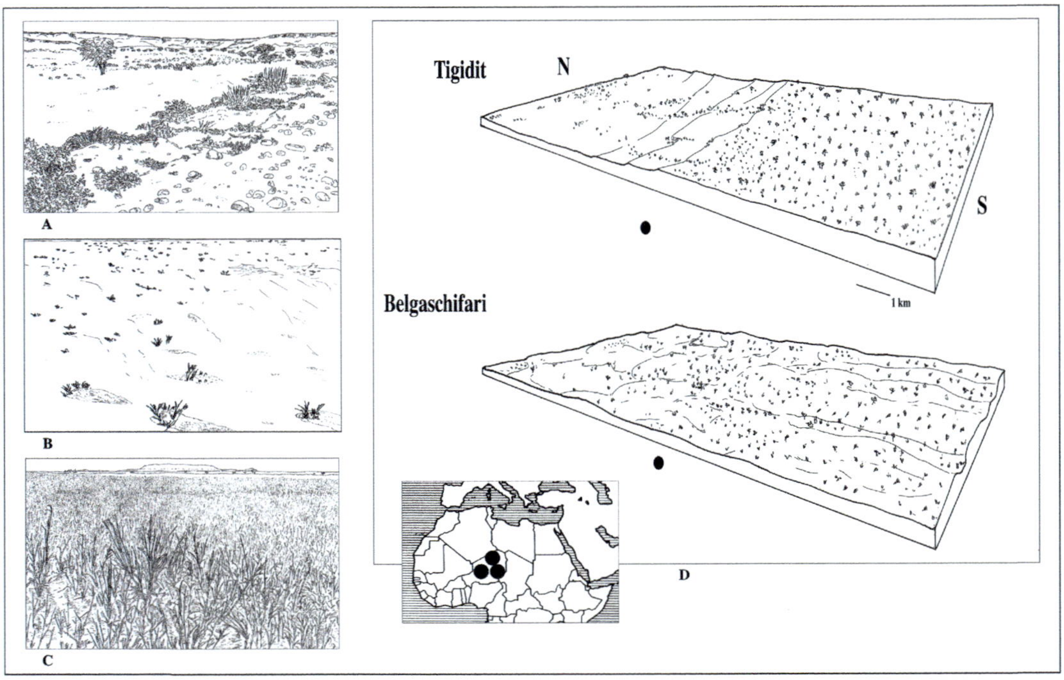

FIGURE 6.5 Aspects of the desert. Note the difference between the contracted vegetation (Wadi Achelouma, northeastern Niger A) and short-lived diffuse achabs which may also exist as wild sorgho fields, for example in northern Niger (C). The southern margin of the Sahara is represented by the vegetation changes at the Tigidit escarpment (16° 25′N/7°55′E) and at the Belgaschifari well (16°2′N/13°16′E) in Niger. The black dots indicate the boundary of the desert (Schulz, 2020, modified).

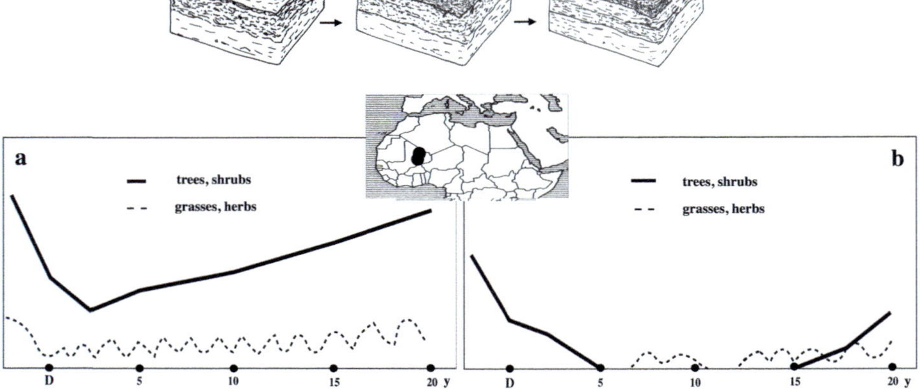

FIGURE 6.6 The development of pellicular soil surface structures in sand (above) and silt/clay (middle).

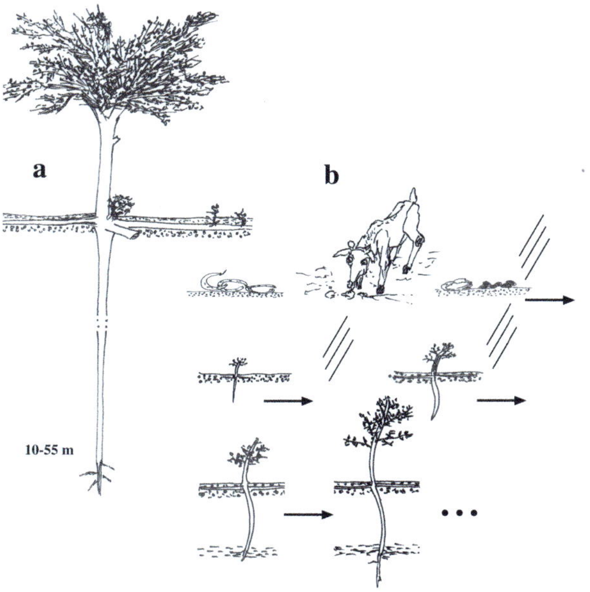

FIGURE 6.7 The Acacia strategy - vegetative (a) and generative (b) features. a: with lateral roots and taproots, root and stem suckers; b: uptake of fruits/seeds by animals, partial fermentation in the digestive system and distribution. Germination after a number of suitable rainfalls. Unique investment in the taproot until a water horizon is reached. From that point on, subaerial growth may start. After Schulz (2020), modified.

The different regeneration time of woody and herbacious vegetation on sand (a) and silt/clay (b) (after Ibrahim and Schulz 2019, modified, see fig. 6.7).

Chance and Unpredictability: Acacia and Achab Strategies of Regeneration

The genus *Acacia* depicts strategies for regeneration in arid regions, which are based on a combination of vegetative and generative features (fig. 6.7). Near-surface roots and taproots utilize regular and aleatoric rainfall, as well as groundwater horizons. The mode of photosynthesis and gas exchange characterizes them as phreatophytes, which are less dependent on the regional climate (Ullmann; 1985) Root- and stem suckers react to damage or drought periods. Trees are resisting to pollarding or to the reduction to gullivers by elephants (Blanco *et al.*, 2015; Munkert, 2009; Grouzis and LeFloc'h, 2011).

After exceptional good years of fruiting, seeds are taken up by animals in great quantities. Fermentation in the digestive tract, migration of animals, and deposition lead to wider distribution of the species. A series of suitable rainfalls are then necessary for germination. During the first years, Acacias invest in their taproot. Only when the groundwater horizon is reached, the subaerial growth really starts, and the tree becomes independent of most climatic variations. This strategy also explains the extension of *Acacia* in the 19th century after the extreme droughts at the end of the 18th century (Schulz *et al.*, 2021).

The short-time rain floras – achabs comparable to the gizu (Wilson, 1978) – develop from seed banks stored in the sandwich layers of fine sand/silt covered by coarse sand (Figure 6.8). Rainfall or dew can be absorbed by the silt/fine sand layer where the seeds may germ, cf. Williams *et al.* (2012). After the short life cycle a new seed bank may be deposited under comparable conditions, but the roots – even dead ones can form a stabilizing root carpet for several months. Thus, achabs are both important pasture resources and an example to demonstrate adaptation to the general aleatoric character of the Saharan climate.

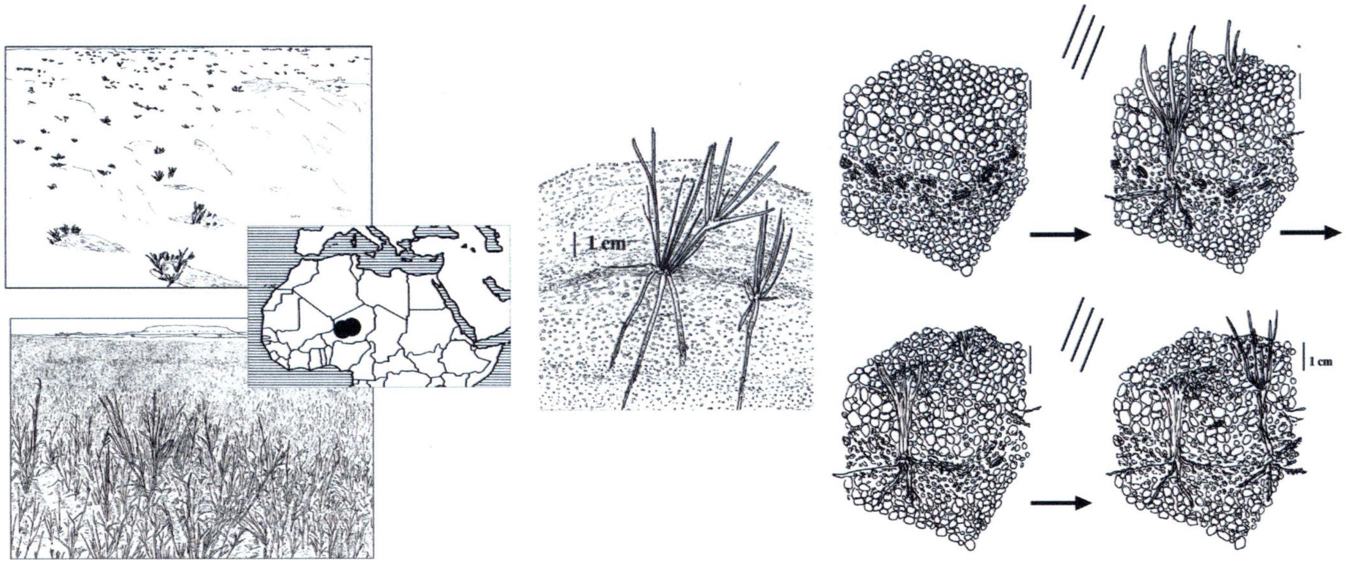

FIGURE 6.8 The dynamic of the achabs. The different size of the achabs plus the development of grasses and therophytes from a seed bank in a sandwich of silt under coarse sand and the formation of a root carpet after repeated precipitation (own draft).

Moreover, Saharan plants in the achabs have a long seed dormancy and most of them show the weedy behavior of colonizers (e.g. *Tribulus terrestris, Aridtida* sp.) and a great genetic plasticity, allowing them to develop several varieties (cf. Mercuri *et al.*, 2018).

Change with Altitude – the High Mountains

High mountains collect rainfall and humidity by dew fall. Thus, they act as water towers for the respective regions and allow the persistence of a water demanding plant cover as altitudinal change (Dubief 1959/1963, Joseph, 1991). The Ahaggar and the Tibesti Mts. depict the differences between desert and semidesert again by the change to diffuse shrub formation from an altitude of about 2000 or 2500 m asl. on. The Air Mts. in northern Niger, however, shows a Sahelian type of altitudinal change with a transition to savannas from about 1800 m asl. on (fig. 6.9).

The Saharan Savanna

Back to the plains, the plant cover changes again at about 16° to 16.5°N from the contracted type of tree-tussock grass-formations to the diffuse one within the same Saharan *Acacia-Maerua-Panicum*-formation. Thus, from the Saharan desert to the Saharan savanna. Savanna as a tropical diffuse tree-grass-formation. Figure 6.9 explains these characteristic traits of desert and savanna in the Sahara. A first step is performed in rocky areas where humidity is stored in fissures such as the Agadem- and Homodji escarpment in eastern Niger. It refers to the type "flowerpot-savanna" Schulz and Pomel, 1994; Schulz *et al.*, 2009). The second step represents the real change from the linear *Acacia-Panicum* wadi vegetation to its diffuse form of an *Acacia-Maerua-Panicum*-savanna at about 160°N.

The external and internal boundaries of the Sahara appear over the whole continent on various substrates. Subsequently, they are climatic ones depending on a combination of regular and aleatoric rainfall, contrary to the simplistic delineations by isohyethes (100 mm to the North and 150 mm to the South; cf. Giessner, 1984).

Sahara *est omnis divisa in partes tres: quarum unum dominat semideserta, aliam deserta et tertiam savanna saharica!* (adapted after Caesar 50/51 BC).

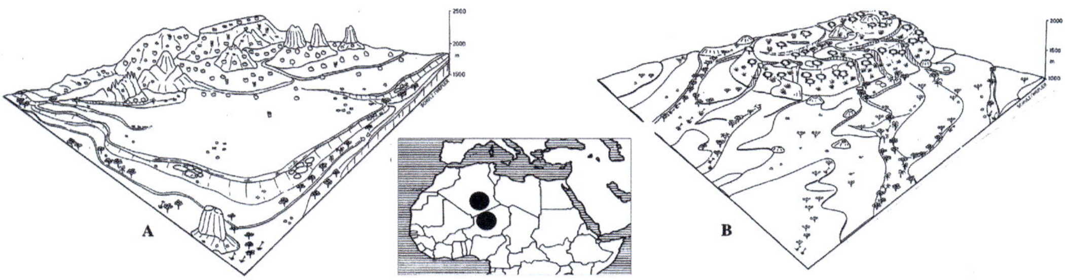

FIGURE 6.9 The general altitudinal change in the Sahara. A: Ahaggar Mts. Algeria standing also for the Tibesti Mts., Chad. The change from the contracted wadi vegetation to a diffuse Artemisia-semi-desert above 2500 m altitude represents the Mediterranean type of change. B: Air Mts. Niger with the passage from contracted wadi vegetation to diffuse Acacia; a diffuse Acacia-Commiphora-Rhus-savannah shows the Sahelian type of altitudinal change (after Schulz, 2020, modified).

The Sahel

Some kilometres to the south (from the desert-savanna transition), the savanna changes again- from the Saharan to the Sahelian type (figs. 6.10, 6.11), the latter being characterized by *Acacia*, *Maerua*, *Commiphora,* and annual grasses such as *Panicum* (see also Barry *et al*., 1986). Figure 6.11 shows a traditional pasture zone. Further south, the Sahel becomes the zone of agriculture where mostly pearl millet and sorghum are cultivated. Thus, the Sahel represents a zone accentuated by cultural landscapes, with all types of degradation and regeneration. The southernmost Sahel shows already the economy of parks, which are a traditional agro-forestry system, in particular characteristic of the Sudanian region/ (see Seignobos, 2017). Thus, the Sahel is a savanna region, and the same remarks as for savanna in general are valid for this zone: "Most disputed, but less understood" (Cole, 1986). In this way, most of the savanna types (cf. the vegetation map figs. 6.2A, 6.2B) are part of different agrarian systems. Combretaceae savannas are best examples, as they are for the most part fallows at different stages of regeneration.

Anyway, despite all the degradation phenomena, these are stable savannas with a tendency for regeneration and soil formation/soil processes in the direction of cambic or luvic soils such as chromic arenosols

FIGURE 6.10 Two types of the Saharan savanna. A Maerua-Capparis-savannah in rocky areas, such as the Agadem escarpment in Eastern Niger. A first transition to savanna is performed when humidity is collected and stored in fissures and cracks. In the plains, the *Acacia-Panicum* wadi formation changes into *Acacia-Panicum* savanna at about 16°N. (after Schulz, 2020, modified).

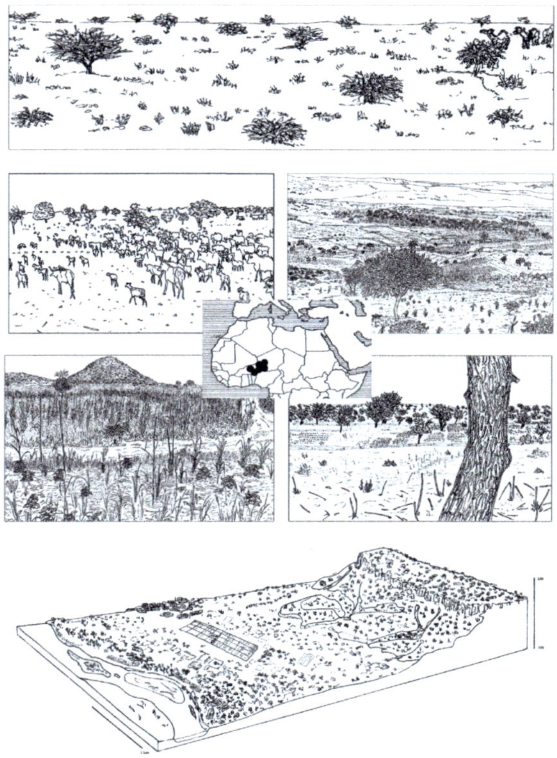

FIGURE 6.11 The system "Sahel": (a) The northernmost Sahelian savanna on the Tigidit plateau. An Acacia-Maerua-Commiphora-Aristida savanna with annual grasses, contrary to the Saharan Acacia-Panicum savanna with tussock grasses. Acacia-Panicion versus Acacia-Aristidion sensuBarry *et al.* (1986). (b) A cattle, goat, and sheep herd in an acacia savanna near Agalak, N-Niger, (c) Acacia-Piliostigma savannah near Keita, Central Niger. An intensively used and degraded agrarian landscape with anti-degradation measures (stone walls). (d) The millet cultivation and remains of Combretaceae savannas near Birni-n-Konni, Central Niger. (e) A Faidherbia albida park with cattle near Niamey, SW-Niger. (f) The Upper Cretaceous cuesta and the Niger alluvial plain at Gaya, SW-Niger. Transition from the Combretaceae savanna and the Sudanian Parkia-Vitellaria-parks (after Schulz, 2020, modified).

for the Saharan or North Sahalian savannas, or chromic luvisols or cambisols for the Combretaceae savannas in the South (Vogg and Meckelein, 1991). Later, we will have to discuss whether this savanna system might be taken as a model for the Holocene developments.

FIRST COURSE: FOOD IN THE SAHARA – A MODERN MODEL FOR THE PAST

This overview concerns the traditional diet and the knowledge about it. It is still preserved – outside the urban areas – and necessary in times of scarcity. This knowledge is particularly important because it can help to improve crops for cultivation in arid regions.

A normal meal in the Sahara and Sahel consists of cereals in the form of porridge or galettes/cakes and a sauce of milk, herbs, or vegetables (fig. 6.12). Animal protein is rare – meat is only eaten on special

FIGURE 6.12 The Saharan meal. (A) The ingredients: Flour and dried herbs, Fachi/Niger. (B) Porridge with milk, Kerr/Ennedi. (C) A gardener in Timia/Niger. Date palms, citrus, and vegetables (D) Collection of Rhus leaves on the Bagzan/Air Mts./ Niger. (E) Goat cheese at Wadi Mekkerene / Air Mts./Niger (Photos A, C, D, E by E. Schulz; B by T. Musch).

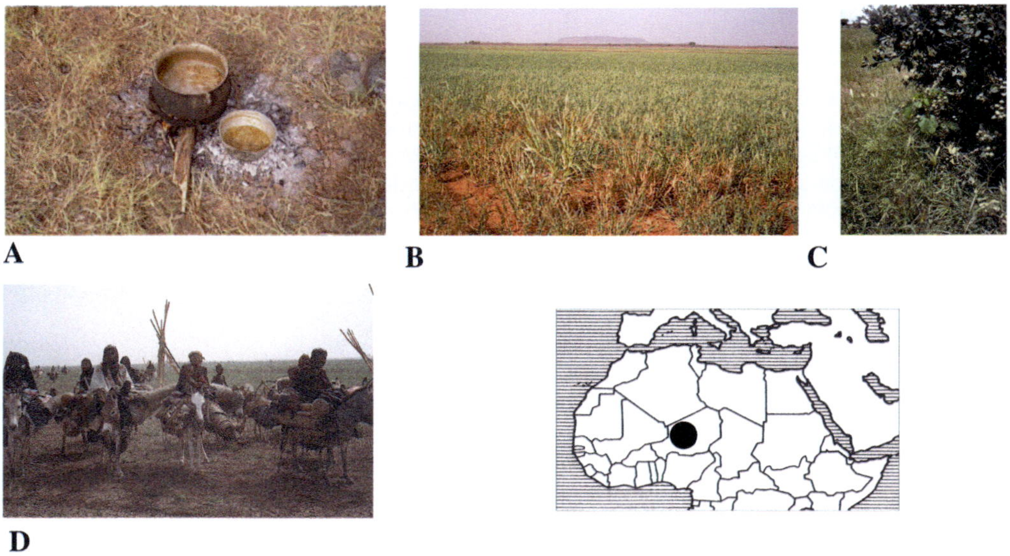

FIGURE 6.13 The Saharan meal for the nomads. (A) Boscia-soup with Sorghum, Azawad/Niger. (B) A wild sorghum field Azawad/Niger. (C) Boscia senegalensis shrub with berries, Azawad/Niger. (D) A group of Fulan nomads in the Azawad/Niger (Photos Schulz, after Schulz and Adamou, 1992).

occasions, and hunting success is not guaranteed (fig. 6.12). In some places, insects or larvae are also eaten. Salt or red pepper are the most common spices. Salt is a necessary component of the diet of humans and animals (Gentilini, 1994); it also represents the traditional wealth of the Sahara and is the basis of trade with the Sahel (Ritter 1980, Spittler 1989). A basic law of the Sahara also applies to food: permanent resources are limited, and chance determines which short-lived additional food sources are available (fig. 6.13).

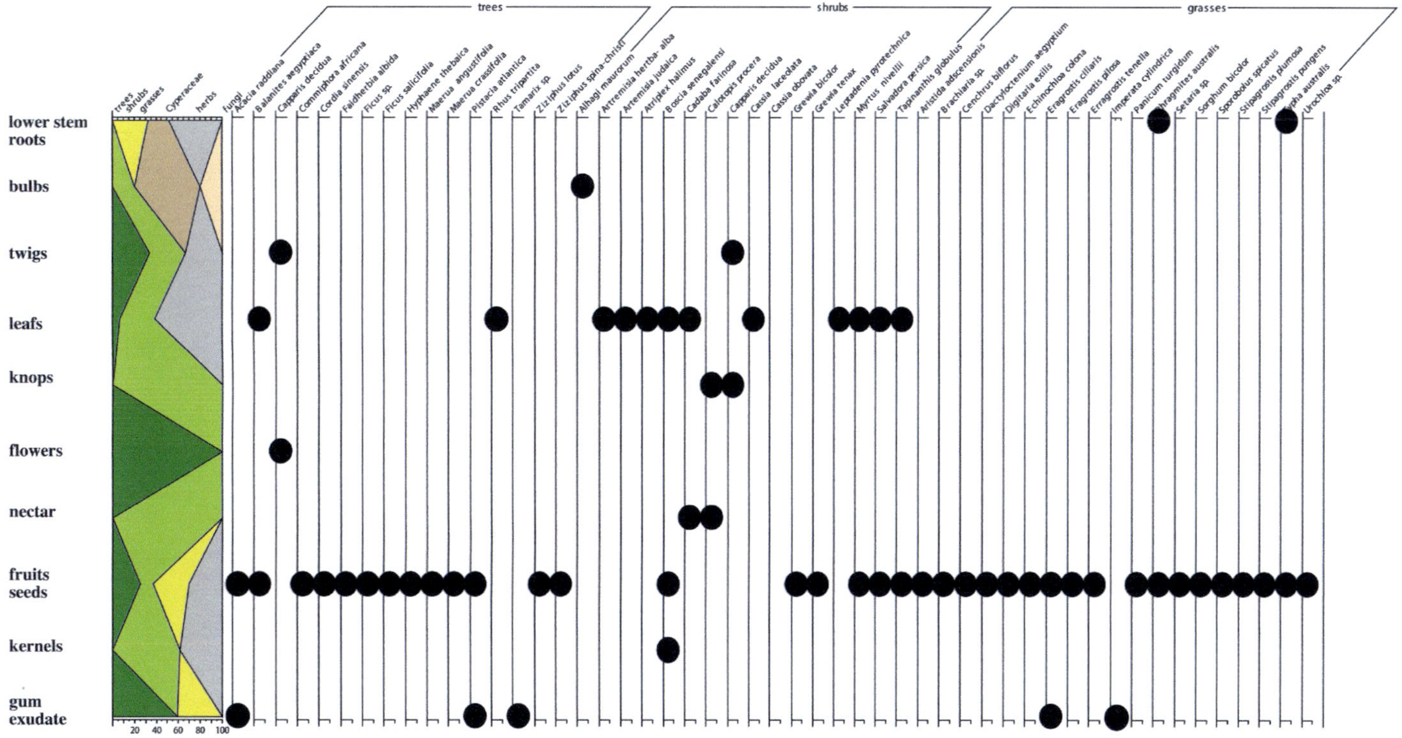

FIGURE 6.14A The Saharan food package. Part of trees, shrubs, and grasses. The diagram shows the great variety of wild Saharan plants used by man. The most important part comes from grasses, which allow obtaining yields of up to 250 kg/ha. (Diagram: E. Schulz).

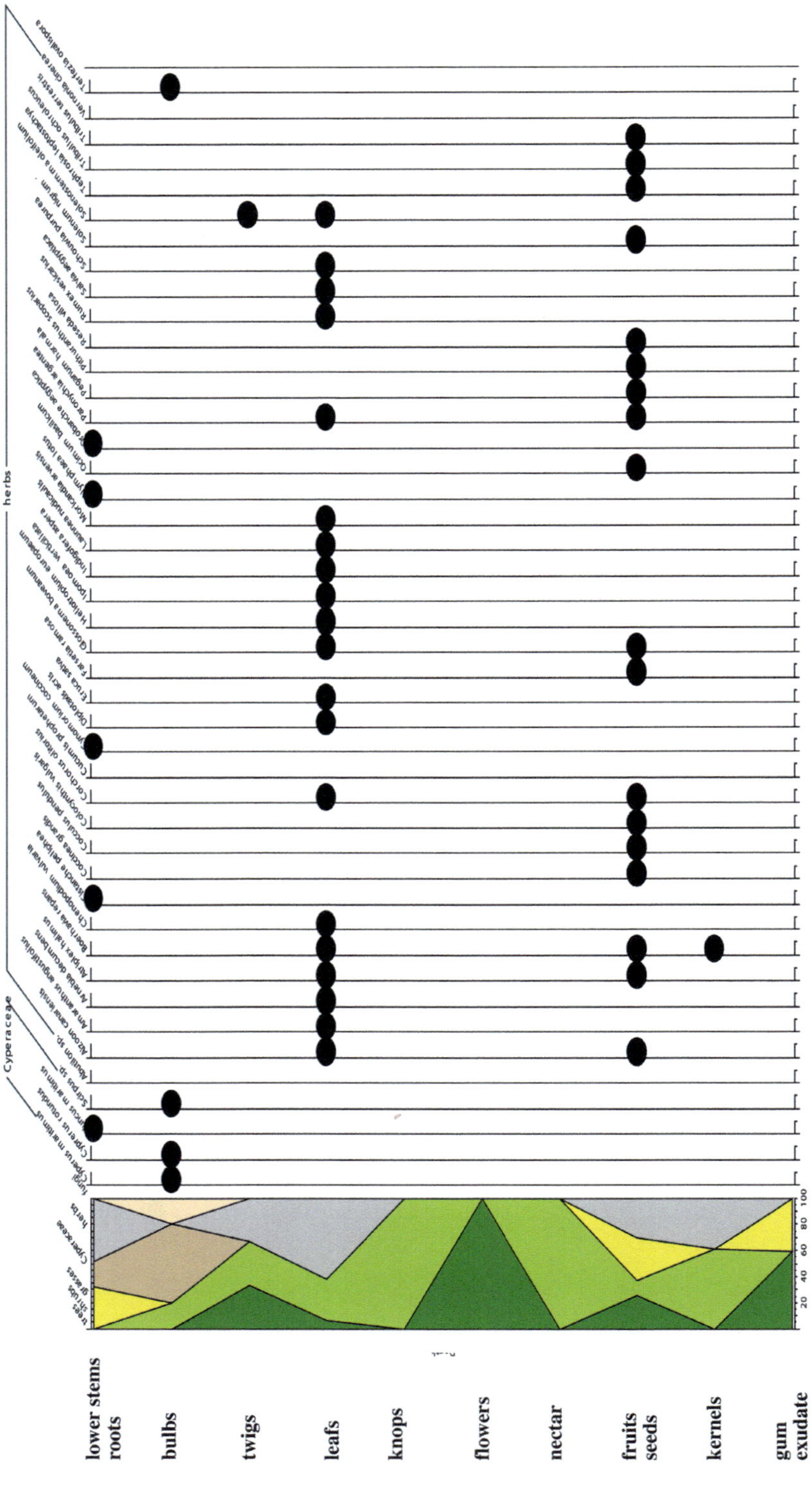

FIGURE 6.14B The Saharan food package. Part of Cyperaceae and herbs. (Diagram: E. Schulz).

The food habits of the Tuareg Kel Ewey in the Air Mts./Niger are simple and clear. The preferred meal is a porridge of millet with some milk, which is seen as the best meal ever. Food from gathering is regarded as uncivilized. However, the Tuareg know about the resources for gathering in times of shortage (Spittler, 1994).

The examples observed among Tubu of northern Chad (Tibesti, Borku, and Ennedi) will give a picture of today's eating habits in the Central Sahara. Today, those Tubu families who continue to eat traditionally prepare meals from mostly local products. At breakfast, for example, these consist of camel's milk, (sour) goat's milk or *dooga*, a type of pancake sweetened with the decoction of Doum fruit (*Hyphaene thebaica*) (cf. Musch, 2020). For lunch or dinner, there is a kind of cooked cereal dish (*tibi tukuli*), which is consumed either with milk or with a sauce (vegetables, small pieces of meat to enhance the flavor). Meat is served on special occasions. The cereals used to make these meals are millet or wheat. In between, dates or various dishes (*munumunu, tinni saw* ...), whose main ingredient is dates, can be served. In the latter, dates are crushed and mixed with wild grains or goat butter, for example. The now intensive trade with Libya and the improvement in the economic situation, which is due in particular to profits from artisanal gold mining, are also changing eating habits. Pasta ("spaghetti"), Maggi flavoring and canned food are now available; canned drinks, sweets, and sweet cookies are served when visitors are received. Meals are often no longer prepared by women or children, but by "cooks" who live with the family. These "cooks", who come from other regions, often only know how to prepare pasta with certain minor variations in the sauce.

The procurement of food is divided into production and collection. Production consists of oasis agriculture and mobile livestock keeping (see Adamou, 1979; Adamou and Morel, 2005). Both sectors depend on the quantity and access to water. Traditional oasis agriculture is irrigated; its main product is dates of many different varieties (cf. Musch, 2018). Figs or citrus fruits are grown in the shade of the date palms. Cereals, alfalfa, and vegetables grow in small, irrigated beds. Salinization is a constant danger, so the amount of water for cultivation must be limited (Musch, 2023; Zabel, 1983). Millet as a staple food cannot be produced in sufficient quantities in the southern Sahara. Some trade with the Sahel and the collection of wild grasses and fruits compensate for the shortage (Schulz and Adamou, 1992; Schulz and Adamou, 1997; Spittler, 1989). Local and nomadic livestock farming represent a second branch of production. Transhumance and permanent nomadism are the means to utilize the random food sources in the vast region.

The Saharan Food Package

A joint survey at the end of the 1980s showed the broad knowledge of the Sahara inhabitants about the plant resources for food and medicine (Schulz and Adamou, 1997). Thanks to the trust the healers had built up over many years with Aboubakar Adamou, it was possible to collect plants and related information directly in the field (table 6.1). It is based on our own investigations (Schulz and Adamou, 1992, 1997) and on the information provided by Bernus (1967, 1980a, 1984), Biagetti (2015), Burkill (1985), Busson (1965), Chapelle (1982), Gast (1968, 2000), Jacks (1964), Klute (1992), Nicolaisen (1963), Nicolaisen and Nicolaisen (1997), Spittler (1989), Tubiana (1969). Grasses and herbs are the most important elements among the plants, but trees and shrubs are also regularly used. Among the grasses, *Cenchrus biflorus, Echinocloa colona, Panicum laetum, P. turgidum, Sorghum aetiopicum,* and *S. bicolor* are the most

TABLE 6.1 The Caloric Uptake in Different Societies (after Leonard 2003, Schulz and Adamou, 1992, 2024, modified).

SOCIETY	*UPTAKE KCAL/D*	*% ENERGY OUT OF PLANTS*
Hunters and gatherers	2,100	33
Animal keepers	1,400	20
Cultivators	2,000	95
Industrial society	2,550	77

important. *Panicum laetum* and *Sorghum aethiopicum* in particular can produce collection yields of u
to 250 kg of grains per ha (Schulz and Adamou, 1992). This underlines the importance of *achab* as
resource.

The Caloric Dimension

The two following tables show the mean uptake of calories in different societies and the caloric value o cultivated and uncultivated grasses/cereals. They underline the importance of gathering plants for presen and past societies. A comparison of the caloric uptake per day for different societies is given by Leonar (2003).

According to Felix-Katz (1980), nomads on NW-Niger take up 2700 Kcal/d mainly through milk Ousseni (2005) is an oral communication estimates that the yearly consumption of cereals is about 200 k per person. The nutrition content of collected plants vs. cultivated plants is given in Table 6.2.

To give an example on the quantity of gathering food, grass can be divided into three groups: A first group of wild cereals is called *ishibin* and englobes small annual grasses like *Aristida adscensionis* *Brachiaria laeta,* or *Stipagrostis plumose.* In this group, yields of up to 40 kg/ha can be obtained. Yield obtained from subterranean formicaries are about 5 kg per location.

A second group englobes annual and permanent grasses like *Panicum turgidum, Cenchrus biflorus* and various *Pennisetum* sp. From these grasses, yields of about 70–100 kg/ha can be obtained (Schul and Adamou, 1992).

A third group with annual grasses and herbs like *Panicum laetum*, *Sorghum aethiopicum*, o *Colocynthis vulgaris* gives yields between 100 and 250 kg/ha. This group represents an important foo resource; nevertheless, in order to leach the bitter taste of the grains, much labor is necessary to cook an dry them preliminarily. To illustrate the importance of these resources: the annual yield of pearl millet an *Sorghum* in the contemporary Sahel ranges around 850 kg/ha but can decrease to less than 500 kg (D Rouv, 2004, Traore *et al.*, 2017). Thus, this third group of grasses (*Panicum laetum*, *Sorghum aethiopum* range already in the yield categories of cultivated plants (Schulz *et al.*, 2021).

All estimations are based on collections per 10 m^2 or 100 m^2 and weighting the grain content.

Hunger is a constant experience in Saharan societies (Bernus, 1980b, Gast, 1968, Nicolaisen an Nicolaisen, 1997; Spittler, 1989). To fight hunger efficiently, managing and preserving knowledge abou the collection of resources and the compensations for food shortages is important (Schulz and Adamou 1992; Spittler, 1989). Nevertheless, in some years, even knowledge about alternative food resources can not prevent hunger and related calamities. An example is provided by the locust invasions of recent years starting from 2003/2005 (Ceccato *et al.*, 2007; Kitz, 2004; see also Skaf *et al.*, 1990).

TABLE 6.2 The Nutrition Content of Some Collected and Cultivated Plants. Note the elevated content of proteins and carbohydrates of the collected plants (after Bergeret and Ribot, 1990; Busson, 1965; Schulz and Adamou, 1992.

SPECIES	*PROTEIN % DRY WEIGHT*	*CARBOHYDRATES % DRY WEIGHT*
Balanites aegyptiaca		
kernels	27.6	20.8
pulp	87.4	5.6
Cenchrus biflorus	21.0	67.2
Dactyloctenium aegyptium	17.7	73.3
Eragrostis ciliaris	20.0	65.6
Pennisetum americanum	10.0	71.0
Zea mays	11.7	80.2

FIGURE 6.15 Traditional medicine as well as alimentation still represents a treasure of knowledge. Field survey with traditional healers in the Air Mts. and procession of collected leaves from the Bagzan, Air Mts./Niger. (After Schulz and Amadou, 1992 modified).

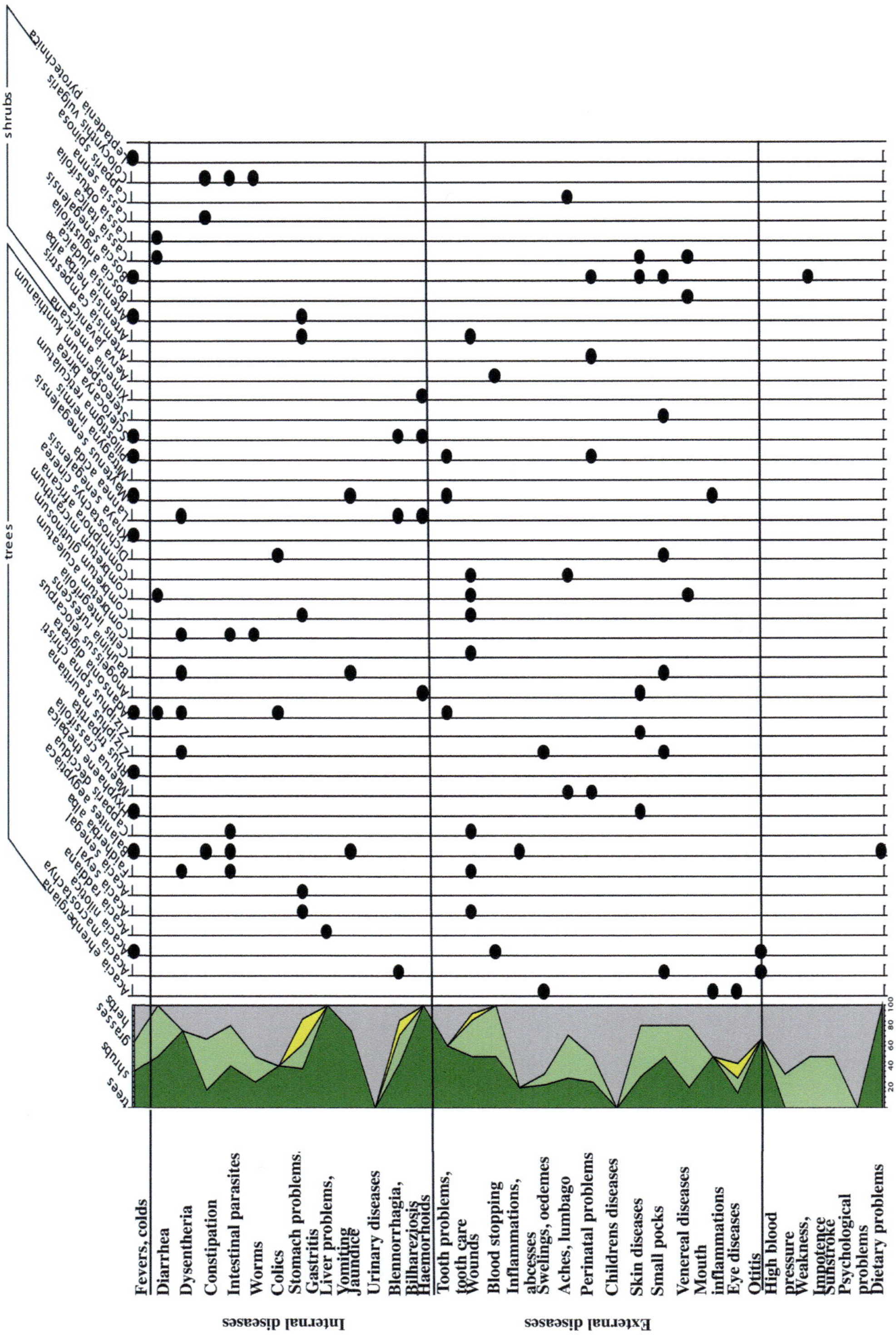

FIGURE 6.16A Saharan medical package. Part of trees and shrubs. (Diagram: E. Schulz).

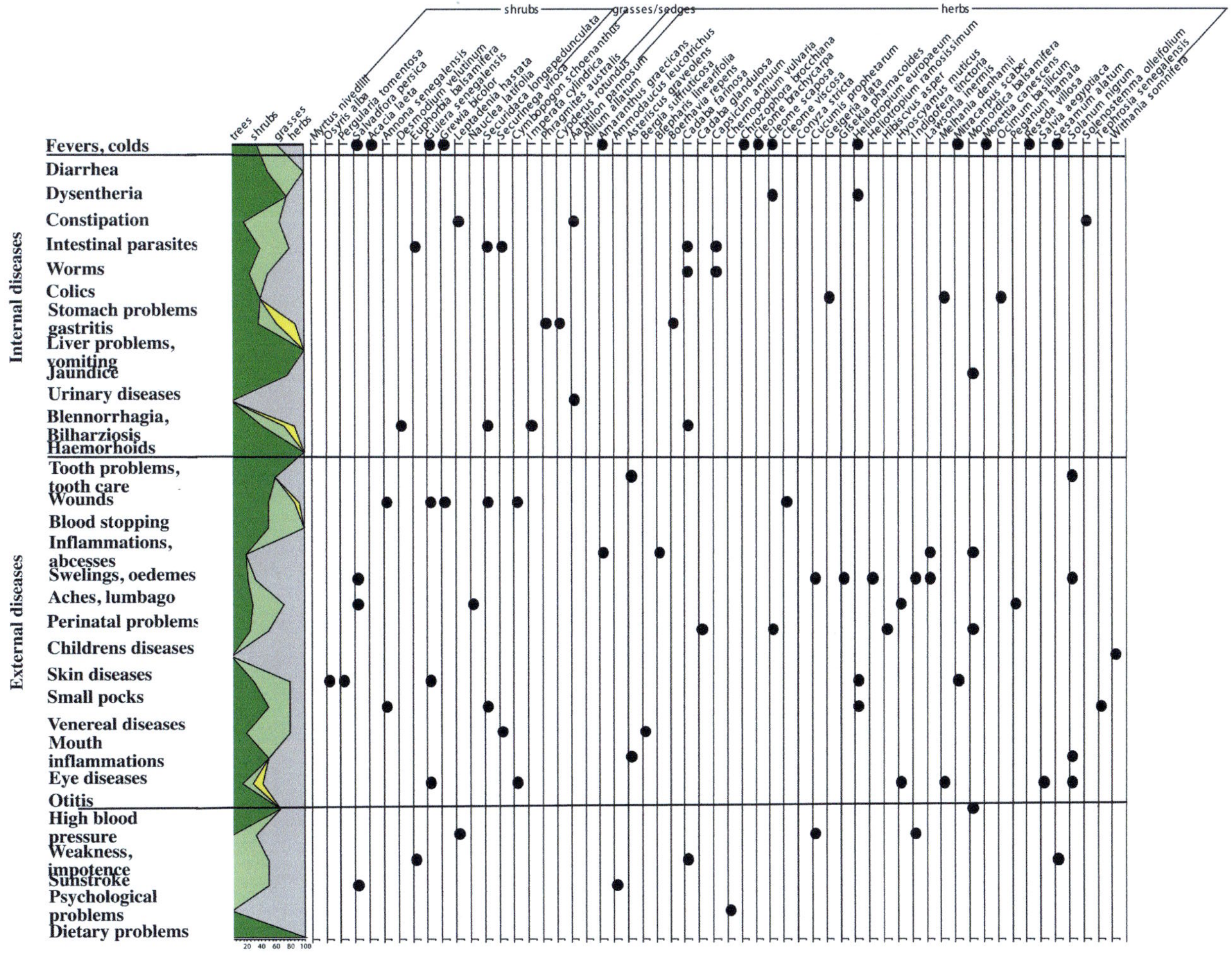

FIGURE 6.16B Saharan medical package b. Part of grasses and herbs. (Diagram: E. Schulz).

The Traditional Medicine in the Central Sahara as a Model for the Holocene

Our re-evaluation of the pollen profiles from the Kawar lakes and the landscape reconstruction for the Holocene also revealed the interest of traditional medicine for the whole region again. Figure 6.11 is based on own field surveys in northern Niger (Schulz and Adamou, 1992) and on respective literature (Adjanohoun *et al.*, 1980, 1981; Arbonnier, 2003; Bernus 1967,1969; Burkill, 1985, Dupire 1957, Gast 2000, Humann, 1933; Ikiri *et al.*, 1984; Maydell, 1992; Nicolaisen, 1963; Spittler, 1989). Our own investigations were conducted in the late 1980s in the Air Mts. and their forelands. Additional information and observations were also collected on other expeditions in the Sahara.

It is obvious that many plants come from the North of the Sahara or from the Sahel, as for example *Artemisia, Myrtus, Combretum micranthum,* and *Guiera senegalensis*; thus, medicine also relies on ancient networks of trade and exchange of the concerned regions. The diagram shows the importance of trees and bushes with bark and leaves as the most used parts. Most medicaments are combinations of several ingredients. Decoctions, infusions, or powders are the most common forms of preparation.

Colds, fevers, and intestinal problems are the most important diseases and infirmities; wounds, inflammations, perinatal problems, or skin problems are among the most relevant external diseases. Note however that the traditional medicine is a holistic one, integrating also an important part of magic and psychology (Hureiki, 2000, Ritter, 1994, Schulz and Adamou, 1997).

Finally, this modern model – food and medicine – will be rediscussed for the questions of Holocene food resources.

SECOND COURSE: LATE PLEISTOCENE AND HOLOCENE VEGETATION AND LANDSCAPES

Reconstructing Saharan vegetation history makes it necessary to consider a variety of approaches, including the analysis of sediments, the reconstruction of long-distance transport of pollen (Schulz, 1990, 1999), the research on insect pollination, as well as the consideration of evidence linked to macro-rests or charred material. In addition, there is knowledge about present vegetation dynamics. At the same time unknown vegetation must also be considered, as well as convergence with today's Sahelian and Sudanese savanna types.

For the Sahara, the northward shift of vegetation belts by several hundred kilometers was a common model of interpretation for a certain period or it was assumed that Mediterranean elements had penetrated the mountains of the central Sahara. In recent years, however, the idea of interweaving of different vegetation units or landscape types has gained acceptance. Saharo-Sahelian savannas were intercalated by Sahelo-Sudanian gallery forests or shrubs, resulting in various landscape mosaics (Dinnies *et al.*, 2021 Lezine *et al.*, 2011; Lezine and Lemmonier 2021; Maley, 1983; Mercuri *et al.*, 2013, 2015; Neumann 1994; Schultze, 1994; Schulz, 1994b; Watrin *et al.*, 2009).

Finally, there is the old insight of ecologists such as H. and E. Walter (1953), which states that organisms living at the edges of their distribution area can move to neighboring habitats if this is necessary to satisfy their basic ecological needs.

The sediment and pollen records from Bilma, as well as those from the Kawar lakes, are well suited both to address the questions raised above and to provide insights into the ecological conditions of humans during the respective periods.

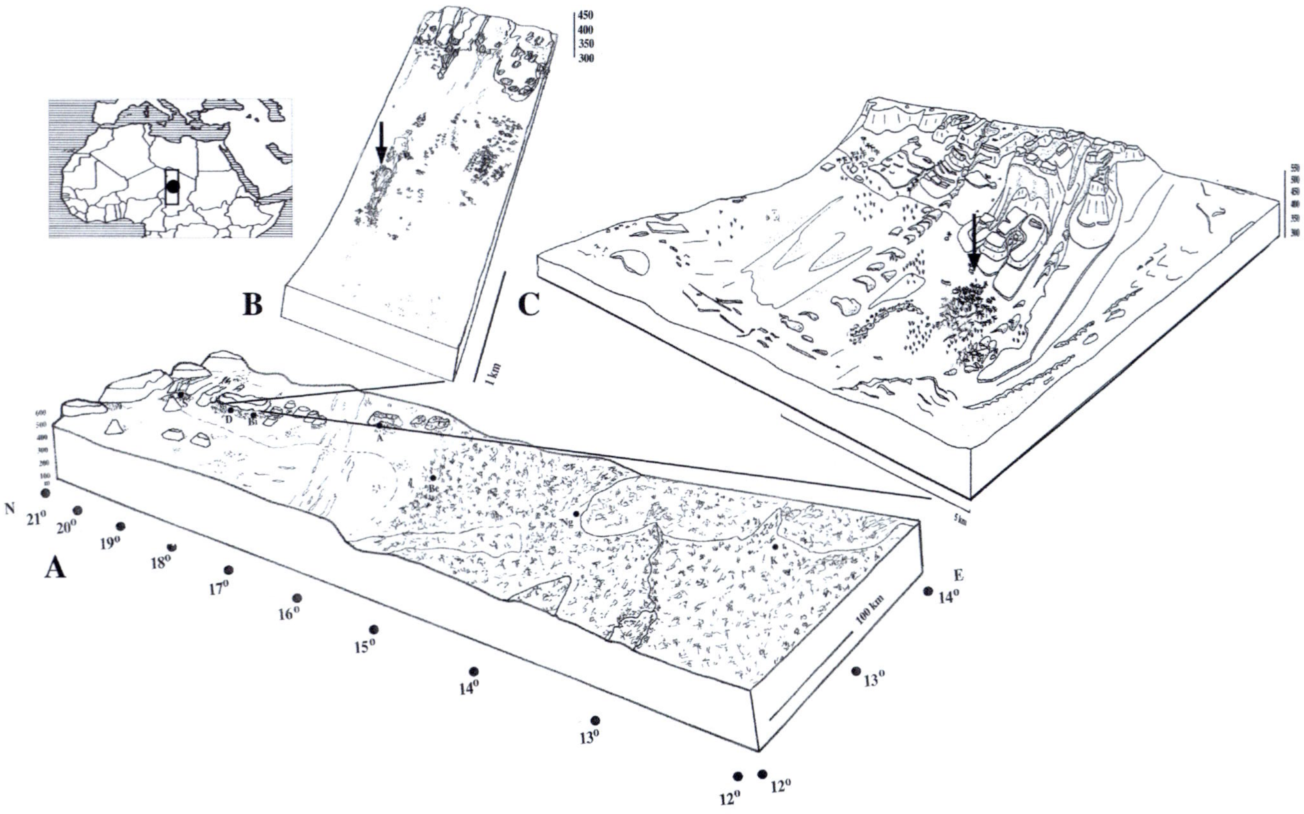

FIGURE 6.17 The present landscape situation of the region between Djado in the North and Lake Chad. The block diagram A shows the savannas in the north of Lake Chad with its sharp boundaries around 18° N and the oasis chain along the Bilma escarpment. B shows the oasis of Achenouma with Lake Ayama. The Bilma oasis is depicted in C. Note the step and dissected escarpment. The arrows indicate the coring points (Schulz *et al.*, 1990 modified).

The Physical Background of the Bilma and Kawar Lake Records

The Kawar is a chain of oases that runs for 160 km in a north-south direction along the Bilma escarpment in NE Niger (fig. 6.17). Geologically, it is part of a sub-basin of the Chad Basin. It is a three-part landscape consisting of a cliff, a foreland depression, and a sandy slope that links to the sandy plains of the Ténéré. The escarpment, reaching up to 450 m, is made of sand and mudstones and originates from the Cretaceous and early Tertiary. It is largely dissected and highly karstified (Sponholz, 1992). Tectonic movements are certainly responsible for the steepness of the escarpment as well as for the deep lake depressions in the foreland (Baumhauer, 1986; Joseph and Aranyossy, 1992; Pachur and Altmann, 2006). The escarpment dominates 150 m the foreland depression, which is characterised by a desert vegetation of scattered *Acacia-*, *Hyphaene-*, *Tamarix,-* and *Phoenix*-trees. Tussock grasses such as *Panicum turgidum*, *Imperata cylindrica*, *Stipagrostis pungens,* and others dominate the grass layer (Boudouresque and Schulz, 1981). To the west, a sand ramp of about 20 m connects the foreland depression with the sandy plains of the Ténéré. The foreland depression hosts a chain of oases which are famous for their date- and salt-production; their inhabitants participated in the trans-Saharan trade (Musch, 2018; Schulz *et al.*, 2020). The groundwater situation is complicated. There are both near-surface groundwater horizons, which feed the numerous shallow wells in the Kawar foreland depression, as well as fossil groundwater which emerges in lakes and spring mounds. This groundwater is mostly of Holocene age (Baumhauer, 1992, 1997; Joseph and Aranyossy, 1992). According to investigations into their isotopes, it is formed in the geological basin of Bilma itself and does not represent a remote inflow from the Tibesti Mountains.

The Sediment Records of Bilma and the Kawar Lakes

The Bilma F3 Sediment Record (18° N/12° 55'E)

The sequence was drilled in 1988 in the plain in front of the southernmost part of the Bilma escarpment (Figure 6.18). It was conceived to secure the fresh water supply for the Bilma oasis next to a spring mound (Joseph and Aranyossy, 1992). The record consists of overlying sands with some peat remains, lacustrine clays with intercalations of cemented sands, and lower sand layers. An artesian groundwater horizon was reached at 42 m depth. It broke out and could not be closed any more. The sequence was sampled at 1 m's intervals. The thickness of the lake sediments is explained as the result of a tectonic subsidence of the southernmost part of the foreland depression of the Bilma-cuesta (Joseph and Aranyossy, 1992). This would also point to an approximately 30 m deep lake in the Holocene, which was around 3 km wide and 100 km long and was probably connected to the more southerly lake of Zoo Baba even though there were several thresholds (see Figure 6.18).

a. Dating

The sequence is chronologically arranged by five dates (7.9 m depth 3740 +/-150, 9.8 m depth 6860 +/-120, 11,3 m 8590 +/- 120, 27 m 23680 +/- 3850, and 34 m 21490 +/- 2940). Dating was done by J.C Fontes in Gif sur Ivette. The lower two dates were on organic matter; the upper dates were done on carbonates. However, in the 1990s, radiocarbon dating was still limited to ages around 20 ka (Fontes and Gasse, 1991, Maley, 2004). Despite of its uncertainties, the sequence covers the upper part of the Late Pleistocene – regionally "Kanemian" and the African Humid Period (AHP) of the Holocene (Joseph and Aranyossy, 1922; Larrasoane *et al.*, 2013).

b. The Bilma-pollen Record

The sediment samples were prepared according to the classical procedure (Faegri and Iversen, 1968), destruction of carbonates with HCl, destruction of silicates with HF and subsequent acetolysis, and storing in glycerol. The samples were counted until the pollen spectra were constant. Determination was based on our own reference collection and the respective literature. All pollen types except ferns and

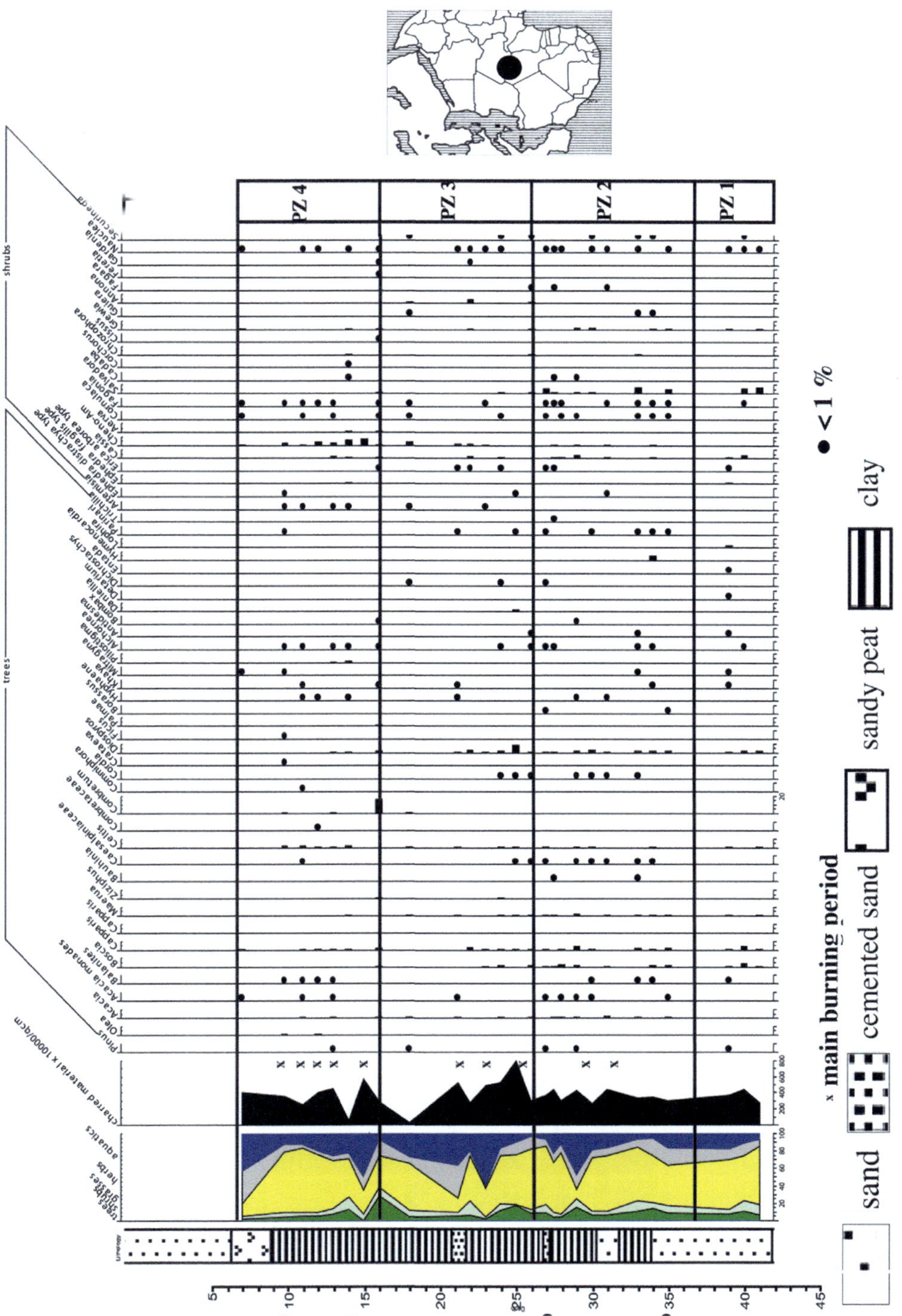

FIGURE 6.18A Pollen-diagram of the Bilma record. Part of trees and shrubs (Diagram: E. Schulz).

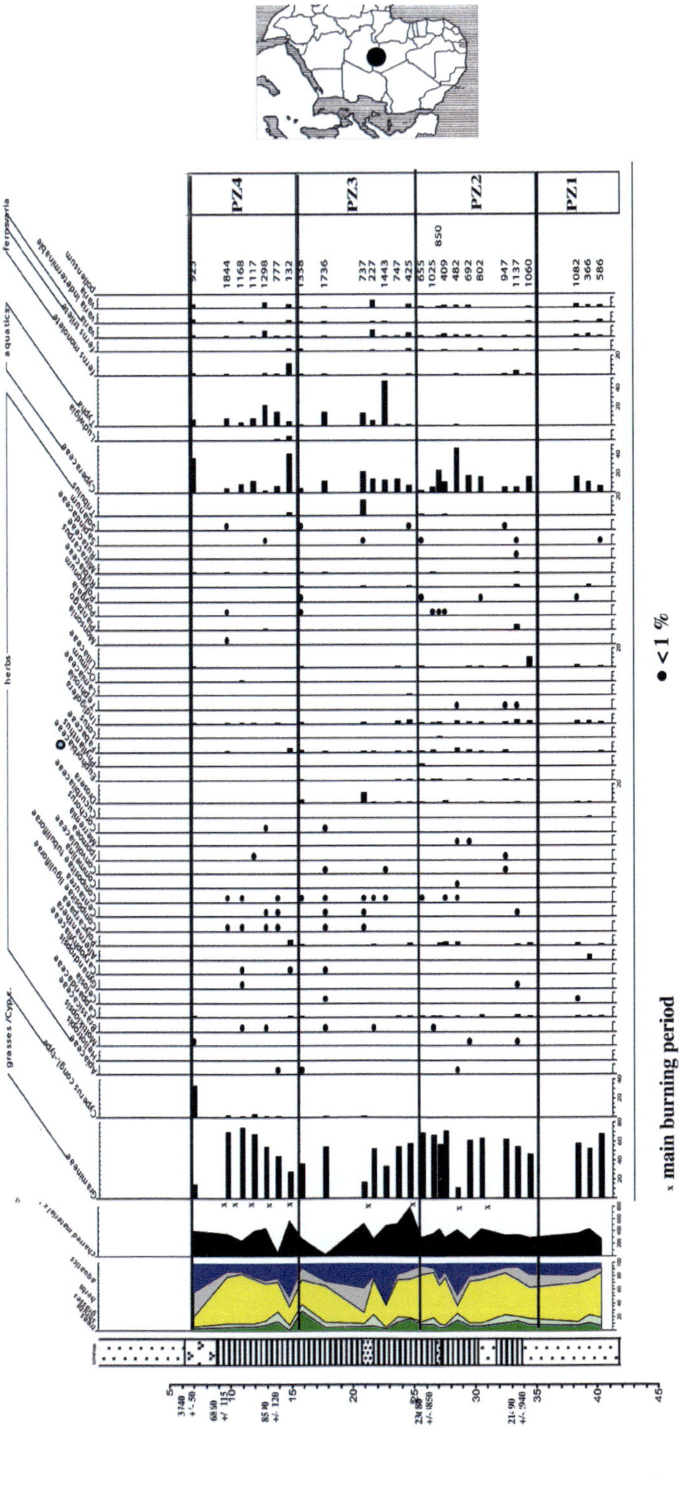

FIGURE 6.18B Pollen-diagram of the Bilma record. Part of grasses, herbs, and aquatics (Diagram: E. Schulz).

spores were included in the pollen sum. First doubts on the ability of the sediments, the type of coring, and the low resolution (about 1 m) could be dispelled by the logic and clear zonation of the diagram, especially by the four aquatic cycles.

The diagram is characterized by the predominance of the elements of an open vegetation. The proportion of trees and shrubs remains almost constantly below 10%. The main diagram shows about four cycles of aquatics after a basic phase of constant values. Pollen types, which reach values less than one percent, are only represented by their presence. Pollen zones are defined by the repartition of spectra and of some elements like *Drosera* which is a real indicator for Sudano-Guinean march microhabitats and microclimate (Burkill, 1985 ff). The following four pollen zones can be distinguished:

PZ1 41-36.5 m

This unit belongs to the dry period of the "Kanemien" which is regarded as the time of the great dune formations and regression of all previous lakes (see Servant, 1983; Maley, 2004). Among the trees and shrubs, two groups can be distinguished:

- The Saharan elements: *Acacia, Balanites, Maerua, Salvadora, and Fagonia*
- The Sahelian/Sudanian elements such as *Mitragyna, Alchornea, Antidesma, Lannea,* or Combretaceae

The presence of *Drosera*, a Sudano-Guinean marsh plant, is also striking; it takes no part in any long-distance transport and so, it shows clearly the presence of appropriate sites under a suitable microclimate (see above). Its presence in the diagram is marked by a dot.

PZ2. 36.5-24.5 m

Grasses and Cyperaceae dominate in this zone. Pollen spectra show the greatest variety of all pollen zones. Parallel to the Saharan types (*Acacia, Balanites, Capparis*, and Maerua), we find other species that are members of Sudanian savannahs, like *Khaya, Alchornaea, Antidesma,* and others. Caesalpiniaceae are constantly present, as well as *Drosera*. Grasses and Cyperaceae reach a first peak. It ends with *Boscia*, Caesalpiniaceae, and *Commiphora*. The Sudanian shrub *Nauclea* is present too.

PZ3 24.5-16.5 m

This unit belongs to the first half of the Holocene humid period. Burning reaches its first maximum and decreases after that. Diversity decreases in this section, particularly that of the Sudanian elements. However, *Nauclea* remains a continuous element. *Erica arborea*-types are part of the long-distance transport coming from North or Northeast (cf. Dinnies *et al.*, 2021). The end of the presence of *Drosera* simultaneously marks the end of the pollen zone. Grasses slowly decline and rise again towards the end. *Typha* joins in this unit and remains present until the end of the profile, indicating the outer reed belts. This shows that the lake surface may have already decreased in size until that moment.

PZ4 16.5-7.0 m

This unit covers the second half of the Holocene humid period. Burning rises again with fresh fires. Diversity of pollen types rises again too. Grasses rise to a maximum. Saharan elements like *Acacia* and *Balanites* are regularly present, as well as the Sahelo-Sudanian elements, like *Celtis, Nauclea,* and *Alchornea. Combretaceae* dominate in this section. Ruderals and sand dwellers (*ChenoAm, Cornulaca, and Fagonia*, but also *Cyperus conglomeratus, Artemisia, Ephedra, Pinus,* and *Olea*) mark the long-distance transport from the North.

- The upper part of the record (6.5-0 m) was very poor in pollen or completely pollen-sterile.
- The aquatic milieus as indicated by diatoms (Amossé, 1925; Baumhauer, 1990, 2004).
- The part of the sequence below 20 m allowed only some findings of *Navicula* sp. or *Campylodiscus clypeus*.
- The Early Holocene layers included periphytic-benthic diatoms like *Campylodiscus clypeus* or planktonic species like *Aulacoseira granulata* or Cy*clotella stelligera*. To an oligotrophic, Si-rich milieu points the dominance of *Campylodiscus clypeus*, *Cyclotella* spp., *Epithemia* spp., *Fragilaria* spp., *Aulacoseira granulata,* or *Rhopalodia gibba*.

In the Late Holocene, *Anonoeoneis sphaerophora* points to alkaline Na-carbonate-bicarbonate-rich waters, but also to Na-chlorites. Thus, the diatoms indicate different oligotrophic freshwater milieus in a heterogenic-meromictic waterbody (Amossé, 1925; Baumhauer, 1990) (figs. 6.20 B, 6.21 A).

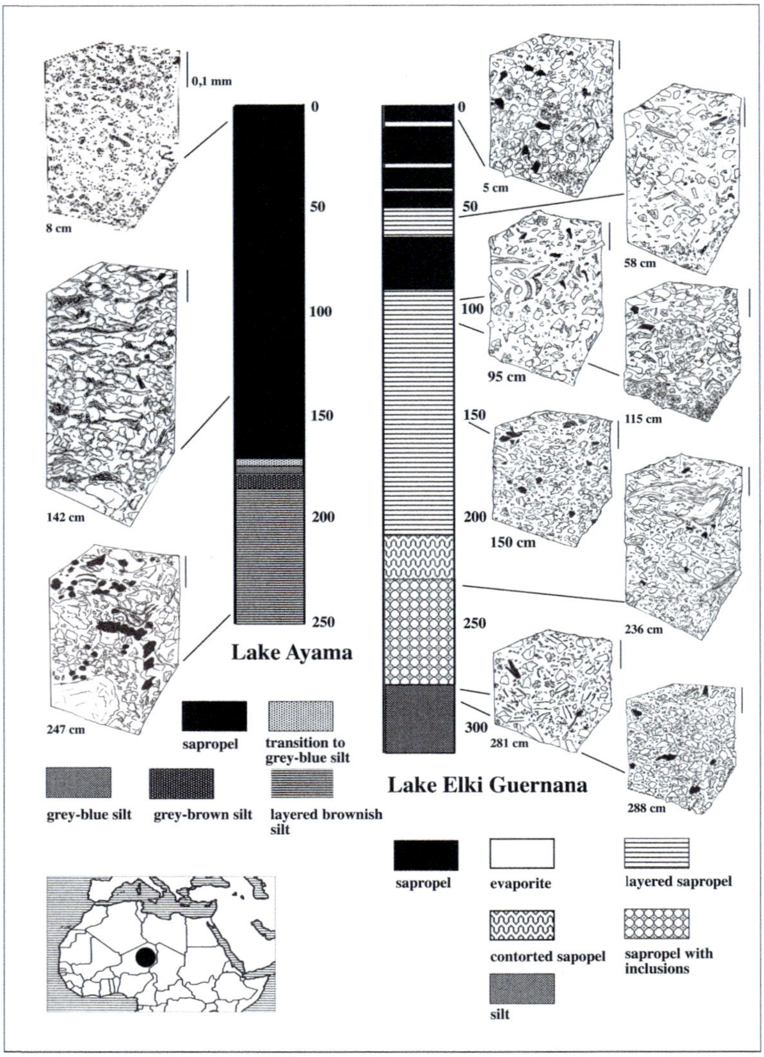

FIGURE 6.19 Sediment structure of the two Kawar Lake records.

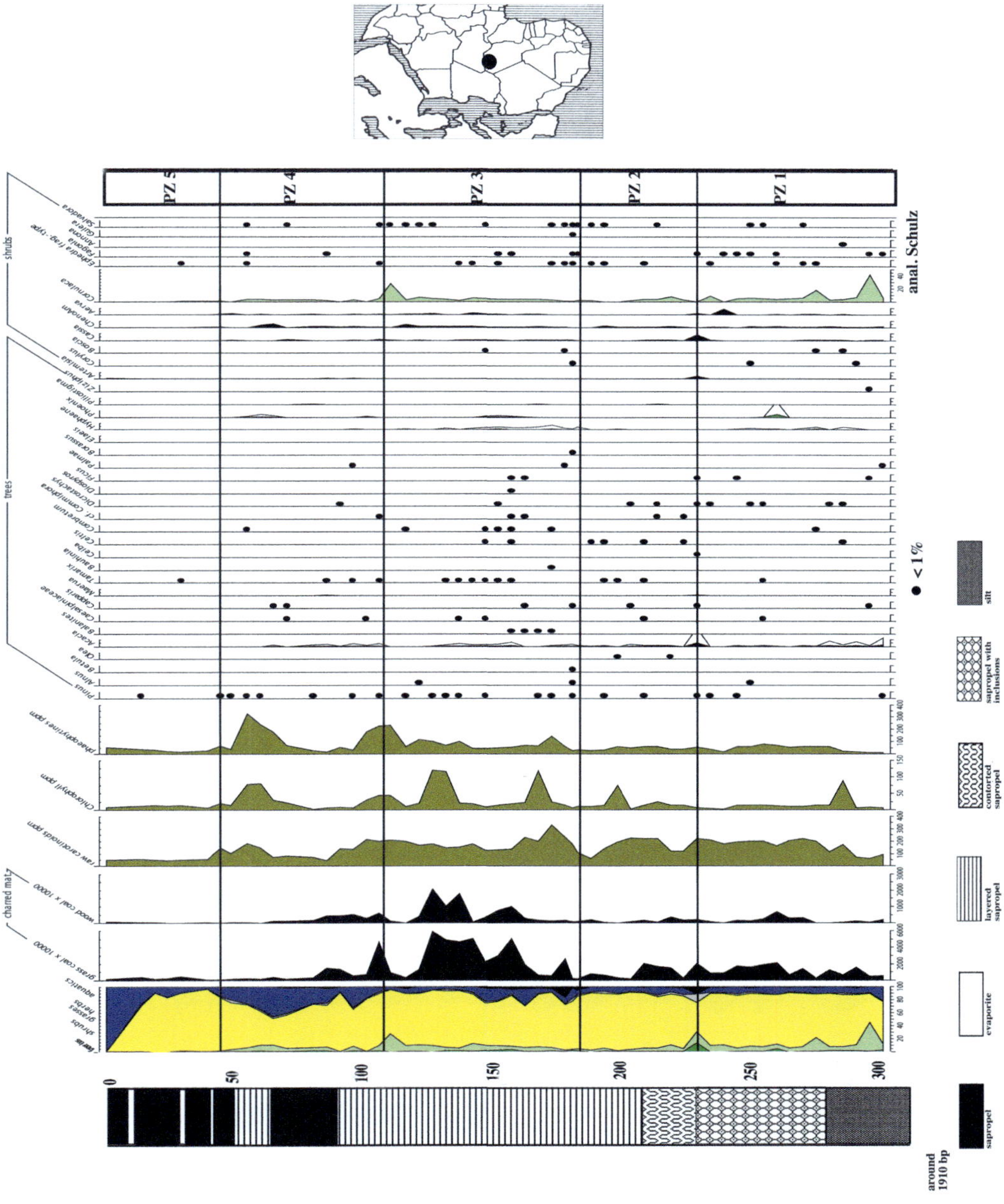

FIGURE 6.20A Pollendiagram of the Lake Arrigui record. Part of charred material, pigments, trees, and shrubs. (Diagram: E. Schulz)

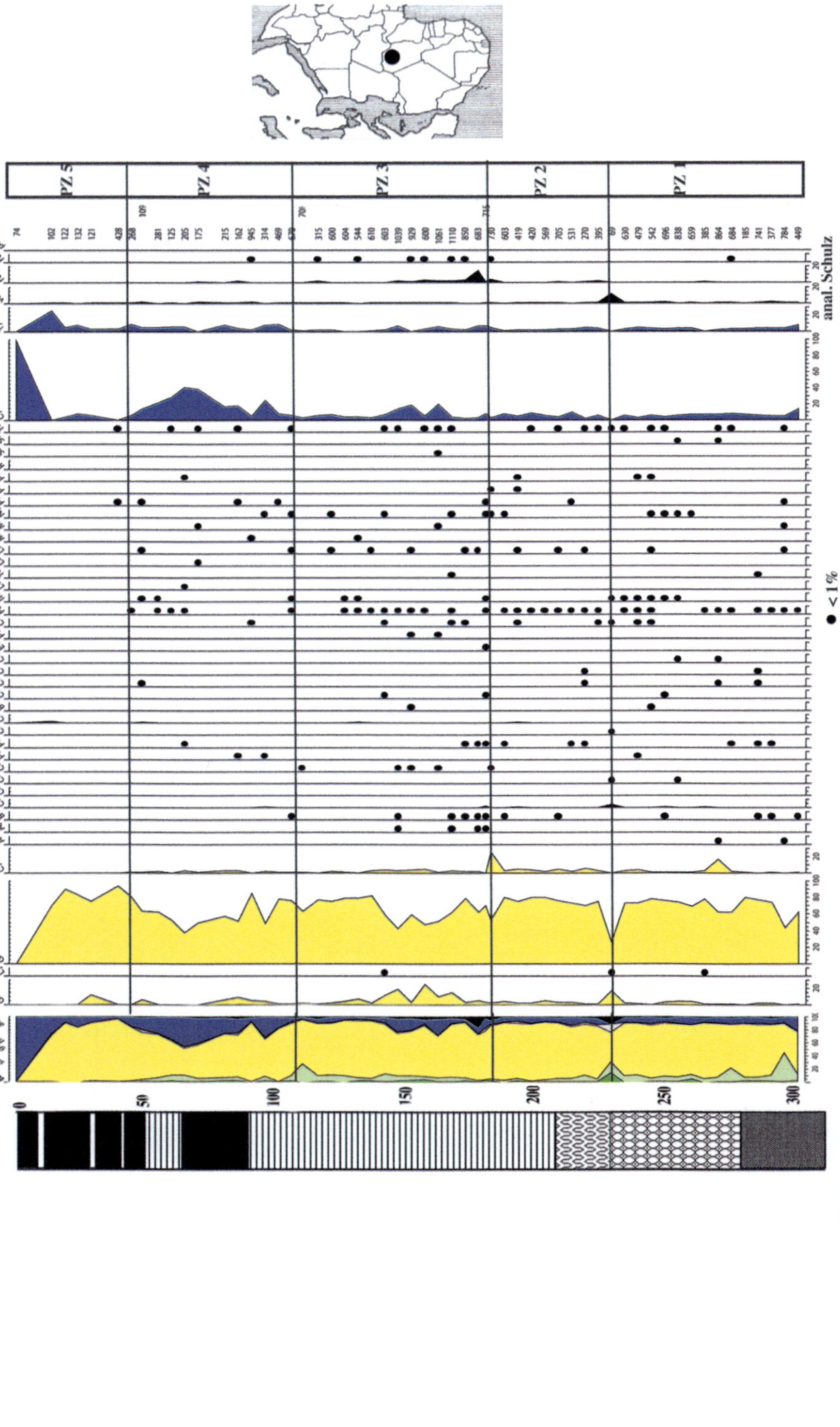

FIGURE 6.20B Pollen diagram of the Lake Arrigui record. Part of grasses, herbs, and aquatics. Sedimentology from Schultze (1994)

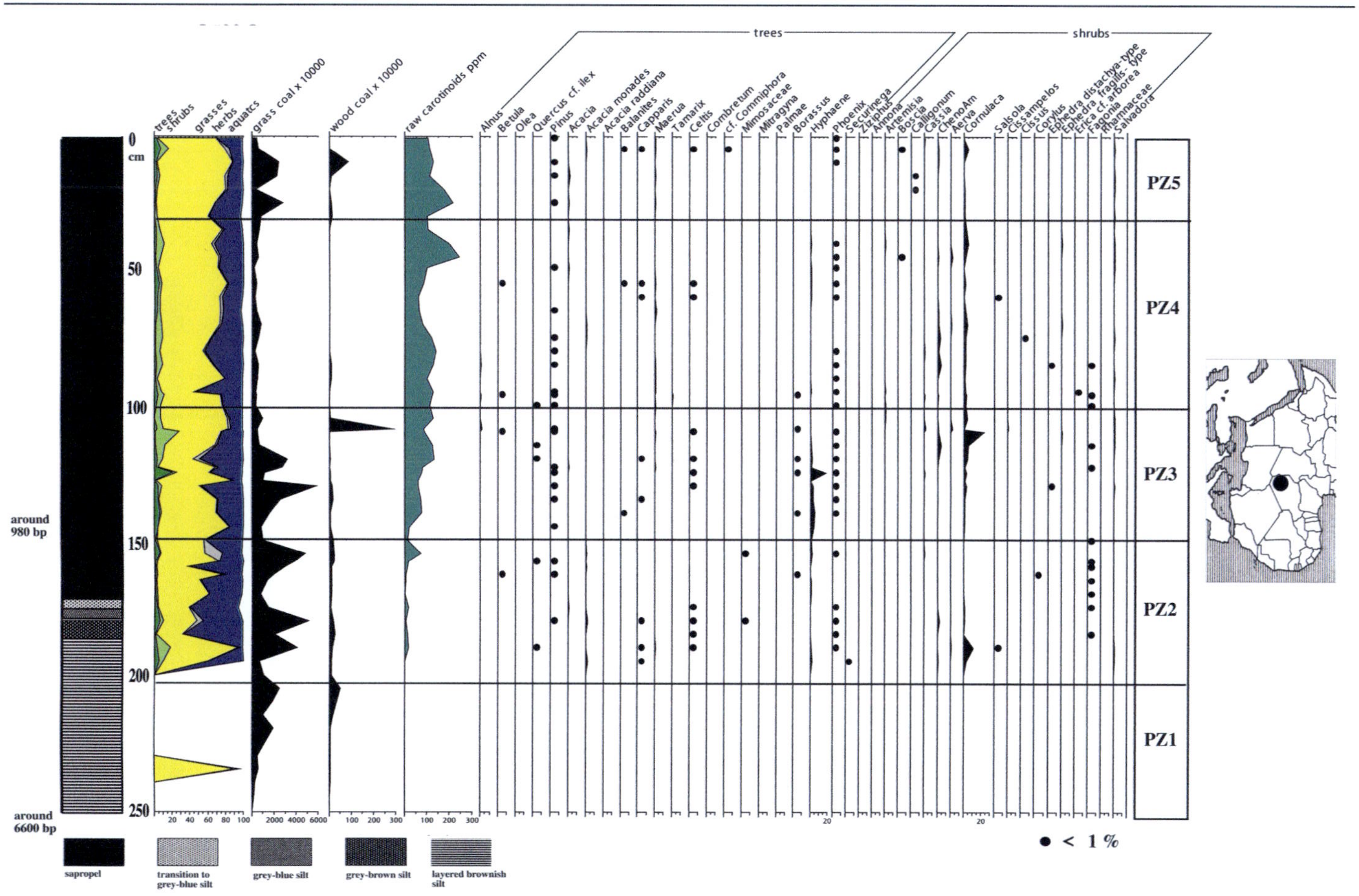

FIGURE 6.21A Pollen diagram of the Lake Ayama. Part of charred material, pigments, trees, and shrubs. Sedimentology from Schultze (1994).

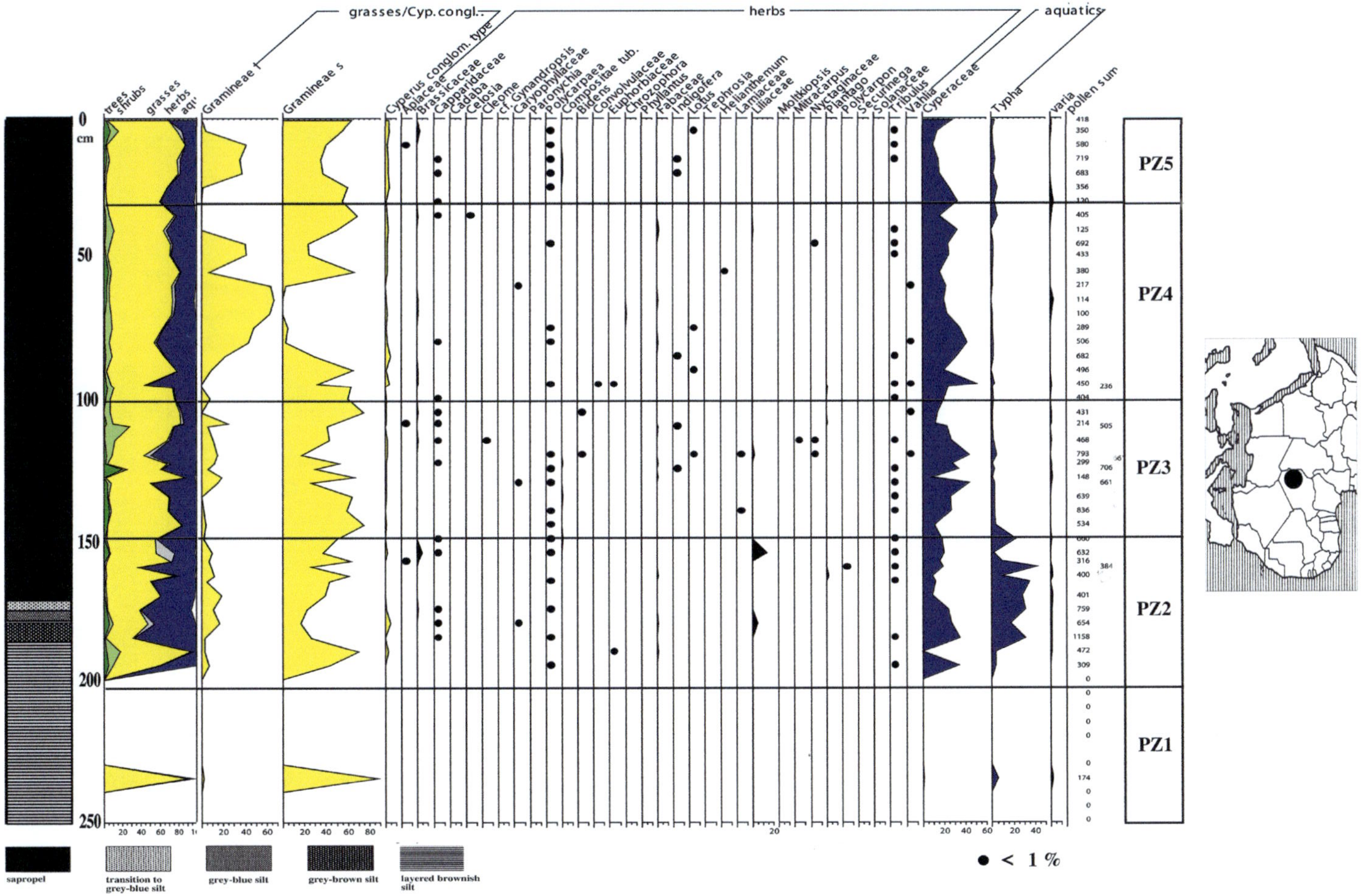

FIGURE 6.21B Pollen diagram of the Lake Ayama record. Part of grasses, herbs, and aquatics. Sedimentology from Schultze (1994).

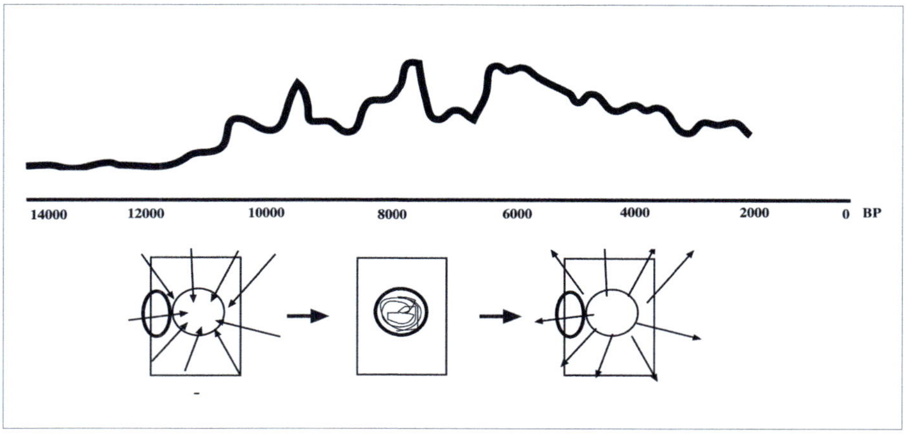

FIGURE 6.22 Estimation and measurement. A simplistic model for the Holocene Central Saharan population. An estimation of the population dynamics and an interpretation of the genetic history of the Holocene population (after Brierley *et al.*, 2018, D'Attonosio *et al.*, 2018; Xue, 2018).

The Kawar Lakes

In 1989, cores were taken in two hypersaline lakes in front of the Bilma cuesta in the two oases of Arrigui and Achenouma (cf. Figure 6.18 B) with help of a modified Cullenberg corer mounted on a raft. The shallow lakes are surrounded by a belt of *Phragmites* and *Typha* and by some Hyphaene stands. The oases in the foreland depression are characterized by *Acacia, Hyphaene,* and *Phoenix* trees. Scattered tussock grasses such as *Panicum turgidum*, *Imperata cylindrica*, *Stipagrostis pungens,* and others dominate the grass layer (Boudouresque and Schulz, 1981). The sediment records were analyzed for their pollen and diatom content as well as for pigments derived from microorganisms. Those are well suited to indicate the trophic state of the lake where carotenoids are fixed to the fine-grained sediments, whereas chlorophylls or pheophytins are part of the littoral milieu (Schultze, 1994).

In contrast to the Bilma sequence, it was possible to consolidate the cores and to obtain thin sections (Pedological Laboratory at Gent University). However, due to the different salt contents, not all parts of the records could be consolidated properly and transformed into thin sections. Figure 6.20 shows the sediment structure from samples, which were chosen according to the pigment content of the respective records.

The Lake Ayama sequence is divided into layered and unlayered structures. The basic sample consists of silt and fine mostly sharp-angled sand. However, layering is not visible. Some charred material is present. Pyrite in "raspberry" agglomerates points to anoxic conditions of the water body.

The middle example shows clear layers. They consist of densely packed silt with charred material (grass coals) and some carbonate nests. Algae layers divide the silt units. The upper sample shows a repetition of densely packed silts and loose layers, both with carbonate needles and grass coal flitters.

These fine layers in the sapropel part of the sequence are event-based. In blown, silt and sand are fixed. The algal layers in the middle sample indicate eutrophic conditions.

The sediments of the Lake Elki Guernama sequence are less prominently structured. They consist of dense, loose silt packages with charred material and carbonate needles. Diatom shells are frequent in the upper part of the record.

Differences between the sediments of the two lakes might be caused by a greater gap in the escarpment, which provoked a stronger input of sand and silt. In all, it points to a stable stratification of the two lakes with the development of a hypolimnion. Probably the depth of the lake was less than 10 m due to the high salt content.

a. Diatoms

The diatoms indicate a presence of hypersaline waters by *Camplylodiscus clypeus* and *Rhopalodia gib berula*, as well as a presence of slightly alkaline-meso-euhaline waters to which points the presence o *Pinnularia borealis*, *Navicula* sp., *Cyclotella meneghiniana,* or *Aulacoseira granulata.*

b. The Time Frame

Even if it is difficult to transfer dating from other profiles, we might use those obtained from forme hand-auger coring at the margins of both lakes concerned (Baumhauer, 1986). So, for the Arrigui Lake it showed an age of 1910 +/-500 bp at 3.47 m depth, and for the Ayama sequence 980 +/-110 bp at 1.43 m depth, as well as 6600 +/-535 bp for 2.54 m depths. Thus, the sapropel layers may have developed during the last 2000 years, whereas the underlying silt may be of Middle Holocene age. However, even in these layers there might be some distortions and lacunae-hiatuses.

c. The Pollen Records: Lake Arrigui (19°04'N /12°53'E)

The sediment record consists of silt and a structured (laminated, contorted) or unstructured sapropel. The water was shallow with a chemocline at a depth of about 24 cm. The basic water layers were quite anoxic

The pollen diagram also depicts the content of pigments and charred material. In general, the dia gram is characterized by the elements of an open vegetation of grasses and some trees (mainly *Acacia Tamarix*, Hyphaene, *and Phoenix*) or shrubs like *Cornulaca* or *Cassia.* Most of the pollen types are only present with values under 1%.

PZ1 and PZ2

The lower zones show regular grass and wood burnings, a rising carotenoid content as well as a content of chlorophylls with some single peaks, and pheophytins in lower values. The pollen spectra are character ized by grasses and trees or shrubs like *Acacia*, *Hyphaene*, *Phoenix,* and *Cornulaca*. *Phoenix* has a first peak in PZ1 but is not present in PZ2. The pollen spectra show open sand vegetation also with ruderals like *Chrozophora* or *Tribulus.* Cyperaceae and *Typha* are only present in lower values. There also is a regular long-distance input coming from the north and including *Pinus*, *Alnus*, *Olea,* and *Ephedra.*

PZ3

This zone shows more diversified pollen spectra mostly through elements of long-distance transport from the North, like *Pinus*, *Alnus*, *Olea,* and *Lygeum*. *Hyphaene* and *Phoenix* may have been present near the lake shores, where trees such as *Celtis and Dichrostachys* may originate from places near the escarpment *Celtis* is, in fact, reported from the Mid-Holocene excavations of the Djado (Vernet and Striedter, 1992) The local vegetation remained unchanged; however, values of Cyperaceae were rising. During this time there were more intense burnings, including those of wood. This may have provoked the release of nutri ent elements with successive rising of chlorophylls and pheophytins.

PZ4

This zone shows less diversified pollen spectra. *Cyperaceae* and *Typha* are rising, pointing probably to a shrinking lake surface. *Acacia*, *Tamarix*, *Balanites,* and *Phoenix* are represented together with the *ChenoAm* and *Cornulaca*. The long-distance input from the North is still present.

PZ5

In this zone, pollen spectra are very poor. Grasses and aquatics are dominant. Probably this was the result of the strong sediment input as it is depicted by the loose sand layers in the sediment structure (see Figure 6.20).

d. The Lake Ayama Record (19°07'N / 12°53'E)
Lake Ayama is a smaller lake some kilometers to the north of Lake Arrigui in a similar landscape. Investigations were the same as for Lake Arrigui, but only carotenoids were analyzed for the pigments. The sediment record consists of unstructured sapropel and silt.

PZ1

The lower part of the profile is almost pollen sterile except for one spectrum with grasses and *Typha*. Coals point to burning.

PZ 2

This zone shows high values of grass coals and some woods. Carotenoids are slowly rising. High values of Cyperaceae and *Typha* point to nearby reed belts. Pollen spectra are poor. There is a long-distance input from the north and from the south. *Acacia*, *Capparis*, *Hyphaene*, and *Phoenix* dominate among trees. *Cornulaca* with grasses depict an open dune vegetation.

PZ3

This is the second zone of intense burning, with a decline in its younger part. Palm trees with *Capparis* characterize the trees and *Cornulaca* has a second peak. Cyperaceae represent almost alone the reed belts. Long-distance input from the north and south are present again.

PZ4

This is a zone with intensive burning but with regular high values of carotenoids. Cyperaceae remain high as in PZ3. *Maerua*, *Capparis*, and *Cornulaca* represent the local vegetation. *Phoenix* is constantly present.

PZ5

That is again a zone of intensive burning of both grasses and wood. Cyperaceae diminish slightly. *Balanites*, *Capparis*, *Phoenix*, and *Cornulaca* represent the local open vegetation.

Correlation of the Kawar Lakes Profiles

Both lake records show very similar traits and may be correlated by the two Cornulaca peaks in PZ2/PZ3 (Ayama) to PZ1 to PZ3 (Arrigui).

Both records show a regular presence of *Phoenix*, confirming thus their young age. In this context, Kaczmarek *et al.* (2023) reported the beginning of the *Phoenix*-oasis economy in southern Fezzan beginning in 3,000 bp. They thus represent the younger part of the Kawar history which is missing in the Bilma record.

Comparison with other records

Four pollen sequences from different sediment types are present in the greater region: Segedim at the northern extension of the Bilma escarpment, the Gobero lake beds Southwest of the Air Mts., the lake beds of the Emi-Koussi crater in the Tibesti Mts., and the laminated Yoa lake beds Southeast of the Tibesti Mts.

- Segedim, Niger. 20°11′N / 150°58′E. Lake beds in front of a cuesta. About 10,410 – 6,250 bp. (Baumhauer *et al.* 2009; Schulz, 1994a). A general predominance of Gramineae, Cheno Am with members of the *Acacia – Panicum* savanna- perhaps with some *Celtis.* A strong long-distance influence from the North and South.
- Gobero, Niger. 17°65′N/9°30′E. Dune depressions, lake beds, and burials at the end of a wadi coming from the Air Mts. About 10000-3800 bp. (Sereno *et al.*, 2008). Predominance of Gramineae and ChenoAm, interpreted as a low diversity savanna with *Ficus, Tamarix,* and *Salvadora.* Gramineae taken as indicators of a Sahelian savanna and ChenoAm as indicators for the Saharan desert.
- Emi Koussi, Tibesti Mts, Chad. 19°47′N / 18°37′E, 3415 m asl. Lake beds in the crater of the Emi Koussi volcano. About 9500-6500 bp (Dinnies *et al.*, 2021) Predominance of Gramineae, *Artemisia,* and Compositae. Importance of the long-distance transport from the North and South. The vegetation is interpreted as a high-altitude *Artemisia* "steppe", perhaps with some *Erica arborea* from the vicinity.
- Lake Yoa, Chad. 18° 03′N/ 20° 31′E. Laminated lake beds Southeast of the Tibesti Mts. About 6000-0 bp. A rich pollen sequence: 6000-4750 bp. An open savanna with *Piliostigma, Celtis, Grewia;* 4700-2700 bp; an open *Acacia-Balanites* savanna; 2700-0 bp. An open Gramineae-ChenoAm desert vegetation with probably some *Artemisia.* From 700 bp on, an increase of *Phoenix* and *Hyphaene* with a rising long-distance influence from the South (*Guiera*).

In all, the Holocene parts of the Bilma and Kawar Lake sequences are comparable to these pollen records from the wider region. Especially, the sequence *Guiera-Phoeni* (of the Lake Yoa record) is also visible in the Kawar Lake records. Thus, a dual vegetation – a Saharan *Acacia-Panicum* savanna s.l. interfingered by Sudano-Sahelian alluvial formations along watercourses and around lakes seems to be the appropriate model for the Holocene – and a background for the reconstruction of landscape types and estimations of pasture and food resources.

THIRD COURSE: MAN AND FOOD IN LATE PLEISTOCENE AND HOLOCENE

There is huge evidence of the presence of man and his different cultures during the Late Pleistocene and Holocene (see f.ex. Banks, 1989; Busche and Heistermann, 1992, Caneva, 1993; Castaneda *et al.*, 2009; Gabriel, 1986; Kröpelin, 2009; Larrasoane *et al.*, 2013; Mercuri *et al.*, 2015; Muzzolini, 1991; Neumann, 1994; Nicoll, 2004; Sereno *et al.*, 2008). We also know about the genetic evolution during the Holocene, as there was a multiple in-migration from all directions, the formation of a common genetic pool, and at the end of the African Humid Period (AHP), again an outmigration in all directions (D'Attonosio *et al.*, 2018; Xue, 2018). These investigations are based on male Y-chromosomes of North African/West African people and their slow mutation rates. They could be traced back to the times of AHP and even before it. The African Humid Period (AHP) was postulated for West and Central Africa with a relatively rapid onset at 14,500 bp and an end at about 5,500 bp. Out of an orbital shift and higher temperatures, it forced an acceleration of the monsoonal regime (De Menocal *et al.*, 2000) and vice versa at the end. Especially the conditions of the end are controversially discussed (see below).

The general estimation of the human population in the Late Pleistocene and Holocene is largely based on the number of archaeological findings and the number of radiocarbon dates at different times; thus, a "summited probability" (see Brierley *et al.*, 2018; Manning and Thimpson, 2014). Moreover, there are calculations on the growth rate of the former population in a scale of 1,1% in 25 years in the period between 14,000 and 11000 bp and from 10,900 to 10,000 bp a growth rate of 7.7% per generation (Manning and Thimpson, 2014) The authors calculate a maximum growth of 63% for the period between 6,700 and 6300 bp. However, we do not have "precise" information about the real population dimension, which would better allow us to evaluate the relation of consumers to the dimension of available food. The genetic investigations (D'Attonosio *et al.*, 2018; Xue, 2018) point to an immigration from various directions into the Sahara during the African Humid Period (AHP) – largely between 14,500 bp and 5,500 bp (see De Menocal *et al.*, 2000), Lezine and Lemonnier (2021), which came, mixed, and left after its end. The authors also detected signs of a population older than the AHP. Thus, we must count on an ancient population in the Central Sahara, which may have survived during the Late Pleistocene or at least in small groups before new people came in several groups and probably conquered a large area down to the Bandiagara chain in Gourma in present NE-Mali (see Huysecom, 2012; Huysecom *et al.*, 2004).

In all, there are two main questions for this period: how many people were present in this area and how many foods they could find for their survival? The first is extremely difficult to imagine – see above as well as it is for the quantity of food resources. Following Jousse (2006), we better should think of the minimum conditions (small game, small fishing, and small gathering) than on the big game of the Ethiopian fauna.

What was the environment they lived in, what was their livelihood? From the Bilma and Kawar records, we can reconstruct the environmental history.

The Landscape Context in Late Pleistocene and Holocene: The Background for Food and Pasture

To understand the background for human activities and food resources, the landscapes of the southern Sahara will be presented graphically for three time slices: the Late Pleistocene/Kanemien around 16,000 bp, the Middle Holocene around 6000 bp, and the period around 2000 bp. The period around 6,000 bp will be that of mapping the food resources.

The Kanemien

For the Bilma sequence, we can state some discrepancies between a dating which announces the Kanemien dry period on the one hand, and the indicators of vegetation including Sudanian elements on the other. The Bilma escarpment was a privileged site because the karst features collected all precipitation coming from North and South; simultaneously, there was a high standing aquifer (Sponholz, 1992). Moreover, in the second part of the Kanemien period, there were lake formations in the Saharan mountains due to cyclonic rainfall, which caused fluvial deposits around the Murzuk basin and in the Air Mts. too (cf. Busche *et al.*, 1979; Maley, 2004). Thus, the special conditions of the Bilma escarpment could have allowed suitable conditions also during the Kanemien period (Figure 6.24). An open grass vegetation supported gathering; in the rocky cuesta there were some tree- and bush-areas, probably belonging to the *Maerua*-type, which were typical for the first steps of savanna formation (Schulz *et al.*, 2009), as also perhaps some *Ficus* trees.

A wide area of open savannas was often dominated by Achabs. South of the Kawar escarpment, in some dune valleys and close to the former lake basin, the vegetation was denser along runnels and depressions. During this period, the outflowing groundwater in front of the cuesta provided sufficient moisture to feed even demanding vegetation elements, while highly arid conditions with areal sand movement prevailed in the area (Maley, 1983). Tributaries of the Lake Chad depression had denser vegetation units including also Sahelian-Sudanian elements. The Lake Chad basin was characterized by dune formation. There was probably no "Lake Chad" itself.

There is a consensus that the people of the Late Pleistocene and Early Holocene were hunters, gatherers, and fishermen (Mercuri *et al.*, 2018). During the "Kanemien" (Late Pleistocene), they lived in a harsh environment, Figure 6.23 for the Kawar region/NE-Niger. Escarpments and rocky areas could collect the scarce rainfall and provided wells and water points. The food spectra might have been less than or comparable to the present day. We must think of small groups, perhaps only ten persons each - necessary for an internal cooperation. This follows the present San people's livelihood in southern Africa (see Lee 1970). It was necessary for them to meet with other groups sometimes during the year for social relations, exchange of experiences, ceremonies, and certainly to avoid inbreeding. However, they were always on the brink of extinction due to diseases, hunger, or hunting accidents.

This situation prevailed up to the beginning of the Holocene.

There is some dissent on the origin of these peoples as simple successors of the epipaleolithics or of a different type (see Hachid, 1998). Out of these problems, the group working in the Acacus Mts / SW-Libya developed a practicable time scheme that relied on the economic activities of the respective people: Early Acacus 11,600–10,200 bp, Late Acacus 10,200–8,200 bp, Early Pastoral 8,200–7,300 bp, Middle Pastoral 7,300–5,900 bp, and Late Pastoral 5,900–4,300 bp (Di Lernia *et al.*, 2012; Fornaciari *et al.*, 2018). Certainly, they did exploit all available resources, and they were also inventive, such as for ceramics, as shown from the Air-Mts (Roset, 2000).

This provided chances of cooking, storing, and transport too. There was a long discussion on an early cultivation of grasses which was regularly proposed for the Mid-Holocene of the Tassili-n-Ajer/D Acacus. It was based on large quantities of wild gathering grasses (*Echinochloa*, *Urochloa*, *Panicum*, and *Cenchrus*). Findings of Earth stores with large numbers of wind-sieved grains were interpreted not only as a stock for future or bad times but also as a surplus of seeds for cultivation. In that way, a regular collection was proposed as "cultivation without domestication" (Dupuy, 2014; Garcea, 2004; Mercuri *et al.*

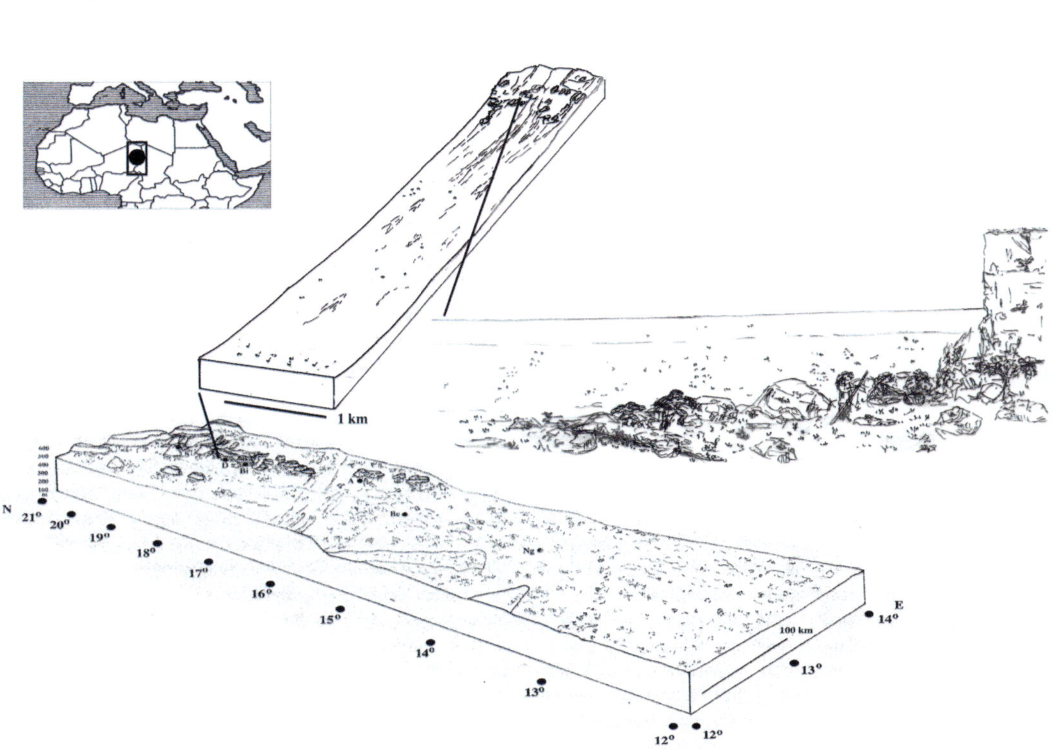

FIGURE 6.23 Block diagrams of a landscape transect from the Djado plateau to the Lake Chad region and of the Achenouma/Kawar region at the time of the Kanemien. It also shows a campsite at the escarpment near Achenouma with some tree and bush vegetation and an outflow of source at the return of some hunters

2007, 2018). Temple and Stojakowsky (2019) presented another interpretation. They describe two different strategies: the one as a rapid consumption of gathered food – from hand into mouth – and the other as foreseeing for the future in preparing for bad times by storing food. That could be a solution to the discussion of early cultivation. However, what were the basic conditions of their lives?

Food in Late Pleistocene and Holocene of the Central and Southern Sahara

An estimation of the food resources for the southern Sahara during the Late Pleistocene and Holocene relies on the former vegetation – within the *Acacia-Panicum* realm – and on its comparison with the present one. In addition, the hydrological situation will have determined the amount of vegetation and possible food. At around 6000 bp there is a maximum of information about the landscape and a fair chance to estimate the food resources. There were consistent and stable savannas with noteworthy soil formation up to chromic arenosols (Grunert, 1988; Mauz and Felix-Henningsen, 2005; Völkl, 1988). Moreover, at that period new people arrived with an additional economy. For the vegetation and landscape types, we can rely on two information sets.

For the first, there is a direct proof for the presence of Sahelian and Sudanian savannas: Charcoals from Fachi / NE-Niger / 19°N, about 7600 bp (cf. Neumann, 1988) pointing to *Terminalia* cf. *macroptera*, *Annona senegalensis*, cf. *Euphorbia/Anacardiaceae*, *Crataeva adansonii*, *Celtis integrifolia*, *Ximenia americana*, *Ficus* sp., *Boscia salicifolia*, *Balanites aegyptiaca*, *Cadaba farinosa*, *Ziziphus* sp., and *Faidherbia albida*. Neumann interprets this flora as the remnants of a hunting fire near the lake shore in front of the Fachi escarpment. An excavation in the Djado-Plateau /NE-Niger at 23°N yielded some fruits of *Celtis integrifolia* of Middle Holocene age (Vernet and Striedter, 1992).

For the second, the Bilma pollen diagram (see above) points to the dual landscape system of a predominant Saharan savanna with the intercalation of Sudano-Sahelian alluvial forest galleries.

In terms of landscape, we must therefore reckon with two units altogether. A Saharan savanna of the "*Acacia-Panicum*" type in the broadest sense and on the lake shores, in valleys and rock fissures with a gallery vegetation of Sahelo-Sudanian trees and shrubs such as *Mitragyna*, *Alchornea*, *Antidesma*, *Borassus*, *Crataeva*, or Combretaceae. This was a period of transition where the lakes closer to the North had already dried out or changed into a sebkha environment such at Segedim oasis at 23°N (see Baumhauer *et al.*, 2009). Anyhow, South of 21°N there were still lakes of different dimension in front of the cuesta.

Figure 6.24 explains this dual landscape between Lake Chad at its high level (see Thiemeyer, 2000) and the savannas up to the Djado Plateau at the time of 6000 bp. This was a period of transition where the lakes more to the North were already dried out or changed into a sebkha environment such at Segedim oasis at 23° N (see Baumhauer *et al.*, 2009). Anyhow, from 21°N there were still lakes of different dimension in front of the cuestas (Busche *et al.*, 1979).

Out of former geomorphology, substrates and hydrology and the repartition of vegetation as well as our knowledge of the present resources in the Acacia-Panicum realm, we may conceive a map of basic food resources for the Holocene – around 6,000 bp- (Figure 6.25) in parallel to the landscape formation (Figure 6.24). Thus, it is possible to recognize three categories of collecting grasses and to quantify them (see the First Course).

A first group consists of the small annual grasses of *Aristida adscensionis, Brachiaria laeta* or *Stipagrostis plumosa*. They may have given yields up to 40 kcal/ha.

A second group encompassed annual and permanent grasses like *Panicum turgidum, Cenchrus biflorus* and various *Pennisetum* sp. These grasses could allow collection of about 70–100 kg/ha.

The third group allows collection yields of between 100 and 250 kg/ha. It consists of annual grasses and herbs like *Panicum laetum*, *Sorghum aethiopicum* or *Colocynthis vulgaris*. Even the latter demands great labor to cook and dry to leach the bitterness; it represents an important food resource. To illustrate the importance of these resources, the annual yield of pearl millet and *Sorghum* in the present Sahel

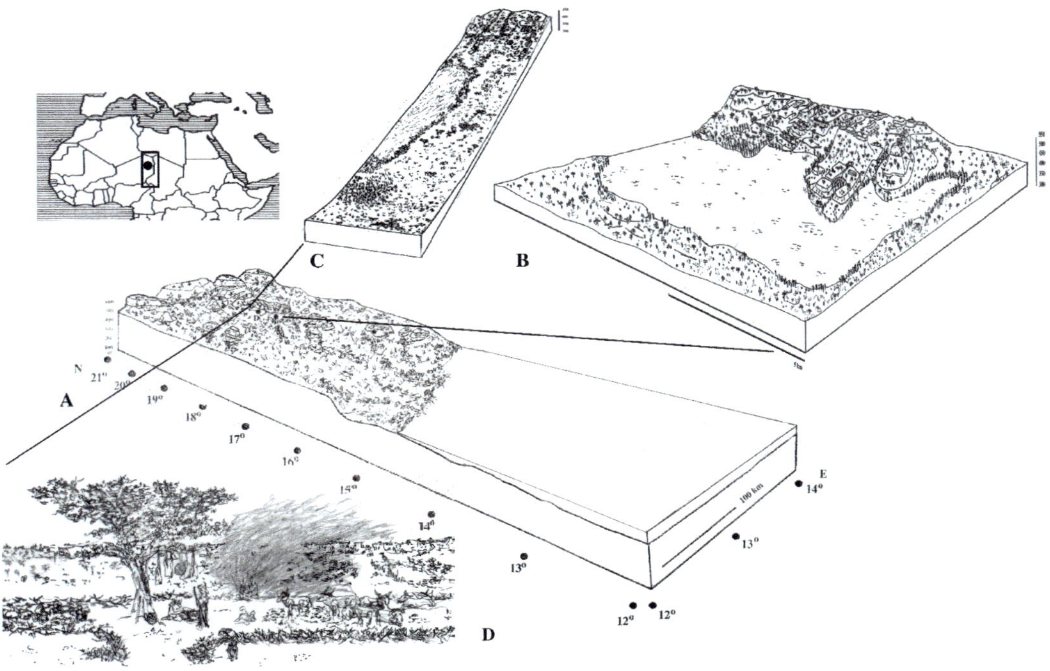

FIGURE 6.24 The landscape situation around 6,000 bp after the arrival of the pastoralists. (a) The Transect Lake Chad - at its high level - to the Djado Plateau with the Saharan savannas and the Sudano-Sahelian alluvial forests. (b) Bilma escarpment with its big lake in the foreland depression. (c) The Kawar at Achenouma showing a shallow lake with a reed belt and troona formation, *Panicum* and *Sorghum* pastures and the Maerua savanna on the escarpment.

ranges around 850 kg/ha but can decrease to less than 500 kg (De Rouv, 2004; Traore *et al.*, 2017). Thus, the third collection group of grasses (*Panicum laetum, Sorghum aethiopum*) range already in the yield categories of cultivated plants.

Grasses were the only category of food, which could be quantified. All other food types are only to estimate. Bushes and trees were regularly exploited as the "Saharan package" indicates.

There also was a "Sahelian package" of food in the surroundings of the former lakes and watercourses (see Figure 6.26). That package consisted of permanent resources and could have helped well in dry seasons, which we must presume anyway.

The Saharan and Sudano-Sahelian packages shed light on the discussions of early cultivation (see above). A similar model is presented as an "entrenched cultivation without domestication" (see Fuller *et al.*, 2021; Winchell *et al.*, 2017). The large "wild millet fields" could have extended over tens of square kilometers and provided noteworthy gathering resources (Wasylikowa and Dahlberg, 1999), even we do not know much about their former regularity.

Anyway, the existence and repartition of rich Achabs (see Figure 6.26) may point to the extended collection resources – without a necessary "entrenched cultivation". They certainly were even more important as food resources during the arid spells in the AHP around 7500 bp- (see Maley, 1983). Anyhow, *sorghum* and pearl millet as fully domesticated plants are found around 6000–5000 bp in NE-Africa (Burgarella *et al.*, 2002) or around 4500 bp (Fuller *et al.*, 2021, Winchell *et al.*, 2017). These findings are mostly based on charred material or ceramic imprints (Amblard and Pernes, 1989). Barich (2019) presented the "shepherd-model": Semi-sedentary people herding small stocks in the southern Sahara had enough time to deal with the large achabs at the end of summer and could collect some promising seeds.

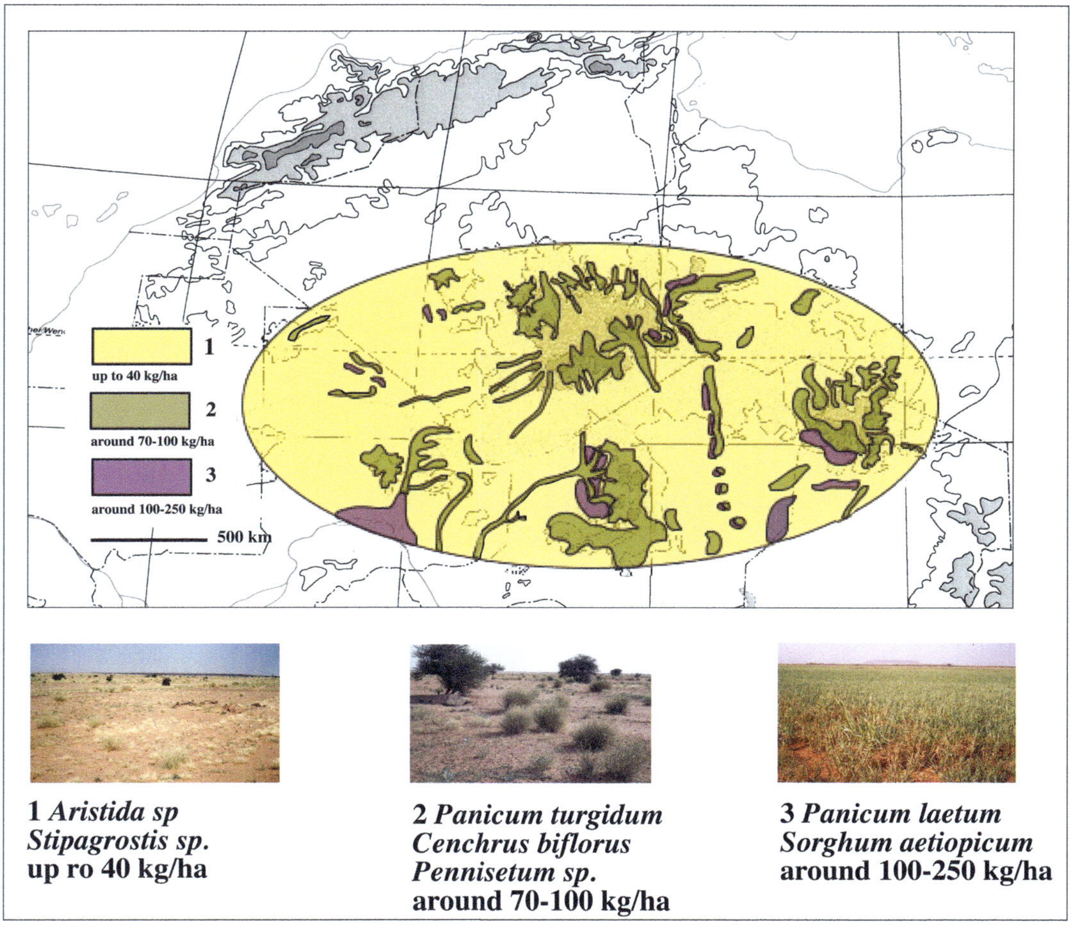

FIGURE 6.25 Map of possible collection yields of grass seeds during the Middle Holocene and the three categories of collecting grasses (Schulz and Adamou, 1997, modified).

As already pointed out above (Starter), hunting and fishing were important food resources too. However, we have to follow Jousse (2006), that small game and small fish certainly represented the bulk of game and fishing but were rarely present in the archaeological records. Insects like locusts played a role as a food resource too (Aumassip *et al.*, 1982–1983, Chapuis *et al.*, 2013). A rare but high caloric forage was honey (Dunne *et al.*, 2021; Musch, 2023).

At this time a new group of people arrived in the region with another economy.

The Additional Model: Animal Keeping and Pastoralism

Around 6,000 bp – the time of maximal resources for gathering and hunting - several groups of animal keepers arrived in the northern Ténére (Smith 1974) apparently from the North-East (Brass, 2013; Caneva, 1993; Dunne *et al.*, 2018; Gauthier, 1984, 1987; Linseele *et al.*, 2014; MacDonald, 1998; Neumann, 1994; Smith, 1974; Vernet, 1993; Wright, 2011). These people came with already domesticated animals: cattle, sheep, goats, and probably donkeys too (Todd *et al.*, 2020).

It is unclear, whether there was really a wave of immigrants – as we do not know about their demographic scale – or an infiltration of small groups (see Di Lernia, 2002) Anyway these were new people. From rock paintings and skeletons, we know that they were black people (Hachid, 1998; Hugot and

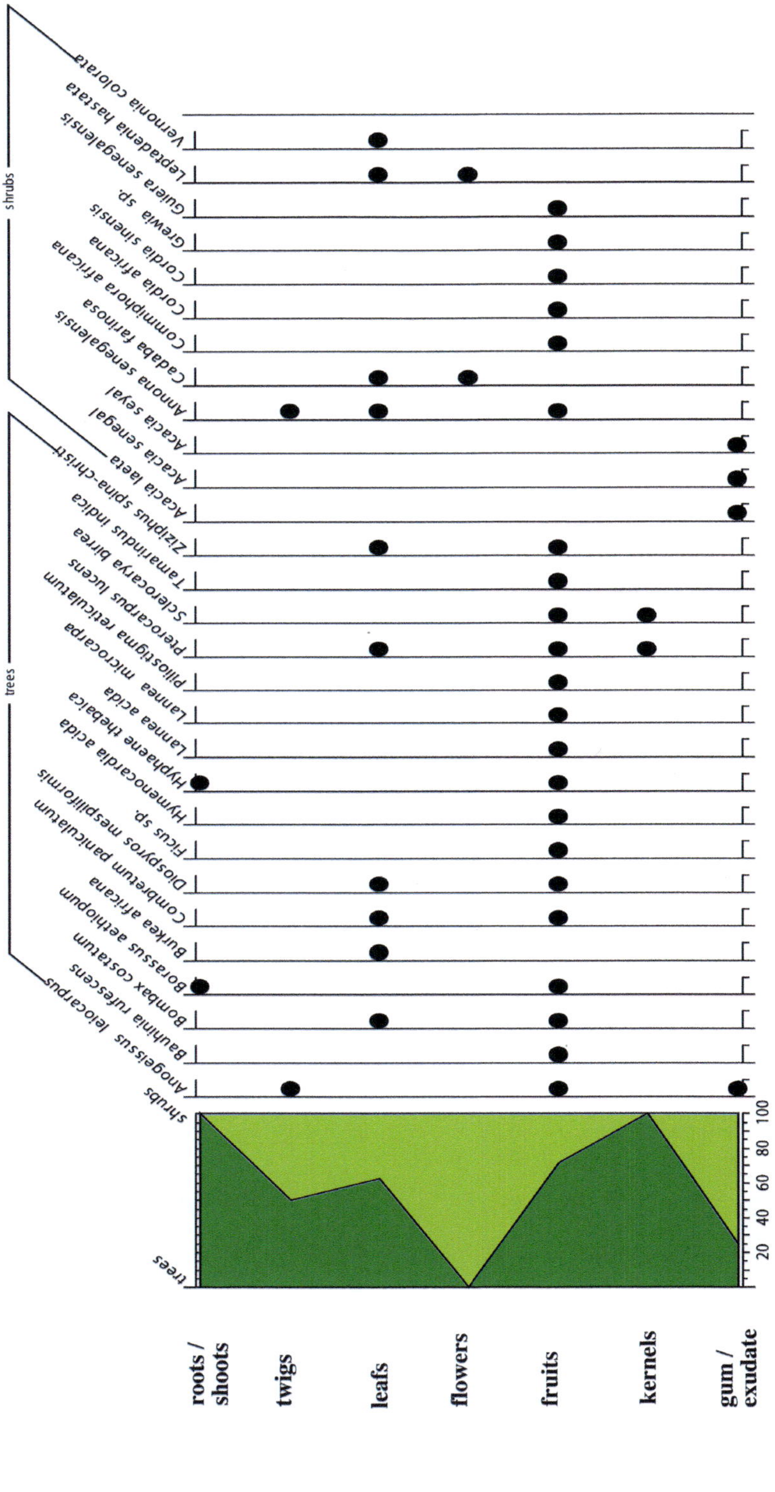

FIGURE 6.26 The additional package of Sudano-Sahelian collecting possibilities in dune valleys and around lakes (Arbonnier, 2003, Bergeret and Ribot, 1990; Busson, 1965; Burkill, 1985 ff; Hahn *et al.*, 2019; Maydell, 1992).

FIGURE 6.27 The pastoralism. Cattle keeping from rock paintings of Tassili n'Ajer near Djanet/Algeria and from Ennedi/Chad and goat keeping as well as donkeys on the Bagzan massif, Air Mts./Niger.

Bruggmann, 1976; Striedter, 1984). Apparently, there were two types of them: One of a greater body size and one of a smaller appearance. The discussion of whether they could be regarded as processors of the present Peulh societies is still unclear (Hachid, 1998).

These people had already a long experience in exploiting the different vegetation units (see Di Lernia *et al.*, 2002, 2013). It was another type of livelihood which allowed an additional alimentary system by food production- after some temptation for animal keeping by the Acacus people in the Early Holocene through corralling of *Ammotragus* f. ex. (Concelliere and Di Lernia, 2014).

However, we have to admit that hunting and gathering were still practiced, as we cannot imagine how many foods could be taken from the domesticated animals. Anyhow, they were able to produce food not only for the moment but also to process and store it (Dunne *et al.*, 2016). It was a noteworthy surplus to the alimentation regime of hunting, gathering and fishing and of storing.

So, the food resources consisted of plants and meat from game, fish, and food from domesticated animals such as milk and milk products – butter, butter oil, or cheese – or blood (Byrnes, 2019; Dunne *et al.*, 2018; Dupuy, 2014; Jousse, 2006; Sadig, 2009). Thus, it was a new kind of additional economy, which also represented a certain security in alimentation, and the caloric input was relatively high (fig. 6.27). It might be that fish were also processed as smoked *tilapia* or oil from them, which is still common in the surroundings of lake Chad (Gardi, 1978; Gartung, 1992; Seignobos, 2017).

Moreover, the number of animals allowed a transport of goods and food, and the development of new storing material such as leather bags – even that could have been the case also by hunter-gatherers before. Plants for tanning and the knowledge about them were present long before (see Nicolaisen and Nicolaisen, 1997).

The pastoralist was bound to their animals – even if we do not know about the size of the different herds. So, they had to organize their life and their groups around the animals. It was necessary to assure a

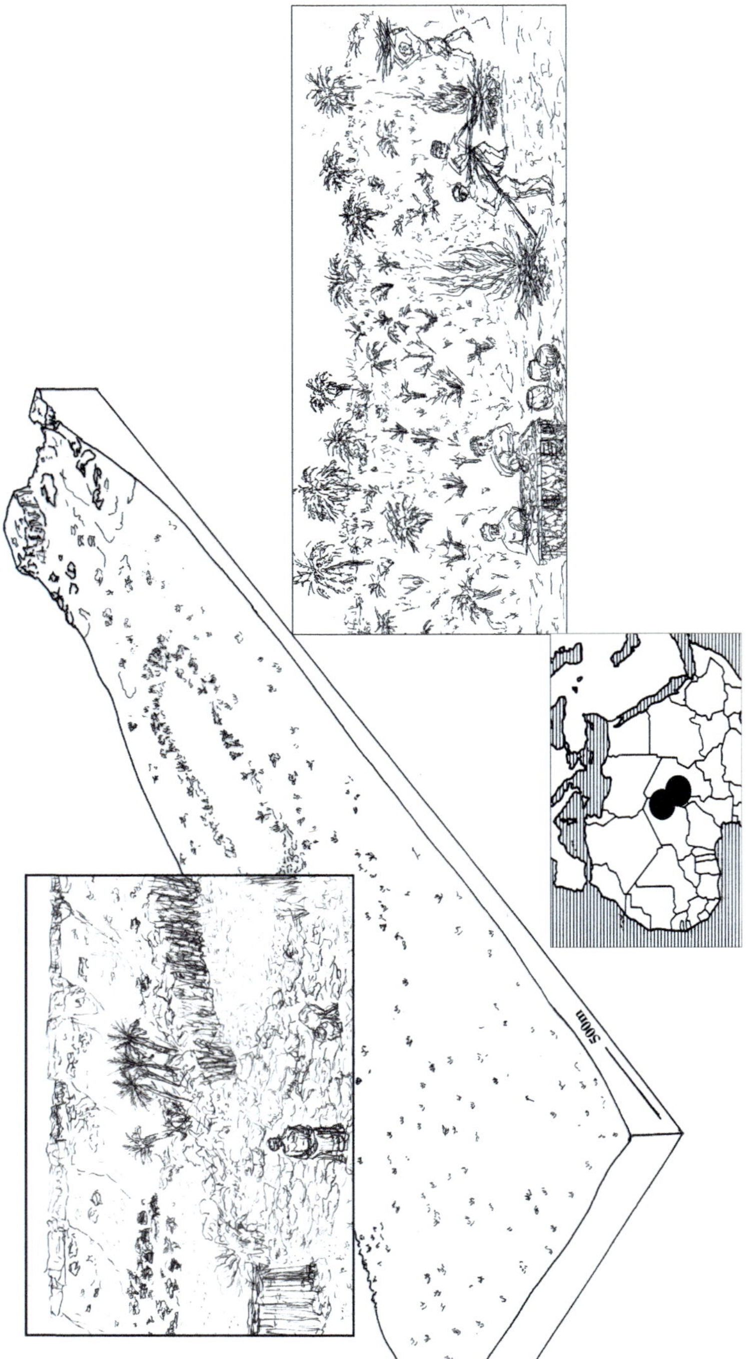

FIGURE 6.28 Troona collection around the Kawar Lakes and salt production by leaching of Capparis decidua -wigs, as it was still practiced in the 19th century on the northwestern shore of Lake Chad.

good and sufficient pasture, access to water, and protection of the animals against predators. Thus, it was necessary to build corrals as milking and protection places (fig. 6.28).

To fulfill all these duties, pastoral groups must have been greater than hunter-gatherers and they must have been socially organized in a more diversified way (Richerson and Boyd, 2001).

Investigations on the cattle herd management of the Peulh-Wodaahbee in SE Niger (Krätli, 2008; Scharaika *et al.*, 2000) presented a model, which could be taken for the Holocene too. They described decade-long cattle herding with a maximum of 50 heads. The herds are permanently accompanied and guided to the different pasture types year-round. At the end of the cold and dry winter period, rough fodder has an important part, which for *Salvadora* persica, for example, can rise to 60% of the pastures (Scharaika *et al.*, 2000). The annual pasture cycle is organized in some middle-distance migrations and afterwards by smaller moves around the camps and the watering points.

Camps follow all movements of the herds. Today, the camps are not enclosed, which certainly is not a measure to wait for the Holocene and its predators (see Jousse, 2006).

Since the herds are permanently kept together, there is a tradition of horizontal and vertical exchange of experiences among the cattle on different pasture and landscape types. Concerning the ovicaprine herds, which are documented for the Holocene (see above), there could be a certain division of the camps as the small stocks will have been guided to the rough parts of the landscape – even mountains – with a different pasture spectrum; see also Holl (2020). The distribution of the so-called "stone places", dense assemblages of stones in dune areas or on hammadas – interpreted by Gabriel (1976, 1986) as camp and fireplaces of cattle pastoralists – indicates a far-reaching mobility to exploit all types of pastures.

Regarding the different tasks and work connected to pastoralism – even with a restricted size of the herds – we must think about 20 persons per camp, however, under the condition of a permanent connection to other groups. This concerned the loan of cows, a planned mating and generally, cooperation and help during calamities.

Coming to the Kawar, we have counted on an organized exploitation of these environments. As dense reeds surrounded the lakes (see Figure 6.24) access to the water was assured by regular burning, as it is documented in the sediment records. Rocky areas offered different pastures, suitable for goats. Clayey areas around the lakes could offer rich pastures at the end of the summer, such as *Sorghum* "fields" or large stands of *Panicum laetum* on sandy parts. Moreover, the lakeshores provided trona and salt necessary for the animals (Gardi, 1978; Gartung, 1992). Other ways, it had to be produced, for example, by leaching of the ash of *Capparis decidua* twigs (Schulz, 2020).

Camps or settlements were certainly outside of the lakes as the numbers of mosquitoes had been annoying or even dangerous. For that task, there were certainly large smoke fires beside the caps, as Figure 6.24 suggests.

There is much reasoning on the magic components of cattle pastoralism, burials, and possible relations of stone monuments for cattle and later on for man (Di Lernia, 2002; Hachid, 1998) up to the apotheosis of a later incorporation of pastoralist in emerging states like the Garamantes (Brooks *et al.*, 2005).

Anyhow and more importantly, these people had to care for their animals. Not only for pasture and water but also for their health.

Animal keeping brought new contact with disease vectors. The tight contact of the animals on the one hand and the permanent contact between man and animal on the other hand brought plenty of chances for an exchange of infections or first types of zoonoses (see Holl, 2020). This could have led to complete losses of smaller or even greater herds and so to the loss of provision of milk or milk products. Finally, it worked as an additional point of the general aleatory background of the Sahara (see above). It also explains that gathering and hunting were still a necessary part of food acquisition for these people. Thus, pastoralists were urged to build up veterinary wisdom and tradition too.

An estimation of former veterinary practices might be drawn from the present experiences in the region (Bernus, 1969; Nicolaisen and Nicolaisen, 1997; Volpato, *et al.*, 2015) and own investigations (Schulz and Adamou, 1992).

Figures 6.29 A, 6.29 B give a repertoire of the employed plants, substances, and different measures such as bleeding, combustion, cauterization, or rustic vaccinations.

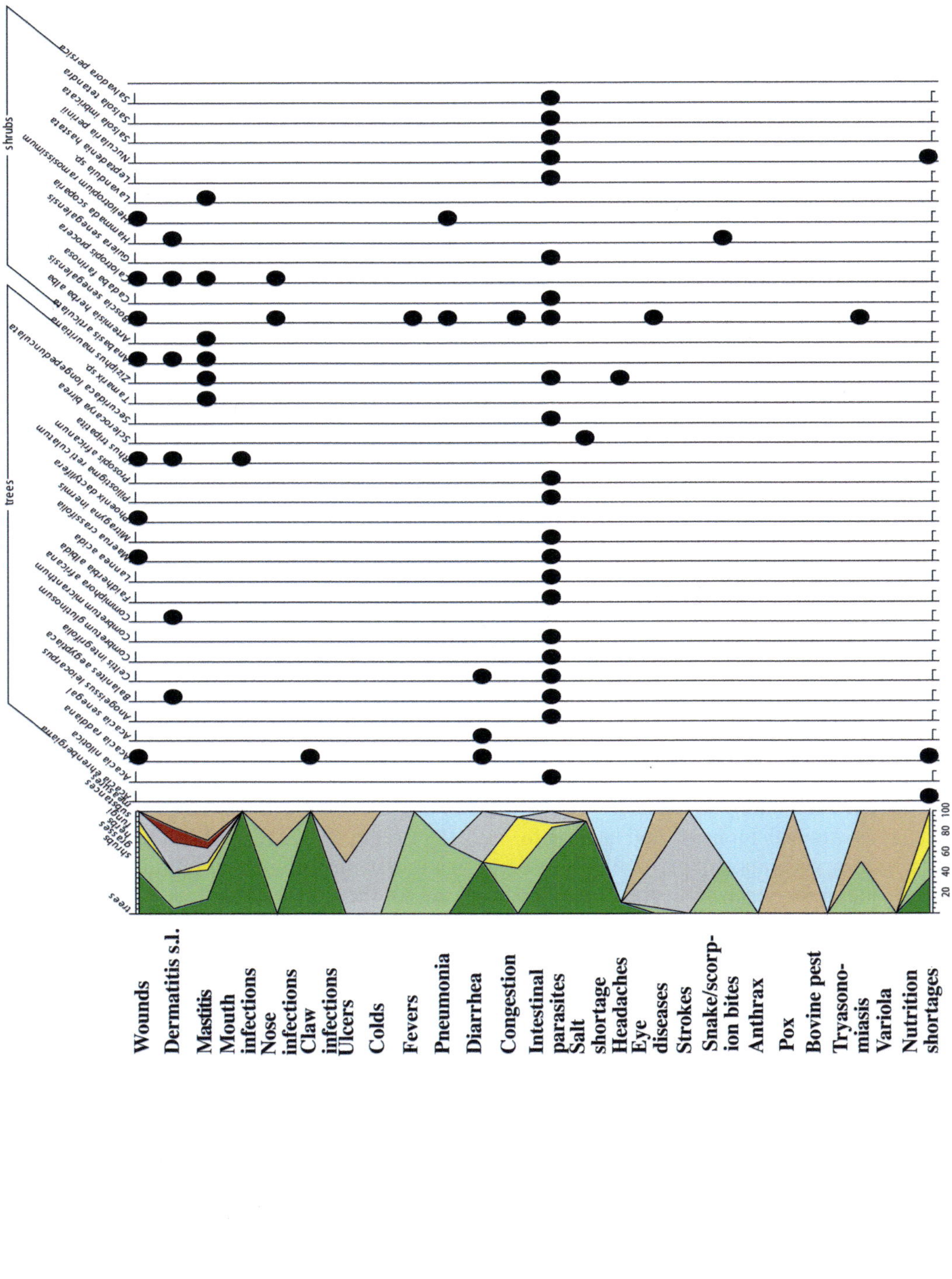

FIGURE 6.29A The Saharan veterinary package part of trees and shrubs

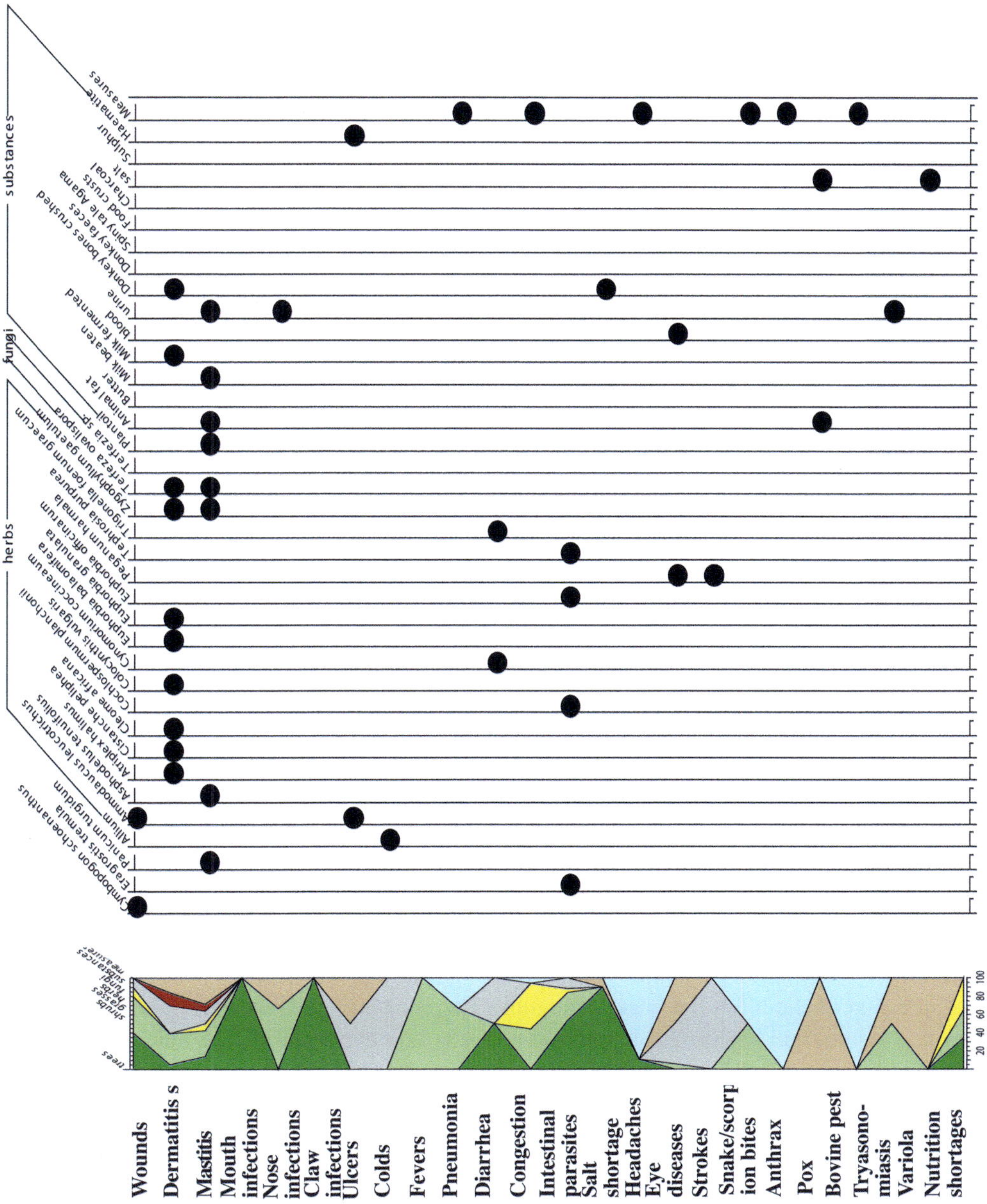

FIGURE 6.29B The Saharan veterinary package. Part of grasses, herbs fungi, and substances.

Compared to the traditional human medicine, the number of herbs, substances, and measures is more elevated.

The repertoire of diseases is large. Wounds, skin problems, mastitis, and intestinal problems are the most treated items. Decoctions of barks and leaves are most employed for treatments.

Figure 6.30 also shows that for infectious diseases the number of treatments is limited. However, the ultimate danger and challenge for cattle keeping was, and still is, trypanosomiasis (Gentilini, 1994; Stich, 2021). Its vectors – the glossinas are bound to Sudanian typed shadowish vegetation along rivers and they can only be fought by fire and eliminating the respective alluvial vegetation.

Herd management implied also to be prepared for shortages of pasture during dry years by changing the movements, e.g., from a regular transhumance to full nomadism or to change the herd's composition or to reduce the numbers of animals (see Krätli 2008; Pamo, 1998, Schareika *et al.*, 2000; Widstrand, 1975).

Moreover, several plants will have been distributed by the animals; *Acacia* seeds are the most prominent among them (see Tews *et al.*, 2004). This has contributed to a certain homogenization of the vegetation units.

Finally, these activities represented an entrance to a systematic land management, which included access to pastures, pathways, and protection sites, e.g., corralling. They also implied several geomorphic features like long trampling passages etc. (Busche and Heistermann, 1992; Zerboni and Nicoll, 2019), and a lot of these practices were connected to fire. Fire and fire management will be treated in the following part of this chapter.

Fire and Landscape Management

Fire was a common phenomenon in the Late Pleistocene and Holocene in the Sahara. In the Bilma record as well as in those of the Kawar Lakes (figs. 6.19, 6.21, 6.22), signs of fire are regularly present. These are charged particles – mostly from grasses and in a lesser amount also from wood. Unfortunately, it was not possible to determine those charged particles in detail. In the lower section of the Bilma-record, during the Kanemian-Period", such particles are pulverized remains of charred material. In the upper (Holocene) section, however, particles are of fresh grass coal and, in some cases, of wood coal. So, what could have been the cause of the fire and its regularity? Savannas are fire-prone landscapes (cf. Balouche, 2009). But under which conditions did burning happen? Which types we may admit for the Saharan savannas? In what way can we explain the regularity of fire documented in the sediment cores?

a. In general, basic conditions for fire are fuel, a suitable environment, and a large energy ignition.
b. Savannas are open grass and tree formations under a seasonal climate. The dry season produces enough dry material, as fuel out of grasses and wood, which can serve as fuel.
c. The question of the lightning energy remains. There are two possibilities: either atmospheric energy through thunderstorms and lightning strikes, or a terrestrial one by man.

The Atmospheric Approach

Fulgurites may serve as markers of climate and landscape development in the Holocene of the Sahara (Sponholz *et al.*, 1993).

Fulgurites (lat. fulgur = lightning) are mineral transformations induced by lightning strike into unconsolidated sediments or into solid rocks. When lightning strikes this material, the mineral components are briefly exposed to extreme temperatures up to 3000 °C (Feldmann, 1988) along the lightning channel, which leads to the transformation of minerals, but above all to the melting of the latter accompanied by the vaporization of all volatile components. In loose sediments, quartz is molten along the lightning

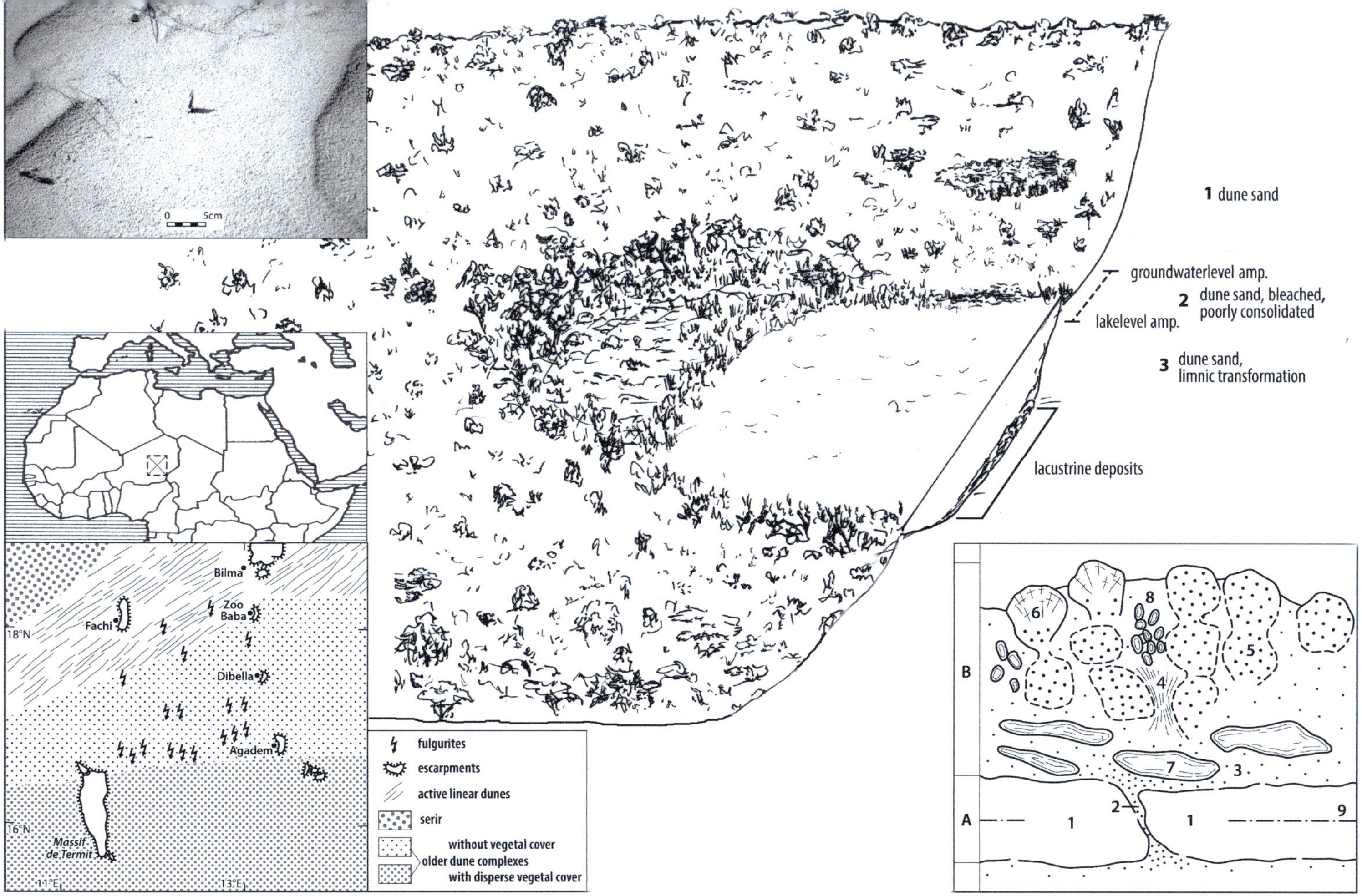

FIGURE 6.30 The reconstructed landscape setting during the Mid-Holocene – seen under the aspect of fulgurites and vegetation. Below right: schematic longitudinal section of a dune sand fulgurite. (a) 1 open lightning channel (9 longitudinal axis) with 2 lechatelierite bridge. (b) fulgurite wall with 3 lechatelierite matrix, 4 semi-crystallized flow mark, 5 cristobalite grains, 6 semi-preserved quartz grains, 7 air bubbles, 8.

channel and re-solidifies as lechatelierite or cristobalite and creates an overall glassy-brittle coating of the lightning channel. In solid rocks, lightning strikes often lead to rock fragments being blasted off and to diffusing mineral transformation in the rock due to high temperatures (fig. 6.30).

Generally, fulgurites can form worldwide with every lightning strike if the conditions along the lightning channel favor the melting of minerals and their immediate re-solidification. However, the complete soaking of the loose material struck by lightning tends to result in impact energy being dissipated as the moisture evaporates. In solid rock, moisture penetration along fissures tends to have a reinforcing effect as the explosive evaporation of the water causes the rock to "burst".

What Is the Significance of Fulgurites for Climate and Landscape Reconstruction in the Sahara?

The occurrence of fulgurites is concentrated in particular localities: most fulgurites occur in a fringe of a few meters wide around paleolake-depressions in dune areas. This position can be explained by high electric potential between completely soaked sand (edge of the depression/near lake water level) and the surrounding high dune bodies, which are dry due to limited capillary water rise in the sand. The possible existence of tall trees in the lake shore area whose roots dissipated the lightning energy is not relevant in this context: a lightning strike to a tree is more likely to "blast" the trunk by exploding moisture in the tree. The energy still arriving in the ground should therefore not have been sufficient to form fulgurites. More likely than a high and dense tree population in the fluctuation range of groundwater and lake water level, the lightning strike seems to have occurred directly in the lake edge area, solely due to the locally highest voltage in this small wet-dry transition rim.

Other authors place the pronounced lake phase with Sahara-wide distribution of lakes in interdune depressions with dune bodies rising in their vicinity in the early- to mid-Holocene (Baumhauer, 1991). This landscape configuration requires strongly contrasting hygric seasons. These correspond to today's trade wind-monsoon dominated seasons. Especially at the beginning of the rainy season in summer, thunderstorms with a high frequency of lightning still occur today and reach approx. 15°-16°N in Eastern Niger.

The concentration of fulgurites in the edge of interdunal depression with traces of bleaching due to the sea phases can also be found elsewhere (Mauritania, Sponholz, 2009) in the Sahara. Despite large regional observation gaps, it can therefore be assumed that the concentration of fulgurite sites shown in Figure 6.30 with an increase from approximately 18° N (rare, but regular findings) to 16°–17° N (high density of sites with an increase in the number of fulgurites per site) represents a zonal phenomenon.

Below left: frequency of fulgurite sites from North to South between about 18°N and 16°N and number of fulgurites per site (schematic).

Middle left: study area in eastern Niger. Above left: fulgurite (vertical structure in the center of the photo (central Mauritania, Sponholz, 2009) at the rim of an interdune depression. Because of their fragile glassy structure, dune sand fulgurites are sensitive to erosion. Usually, only such small remains are found.

Rock fulgurites occur in several places in the central Sahara, too. Analytically proven on the top of Emi Bao/near Segedin/Niger (topographic situation: 20°13′15″ N/12°58′58″ E, 588 m asl. Stratigraphical situation: formation de Segedin/Emi Bao (Mesozoic) according to Faure (1966); Upper Senonian, (BRGM, 1966), northernmost part of the Kawar cuesta.

While the rock fulgurite on the Emi Bao is not dated and has formed at "any time" post-sedimentary in the Upper Cretaceous sandstones, the rock fulgurite in the Enneri Achelouma (Figure 6.31, photo Sponholz, 2009) is dated to about 1995 according to a witness (A.K. Dizi, oral comm. 2006).

As both rock fulgurites are specimens that were stroked as the highest point in the area, they are not considered here in the context of atmospheric versus anthropogenic Holocene fires.

FIGURE 6.31 Rock fulgurite- on a tributary's terrace of the Enneri Achelouma (topographic situation: 22°18'42"N, 12°40'38,9" E, 672 m asl). Stratigraphic situation: Post Tassilien to Grès de Nubie -Nubian Sandstone- (Late Paleozoic to Early Mesozoic, after BRGM 1966. (Sponholz, 2009).

The Terrestrial Approach

The regularity of grass-coal in the Bilma and Kawar records makes lightning fires as an explanation too simple. This can probably be explained by repeated hunting fires on the one hand and by pasture fires in the grass areas on the other. However, the charcoal flora of Fachi-Dogonbolo shows that there were also savanna wood fires. Neumann (1989) proposed hunting fires and subsequent bushfire as a case for the burning of this rich Sudanian alluvial forest at the Fachi lake shores. Huysecom *et al.* (2004) and Ozaine *et al.* (2009) explained well regularity of charcoals in a sediment record is a clear sign of human activity.

Another use of fire was certainly the access to water points or lakes. There were many lakes and waterholes along the cuestas and on the mountain edges, as well as in the dune valleys in the South. The problem, however, was to open access to the water. This was most easily done by burning the *Typha*-reed or Cyperaceae belt, as is still common today (Ballouche, 2019, Ibrahim and Schulz, 2019).

Figure 6.32 depicts the main types of intended fires. Above it describes the reed fire around the former Segedim lake, and below it shows a hunting fire in the grassland in front of the Bilma cuesta which also stands for pasture fires in order to provoke or improve the resprouting of grasses.

Domestic fires englobed heating and cooking but also the smoke fires against mosquitos (see fig. 6.24).

A real danger for the cattle keepers, however, was trypanosomiasis, which is transmitted by Glossina/Tsetse flies. It stood for their animals and themselves. They demand shady areas with temperatures between 20° and 28° C and humidity between 50 to 80% (Lambrecht, 1964). So, they settle along rivers and/or brooks in the South Sahelian-Sudanian area (Gentilini, 1995). Gifford-Gonzalez (2000) developed the idea of a long-retarded movement of Holocene pastoralists due to the danger by Tsetse flies, and Sadig (2009) reported this borderline at about 18°N for these circumstances. Even though it is difficult

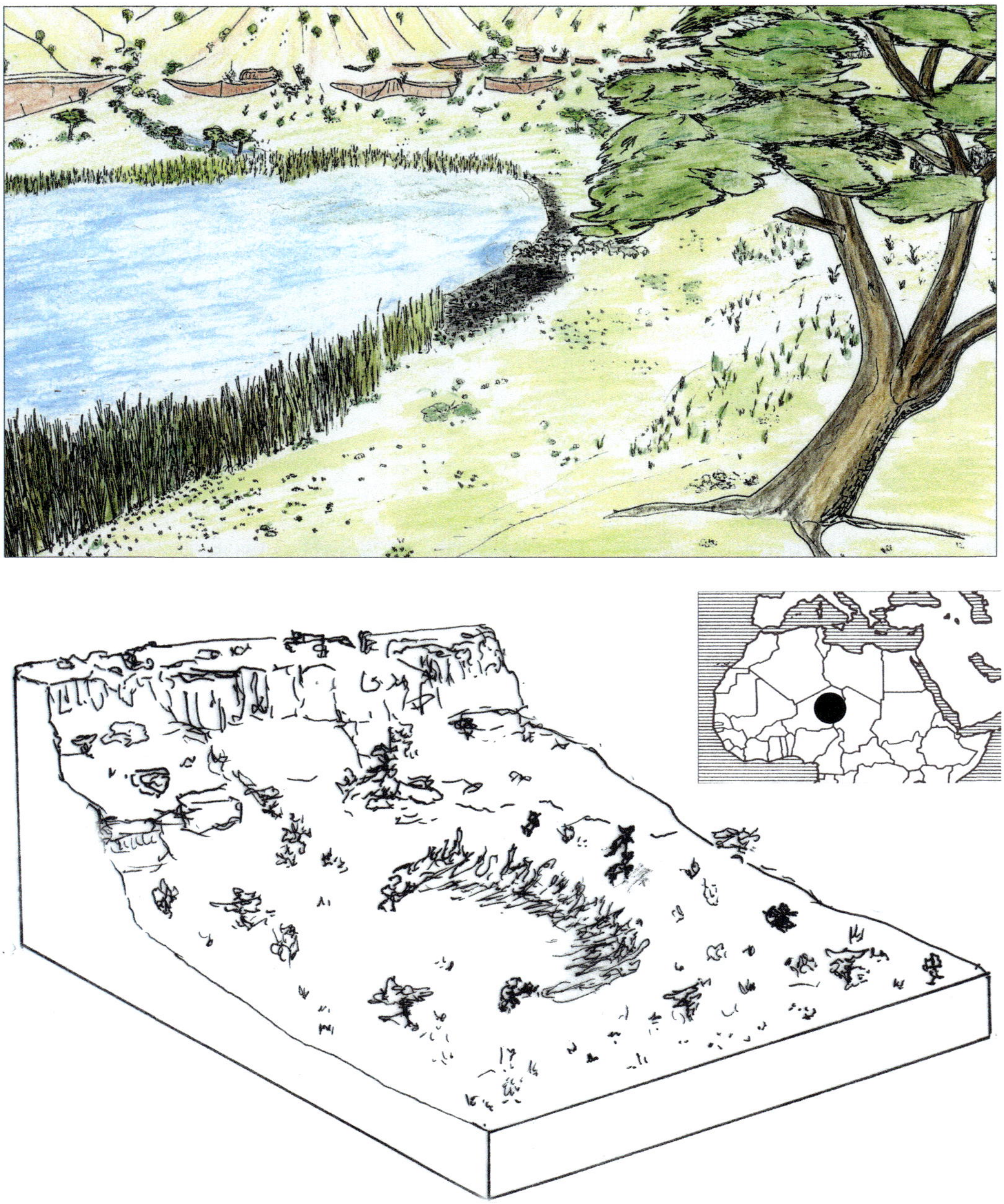

FIGURE 6.32 Fire management in the Holocene of the Central Sahara. Access to water by fire Segedim (above) and a hunting fire in front of the Bilma escarpment (below). This also stands for pasture fires in order to provoke a resprouting of grasses. (Draft: E. Schulz).

to transfer these borderlines from East Africa to the Sahara, it describes an intriguing problem for all cattle keepers, and for about 6000 bp we will have to count it in the Djado-Lake Chad transect too (fig. 6.33). The model of a Sudanian-type alluvial vegetation in the dune valleys and along the southern part of the cuestas indicates favorable conditions for *Glossina* (see also the charcoal flora of Fachi-Dogonbolo). People at that time were well able to treat fevers, but trypanosomiasis was beyond their healing capacities. Today, West African herders work with trypanosome resistant races of taurine and Zebu cattle (Bouye,

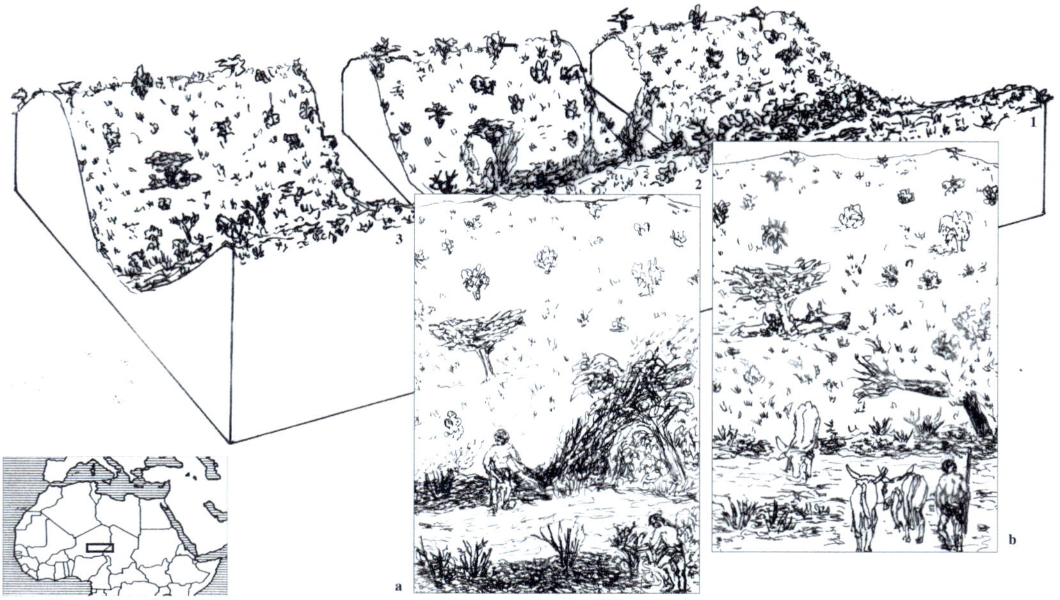

FIGURE 6.33 The fight against Tsetse flies south of the Kawar—The Sudanian alluvial savannas or forests were a danger for man and cattle (1). The adapted measure (2) was a system of burning grasses, herbs, and bushes in order to kill the Glossinae (a) and afterwards cutting bushes and trees to eliminate shadow places definitely (b). As a result, they obtained water and useful savanna with shadow trees for the animals (3) (Diagram: E. Schulz).

2015), but in Holocene, times had only one measure- fire! The only way to protect cattle and men was a systematic clearing of alluvial thickets or shady bushes. This was manageable by a burn and slash action. It would go along with access to water and a further opening of the savannas. However, it would also imply a kind of protection of useful/shadow trees.

A Second Addition to the Saharan Alimentation

Around 2500 bp another food supply arrived in the Central Sahara. It was the introduction of the domesticated date palm (*Phoenix dactylifera*) and the subsequent development of an oasis culture in the Central Sahara. Kaczmarek *et al.* (2023) reported the introduction of date palm cultivation and the development of an irrigated oasis culture of the Garamantes state in the southern Fezzan from 3000 bp on. It was an integrated system of cereals (wheat, barley) and tree cultivation (date palm, fruit trees). For a long time, it was taken as certain that the date palm was domesticated in the Arabian Peninsula and spread from that region to the West. Genetic investigations, however, postulate two different source areas of ancient presence and later domestication. Parallel to the Near East, there should have been an area in North Africa (Zehdi-Azouzi *et al.*, 2015). That model could explain an early presence in Saharan pollen records such as Gobero, Ténéré/Niger for about 10,000 bp (Sereno *et al.*, 2008) or Lake Yoa in northern Chad (Lezine *et al.*, 2011) from about 1,700 bp on with cultivation from about 700 bp on. Looking at the Kawar Lake pollen diagrams (see above), we have a presence of *Phoenix* pollen from about 2,000 bp on, however, with no signs of cereal cultivation. That will point to the introduction of some palm groves in the vicinity of the lakes, as *Phoenix* needs groundwater and is tolerant to salty environments (see Purseglove, 1972). Finally, these palm groves were certainly integrated into the annual migration cycles of the pastoralists when they came at the end of summer for the collection of wild cereals, troona, and from that time on, dates. The century-long abandonment of the Segedim village in NE Niger, however, with regular visits for

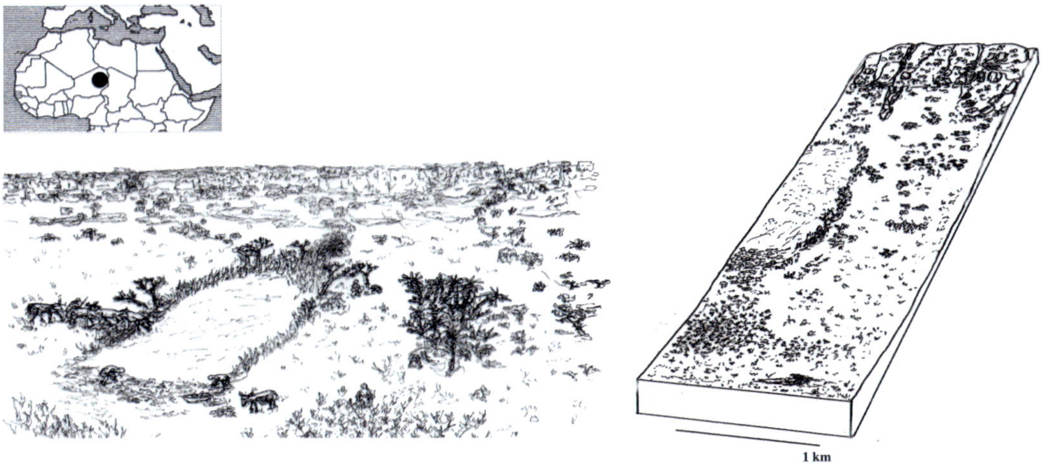

FIGURE 6.34 Exploitation by nomads at Achenouma/Kawar around 2000 bp. Troona, wild cereals, and dates from the palm grove are collected during regular visits. Note the still existing Maerua-savanna on the escarpment, the tree lines directed to the lake, and the burning for access to the lake for cattle. (Diagram: E Schulz).

date collection (Carl and Petit, 1954) may give a model for this economy. Figure 6.34 explains the regular exploitation of the Kawar oasis for troona, wild cereals, water, and dates. Moreover, it points to the importance and probably origin of tree cultivation in the development of agricultural systems (see Krings, 2006; Maydell, 1992; Seignobos, 2017).

During the last two millennia, the Kawar developed into a veritable crossroad between the Tibesti and the Air Mts. by the permanent concurrence of Tubu-Teda and Tuareg and the North and South between Fezzan and the Lake Chad region. The Garamantes were the first to conduct trade by horses with the South. The introduction of the camel supported a continuous commercial exchange between the Mediterranean and the Sudan (cf. Magnavita *et al.*, 2019). The permanent lakes and wells of the Kawar attracted and assured the caravan trade (Austen, 2012). It is explained by Figure 6.35.

The sources on the Kawar between the 9th and 15th centuries are mostly Arab chronicles, for which the caravan trade is said to have been of particular interest. The inhabitants of the Kawar were heavily involved in this trade and became rich through it; women, children, the elderly, and slaves maintained the oases and were involved in horticulture. The salt lakes certainly played an important role at that time. However, between the 16th and 18th centuries, these are said to have been exploited mainly by Tuareg. The caravan trade continued to be of great importance to the inhabitants of the Kawar.

Population density figures for the Kawar were documented in the 19th century by travellers such as Rohlfs and Nachtigall. The population fluctuated considerably, and its density was always highest at the most favorable time of year, i.e., during salt mining and the date harvest (June to August). For seven settlements in the Kawar, the total number of houses may have been 1,140 and the total number of inhabitants 5,200. There were repeated raids by Tuareg or Weled Sliman groups. Travellers of the time were struck by the large number of date palms and the almost complete lack of grain cultivation.

During the last decades, the Kawar was still an important relay for the trade and traffic between Libya and West Africa. This included migration too. From the 1990s, the informal trade and smuggling became more and more important, and the oasis found a growing income source in growing alfalfa for goat and sheep herds as a meat source for this commerce (see Brachet, 2011; Schulz *et al.*, 2009).

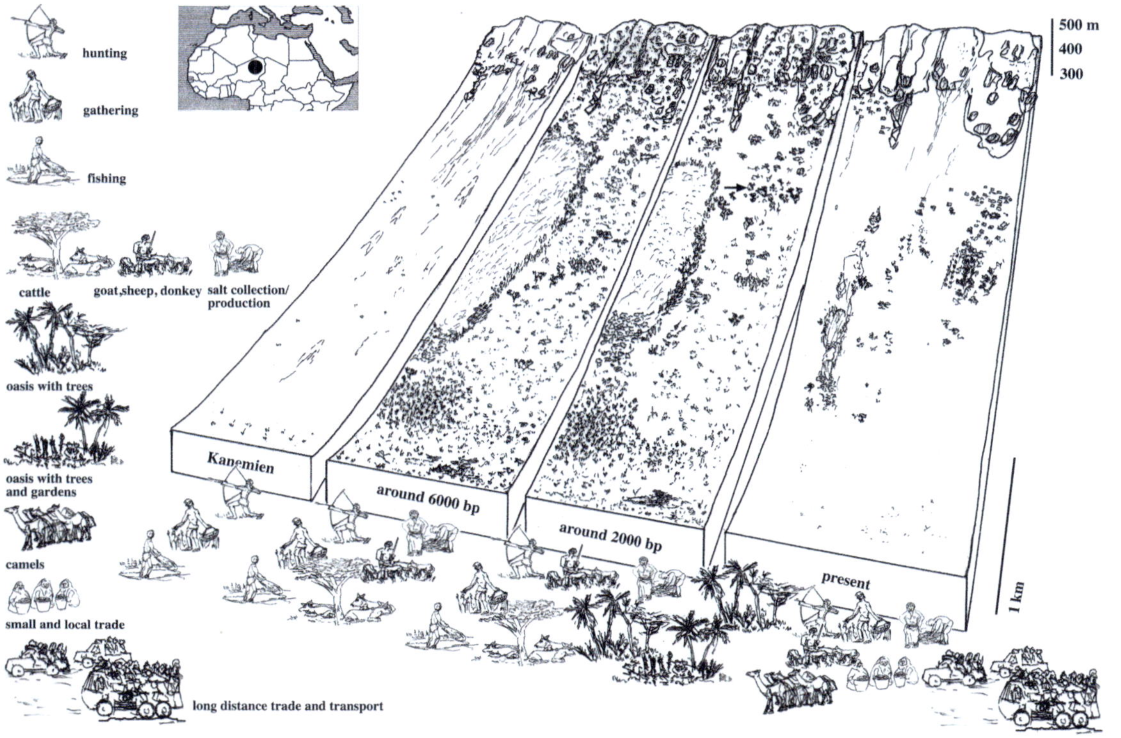

FIGURE 6.35 The economic history of the Kawar region/NE-Niger from the Late Pleistocene on. A model for the Central Sahara Note the persistency of the extractive economy (hunting, gathering, and fishing) as a base of human life and the role of trade and transport (Draft: E. Schulz).

DESSERT: ADAPTATION, FOOD STRATEGY IN SAHARA AND SAHEL AND FIRST TRACES OF CULTURAL LANDSCAPES

The Bilma and Kawar Lake pollen records show clearly the intermingling of Sudano-Sahelian alluvial vegetation with the Saharan *Acacia-Panicum*-savannas during the late Pleistocene up to the Middle Holocene. This process started already within the Kanemien dry period. Based on present knowledge about food resources and traditional use of the contemporary *Acacia-Panicum*-complex allowed us to model the resource situation of the Late Pleistocene and Holocene. In this way, we could obtain an estimation of the Holocene food resources and could also quantify the importance of gathering wild cereals. Combining evidence from archaeology, ethnology, the analysis of rock art, as well as sedimentology, geomorphology, and palynology, we could develop a landscape- and vegetation-history of the Central Sahara with particular interest in the human impact. It became clear that an "extractive" economy of hunting, gathering and fishing persisted over many millennia, even parallel to "alternative" strategies of food production either through animal husbandry or early oasis culture based on date cultivation. However, it is still unclear how many "consumers" were present in the area to exploit the different resources.

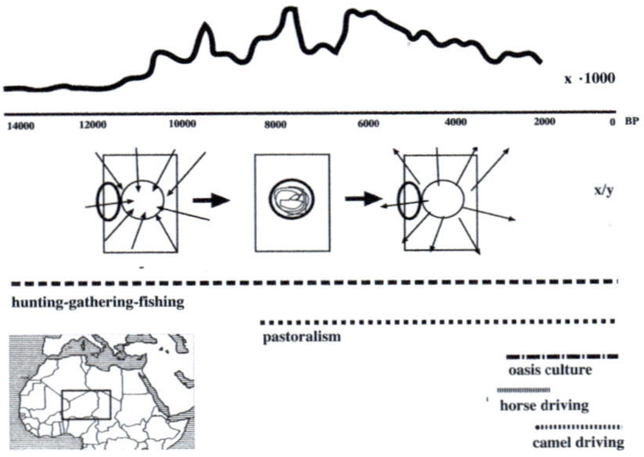

FIGURE 6.36 Persistency, change, and uncertainty in human exploitation of the Saharan resources during Late Pleistocene and Holocene against the background of the unknown number of consumers. (after Brierley *et al.* 2018, D'Attonosio *et al.*, 2018, Striedter, 1984, and Xue, 2018). (Diagram: E. Schulz).

This general reflection resumes the economic history based on gathering, hunting, and fishing with the addition of pastoralism and later on oasis culture (fig. 6.37).

Two points are to be recognized: firstly, the importance of the aleatoric character of climate, landscape, and resources, and secondly the necessary adaptation to this situation. Achabs may assure collection yields up to 250 kg/ha, but they rely on the variable rainfall regime and the respective seed banks. Transferred to Late Pleistocene and the Holocene of the southern Sahara – to the transect Djado-Lake Chad – we must admit that the several arid spells underlined the aleatoric part of the resources in those periods. As the permanent vegetation and resources became scarce, it played a role for the two economic models – extractive and productive – of that period. Mobility was one of the most important survival practices of the time. The importance of aleatoric resources demands flexibility. Thus, exploitation of the few permanent resources – oasis culture or not – in combination with semi- or full nomadism or trade is the adaptive economic model cf. Manning *et al.* (2023).

The extension of the *Sorghum* or *Panicum laetum* achabs explains well the existence of the large grinding stones. Astonishingly, the grinding could have been very effective – in a short time. Nixon-Darcus *et al.* (2024) showed that with the help of large grind stones, one kg of flour could be prepared in about 15 minutes.

The biggest question is still on how to estimate man's capacity to manage environments and to change them into human-made systems or into cultural landscapes.

In sediment and pollen records, evidence for regular burning could be found throughout millennia. Despite the influence of atmospheric ignition, this regularity strongly indicates a continuous human influence. Burning was applied in different kinds of resource management strategies, for example, hunting fires, to promote regeneration of pastures and sanitary fires. The wide-ranging use of fire against the Tsetse fly as a vector of ngana and sleeping sickness led to a definite landscape change. In this way, bushes along water courses and around lakes were eliminated in a two-step procedure, including both burning against glossina and slashing of woody vegetation to create free access to water. Regularity in burning was necessary to prevent the regrowth of the Tsetse fly's habitat. In this way, burning led to a definite change in landscape.

In fact, pastoralists have been blamed, on the one hand, for deteriorating the landscape by bad pasture management which led to overgrazing and thus to a growing albedo, the latter having the same effect as the change of orbital parameters – higher temperatures and dryness (Wright *et al.*, 2011). On the other hand, wise pasture management by the herders which avoided overgrazing could have avoided a greater albedo and thus could have retarded the effects of the change of orbital parameters (Brierley *et al.*, 2018).

FIGURE 6.37 A model for the mixed economy of the Saharan Holocene (gathering, hunting, fishing, and animal keeping). (A) wild sorghum field in the western foreland of the Air Mts./Niger allowing gathering yields up to 250 kg/ha, (B) Panicum achab in the forelands of the Ennedi Mts./Chad allowing gathering yields up to 70 kg/ha, (C) gathering with the swinging basket, (D) "sleeping mills" in the Ennedi Mts./Chad, (E) wind sieving in the Ennedi Mts./Chad, (F) grain stores in the Ennedi Mts./Chad, (G) gathering of Salvadora berries in the Djado/NE Niger, (H) hunting success: Dorcas gazella, (I) a mixed herd of cattle, sheep, and goats Abalak/Central Niger, (J) fossil Tilapia bones in the Djado/NE Niger, (K) rock painting of a fisherman and his wife in the Ennedi Mts./Chad. Assemblage Schulz 2025.

However, for both models, wise land management and restricted albedo development or destructive land management with a growth of open surfaces, there is an unspoken acceptance of a domesticated domestic landscape system, thus cultural landscapes. In this context, the "Wodaabe herding system" (Krätli, 2008; Scharaika *et al.*, 2000) represents an adapted model and points against a rapid end of the AHP (cf. Amaral *et al.*, 2013, Li *et al.*, 2022).

In all, changing orbital parameters and the anti-glossina measures of burning eliminated the Sahelo-Sudanian intrusions into Saharan savannas and led to a homogenization of the landscapes in the southern Saharan savannas. The plasticity and the regeneration potential of the *Acacia-* vegetation types resulted first in an extension of the *Acacia-Panicum*-savannas and later in the development of the Sahelian *Acacia*-dominated savannas. Increasing dryness and a southward migration of the pastoral groups happened then. Later, with continuing dryness, a shrinking of the *Acacia-Panicum*-savannas led to the contracted-linear vegetation formations of the future desert. At the same time, the part of the ever-present aleatoric resources became more and more important among food resources. Finally, the definite and man-made change of the southern Saharan environment during the Late Holocene shows both: the ability of early humans to modify and create landscapes, as well as the emergence of savannas as cultural landscapes, which are called "Sahel" today, cf. Pomel *et al.* (1994).

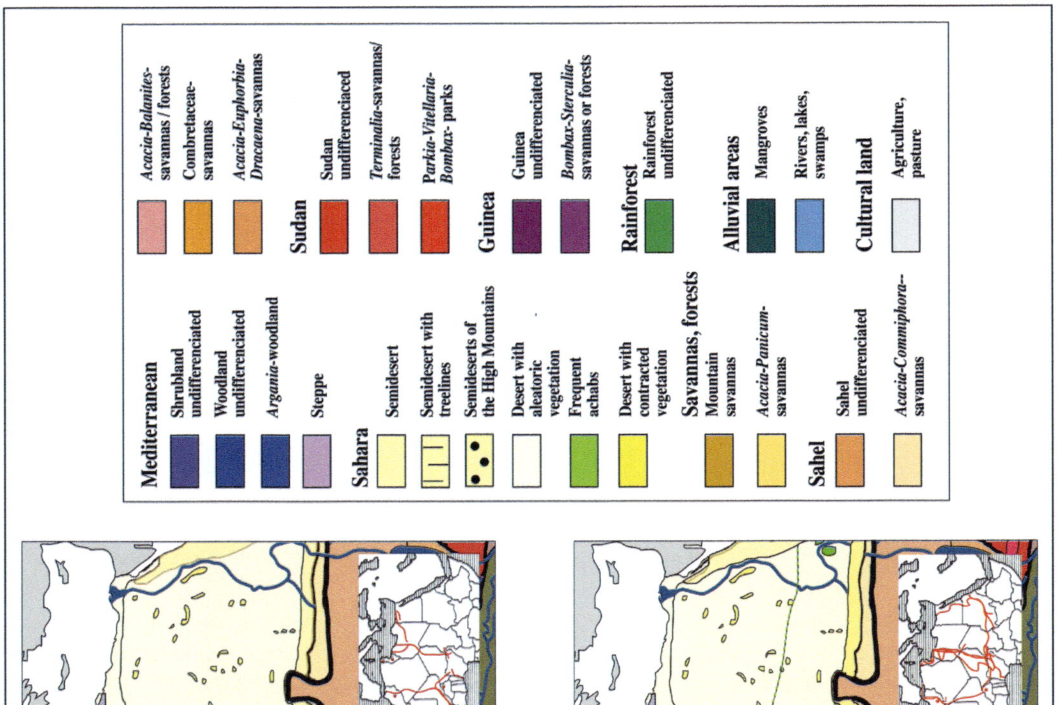

FIGURE 6.38 The Central Sahara in the 19th century (Schulz, 2020, modified).

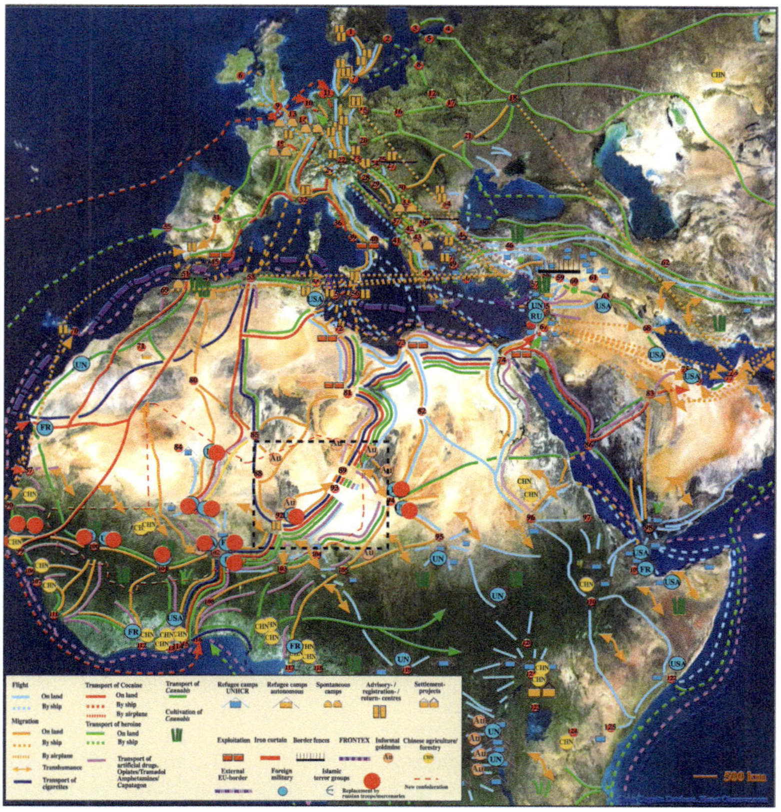

FIGURE 6.39 The mosaic of drugs, flight, migration, and military interferences. The background for the general civil insecurity (Schulz *et al.*, 2009, modified).

However, the question of early cultivation of non-domesticated plants- entrenched or not (cf. Fuller *et al.*, 2021, Neumann 2003; Wasylikowa and Dahlberg, 1999) may simply be answered: that apparently, it was not necessary.

A reconstructed of the Saharan vegetation during the 19th century explains the long regeneration phase after the severe droughts of the 18th century (Schulz, 2020). It is based on the precise reports of several explorers on the same caravan routes. The maps depict two points: the development of a large belt of *Acacia*-dominated savannas out of the regeneration and colonizing potential of the genus *Acacia* and the importance of the achabs, which allowed the enormous development of the pasture economy – especially that of camels.

POSTSCRIPT

An historical window of access and security allowed fieldwork for 30 years – with some interruptions – from the late 1970s on. Research for the traditional use of plants was possible in the late 1980s to reveal and to save that knowledge. It was done in a state of confidence. Today, the importance and value of medicinal and food plants are largely recognized by the Saharan people. The transfer to the next generation is assured. However, this is done within the families or to disciples having the state of confidence. A transfer to outsiders, however, will be hardly possible.

The last years saw a rising unpredictability of rainfall in the region, however with large inundations too. This will be a general feature in the near future (see IPCC, 2023). Thus, Achabs and pastures and subsequent animal breeding might well develop in the future.

However, the growing insecurity in the region interdicts classical fieldwork or any coring-without military protection-, and a general presence of landmines will make this situation persistent. Figure 6.39 depicts the Central Sahara as an epicenter of the drugs, weapons, and migration/slave trade as well as the different interferences of military and terror groups.

REFERENCES

Adamou, A., 1979, Agadez et sa région. *Etudes nigériennes*, **44**, pp. 1–358.

Adamou, A. and Morel, A., 2005, *Niger, Agadez et les montagnes de l'Air. Aux portes du Sahara.* (Paris: Editions de la Boussole).

Adjanohoun, E.J., Ahyi, A.M., Ake, K., Assi, L., Dan Dicko, l., Daoud, H., Delmas, M.S.W., Souza, A.S., Garba, M., Gouinko, S., Kayonga, A., N'golo, D., Raynal, L., and Saadou, M., 1980, *Médecine traditionnelle et Pharmacopée. Contribution aux études ethnobotaniques et floristiques au Niger.* (Paris: ACCT).

Adjanohoun, E.J., Ake-Assi, L. Floret, J.J., Guinko, S., Koumare, M., Ahyi, A., M., R., and Raynal, J., 1981, *Médecine traditionelle et pharmacopee. Contributions aux etudes ethobotaniques et floristiques au Mali.* (Paris: ACC).

Aguilar J. 1987, Soil Forming processes in desert soils. In: *Eremology*, edited by De Boodt and Hartmann, R. (Gent: Rijksuniversiteit Gent), pp. 83–94.

Amaral, P.C., Vincens, A., Gulot, J., Buchet, G., Deschamps, P., Doumnang, J.C., and Sylvestre, C., 2013, Palynological evidence for gradual vegetation and climate changes during the African Humid periods termination at 13° N from a Mega. Chad sedimentary sequence. Climate *of the Past*, **9**, pp. 223–241.

Amblard, S., and Pernes, J., 1989, The identification of cultivated millet (*Pennisetum*) amongst plant impressions on pottery from Oued Chebbi, (Dhar Oualata, Mauretania). *The African Archaeological Revue*, **7**, pp. 117–126.

Amossé, A. 1925, Diatomées des Oasis du Kaouar, de Djado et d'Agram (Sahara oriental). *Bulletin Muséum national d'histoire naturelle*, **31**, pp. 104–111.

Arbonnier, M. 2003, *Arbres, arbustes et lianes des zones sèches de l'Afrique de l'Ouest.* (Pont sur Yonne: CIRAD. MNHN).

Aumassip, G., Betrouni, M., and Hadu, S. 1982–1983, Une structure de cuisson des sauterelles dans les dépôts archéologiques de Tin Hanakaten (Tassili-n-Ajer, Algerie), *Libyca*, **30**, pp. 199–202.

Austen, R.A., 2012, *Sahara.* Tausend Jahre Austausch von Ideen und Waren (Berlin: Wagenbach).

Ballouche, A., 2019, Fire prone landscapes in the West African savanna. Present day dynamics and historical data. *Frankfurter Archäologische Studien*, **37**, pp. 17–30.

Banks, K., M., 1989, The appearance and spread of cattle keeping in Saharan North Africa. In: *Environmental changes and human culture in the Nile basin and northern Africa until the second millennium BC*, edited by Krzyzaniak, l., Kobusievicz, M., and Alexander, J. (Poznan: Archaeological Institute), pp. 57–59.

Barich, B., 2019, Herders-Foragers and low-level food production. Some insights into the early food production in northern Africa. In: *Archaeology of Food*, edited by Baldi, M., Dan, R., Delle Donne, M., Lucarini, G., and Mutri, G. (Rome: Science e Lettere), pp. 75–106.

Barry, J.P., Celles, J.C., and Musso, J., 1986, Le problème des divisions bioclimatiques et floristiques du Sahara. Note V : du Sahara au Sahel, Un essai de définition de cette marche africaine aux alentours. *Ecologia Méditerranea*, **XII**, pp. 187–235.

Baumhauer, R. 1986, Zur jungquartären Seenentwicklung im Bereich der Stufe von Bilma (NE Niger) *Würzburger geographische Arbeiten*, **65,** p. 235.

Baumhauer, R., 1990, *Paläoseen in der südlichen Sahara.* Habilitation Thesis. (Würzburg: University of Würzburg).

Baumhauer, R., 1991, Palaeolakes in the South-Central Sahara – problems of palaeoclimatological interpretation. *Hydrobiologica*, **214**, pp. 347–357.

Baumhauer, R, 1992, Radiocarbon- und Grundwassenisotopendaten aus NE-Niger. *Würzburger Geographische Arbeiten*, **84**, pp. 235–246.

Baumhauer, R., 1997, Zur Grundwassersituation im Becken von Bilma, zentrale Sahara. *Würzburger Geographische Arbeiten*, **92**, pp. 131–146.

Baumhauer, R. 2004, Die spätpleistozänen und holozänen Paläoseen in der zentralen Sahara- neue Ergebnisse aus der Ténéré, dem Erg de Ténéré und dem Erg de Fachi-Bilma, NE-Niger. *Die Erde*, **135**, 3, 4, pp. 289–313.

Baumhauer, R., Brauneck, J., Spronholz, B., Schulz, E., Faran Maiga, O., Ibrahim, S., and Pomel, S., 2009. Holocene palaeoenvironmental change in Central Sahara inferred from Seggedim scarp foot depression (NE-Niger). *Palaeoecology of Africa*, **29**, pp. 39–62.

Bergeret, A., and Ribot, J.C., 1990, *L'arbre nourricier en pays sahélien.* (Paris: Edition de la maison des sciences de l'homme).

Bernus, E., 1967, Cueillette et exploitation des ressources spontanées du Sahel nigérien par les Tamasheq. *Cahier. ORSTOM, sér Sc. Hum*, **IV**, I, pp. 31–52.

Bernus, E., 1969, Maladies humaines et animals chez les Tuareg saheliens. *Journal de la societe des Africanistes*, **XXXIX**, 1, pp. 111–137.

Bernus, E, 1980a, L'arbre dans le Nomads land. *Cahiers ORSTOM Sér Sci.Hum*, **XVII**, 3, 4, pp. 171–176.

Bernus, E., 1980b, Famines et sécheresses chez les Touaregs sahéliens. *Africa*, **50**, 1, pp. 1–7.

Bernus, E. 1984, La végétation. In: La région d'Ingall – Teggida-n –Tessemt. Programme archaeologies. *Etudes Nigériennes*, **48**, 1, pp. 107–122.

Biagetti, S., 2015, Desert pastoralists: the Kel Tadrart Tuareg from Southwest Libya. *Pastoralism, Research, Policy and Practice*, **5**, 6.

Blanco, J., Genin, D., and Carrièrere, S., 2015, The influence of Saharan agropastoralists on the structure and dynamics of *Acacia*-stands. *Agriculture, Ecosystem and Environment*, **213**, pp. 21–31.

Boudouresque, E., and Schulz, E., 1981, The flora and vegetation of NE-Niger (Djado, Kawar,Ténéré). *Willdenowia*, **11**, pp. 363–394.

Bouye, F. E., 2015, Risque trypanosomien et innovations. Le cas de éleveurs d'Afrique de l'Ouest. Thèse Université Montpellier II, p. 385.

Brachet, J., 2011, L'espace feulleté et segmenté d'une oasis saharienne. Rencontre et évitement a Dirkou (Niger). *Mappemonde*, **103**. http://mappemonde.mgm.fr/num31,articles/art11.302.html

Brass, M., 2013, Revisiting a hoary chestnut. The nature of early cattle domestication in North-East Africa. *Sahara*, **1**, 24, pp. 66–70.

Brierley, C., Manning, K., and Maslin, M., 2018, Pastoralism may have delayed the end of the green Sahara. *Nature Communication*. https://doi.org/1071038/s41467-018-06321-y

Brooks, N., Chiapello, Di Lernia, S., Drake, N., Legrans, M., Moulin, C., and Prospero, J., 2005, The climate-environment-society nexus in the Sahara from prehistoric times to the present day. *The Journal of North African Studies*, **10**, 3–4, pp. 253–292.

Burgarella, C., Dergier, A., Clemin, S., David, J., Dew, M., and Pol, D., 2002, The road to *Sorghum* domestication: evidence from nucleotide diversity and gene expression patterns, *Frontier Plant Science*, **12**, pp. 20–27.

Burkill, H.M., 1985 ff, *The useful plants of West Tropical Africa.* (Kew: The Royal Botanical Gardens).

Busche, D., and Heistermann, C. 1992, Wechselbeziehungen zwischen geomorphologischer und prähistorischer Forschung in der Sahara von Ost-Niger. *Würzburger Geographische Arbeiten*, **84**, pp. 169–200.

Busche, D., Grunert, J., and Hagedorn, H., 1979, *Der nördliche Schichtstufenrand des Murzukbeckens (Zentral-Sahara) als Beispiel für das Gefügemuster des ariden Formenschatzes.* (Göttingen: Festschrift Deutscher Geographentag), pp. 45–65.

Busson, F., 1965, *Plantes alimentaires de Ouest africain.* (Aix-Marseille: Thèse Fac. Sciences).

Byrnes, A. 2019. *Modelling early food production in the Mid Holocene of the Eastern Sahara- A sustainable rural livelihood approach.* London: Thesis UCL. https://pol-study.com

Caesar, G., J., 50/51 BC, *Bellum Gallicum.* (Roma: scripturae SPQR).

Caneva, J. 1993, The pastoral Middle Nile: local developments and Saharan contacts. In: *Environmental changes and human culture in the Nile basin and northern Africa until the second millennium BC.*, edited by Krzyzaniak, l., Kobusievicz, M. and Alexander, J. (Poznan: Arcchaeological Institute), pp. 405–411.

Carl, L., and Petit, J. 1954, *La ville de sel.* Du Hoggar au Tibesti (Paris: Rene Juillard).

Castaneda J.S., Mulitzka, S., Schefuß, E., Lopos dos Santos, D.A., Schnighe-Damsie, J.S., and Schouten, S., 2009, Wet phases in the Sahara/Sahel region and human migration patterns in North Africa. *The Proceedings of the National Academy of Sciences*, **106**, 48, pp. 159–2003.

Ceccato, P., Cressman, K, Gianni, A., and Trazaska, S., 2007. The desert locust upsurge in West Africa (2003-2005): Information on the desert locust early warning system and the prospect for seasonal climate forecasting *International Journal of Pest Management*, **53**, 1, pp. 7–73.

Chapelle, J., 1982, *Nomades noirs du Sahara.* (Paris: Harmattan).

Chapuis, M., P. Plantamp, C., Blondin, L., Foucart, A., and Bazele, P, 2013, *On the historical biogeography of the desert locust Schistocerca gregaria.* (Paris: Publications du CIRAD).

Cole, M., 1986, *The savannas*: Biogeography and Geobotany. (London: Academic Press).

Concelliere, E., and Di Lernia, S., 2014, Re-entering the central Sahara at the onset of the Holocene: A territorial approach to Early Acacus hunter-gatherers (SW-Libya). *Quaternary International*, **32**, pp. 43–62.

D'Attonosio, E. Trombetta, B, Bonito, M., Finocchio, A., Di Vito, G., Seghizzi, M., Romano, R., Russo, G. Paganotti, G.M., Watson, E., Coppas, A., Agnanostous, P., Dgoujon, J.-M., Moal, P., Selitto, D., Noveletto, A. and Criciani, F., 2018, The peopling of the last Green Sahara revealed by high coverage resequencing trans Saharan patrilineages. *Genome Biology*, **19**, pp. 20–26.

De Menocal, P., Ortiz, J., Guilderson, T., Adkins, J., Santhein, M., Baker, L., and Yarusinsky, M., 2000, Abrup onset and termination of the African Humid Period: rapid climate responses to gradual insolation forcing *Quaternary Science Reviews*, **19**, pp. 347–361.

De Rouv, A., 2004, Improving yields and reducing risks in pearl millet farming in the African Sahel. *Agricultural Systems*, **81**, 1, pp. 73–93.

Di Lernia, S., 2002, Dry climate events and cultural trajectories adjusting Middle Holocene pastoral economy of the Libya Sahara. In: *Droughts, food and culture*, edited by Hassan, F.A. (Dordrecht: Kluver), pp. 225–250.

Di Lernia, S., Tafuri, M.A., Gallianaro, M, Albaique, T., Balasoe, M., Cavorsdi, L, Fullagar, P., D., Mercuri,A., M Monaco, A., Perego, A., and Zerbori, A., 2013, Inside the "African cattle complex". Animal burrials in the Holocene Central Sahara. *PLoS One*, *8*, 2, e56879.

Di Lernia, S., Massamba N'Siala, L., and Zerboni, A., 2012, Saharan waterscapes, Traditional knowledge and historical depth of water management in the Acacus Mts. (SW Libya). In: *Changing deserts. Integrating people and the environment*, edited by Mol, L. and Sternberg, T. (Cambridge: The White Horse Press), pp. 101–128.

Dinnies, M., Schimmel, L., and Hoeltzmann, P., 2021, Holocene high-altitude dynamics on Emi Koussi, Tibest (Chad, Central Sahara). *Palaeoecology of Africa*, **35**, pp. 27–45.

Dubief, J., 1959/1963, Le climat du Sahara. *Travaux de l'Institut des Recherches Sahariennes*, **1,** 2, pp. 1–275.

Dunne, J., Di Lernia, S., Chlodnicki, M., Kherbouche, F., and Evershed, R.E., 2018, Timing and place of dairying inception and animal husbanbry practices across Holocene North Africa. *Quaternary International*, **471**, pp. 147–159.

Dunne, J., Höhn. A., Franke, G, Neumann, K., Breunig, P. Gillard, T., Walton-Doyle, and Evershed, R.P., 2021 Honey collecting in prehistoric West Africa from 3500 years ago. *Nature Communication*. https//doi.org/10.1038/s41467-22425-4.

Dunne, J., Mercuri, A.M., Evershed, R.P., Bruni, S., and Di Lernia, S., 2016, Earliest direct evidence of plant processing in prehistoric Saharan potters. *Nature Plants*, **3**, p. 16194.

Dupire, M., 1957, Pharmacopée peul du Niger et du Cameroun. *Bulletin IFAN*, **XIX**, ser B, pp. 382–417.

Dupuy, C.H., 2014, Des céréales et du lait au Sahara et au Sahel de l'épipaléolithique à l'Age des métaux. *Afriques* **5**, pp. 1–43.

EBI, 2023, *Ethiopian Center of origin and diversity.* (Addis Ababa), www.Ebi.Gov.ET (correct Link: https://ebi.gov.et/

Faegri, K., and Iversen, J., 1968, *Textbook of pollen analysis.* (Copenhagen: Munksgaard).

FAO, 1988**,** Traditional food plants. *FAO Food and Nutrition Pap*er, **42**, pp. 1–593.

FAO, 2011, FAO says international crop key facing climate change. *FAO-News*, pp. 1–2.

FAO, 2023a, *An international treaty on plant genetic resources for food and agriculture.* (Rome: FAO).

FAO, 2023b, FAO and Europe support relocation of Ukraine's national seed collection site. *FAO News*, April 26.

Feldmann, V., 1988, Comparative characteristics of impactite, tektite and fulgurite glasses. In: *International conference on natural glasses*, edited by Konta, I. (Prague), pp. 215–220.

Felix-Katz, J., 1980, Analyse éco-énergique d'un élevage nomade (Touareg) au Niger, dans la région de l'Azawad *Annales de Géographie*, **491**, pp. 57–72.

Fontes, J.Ch., and Gasse, F., 1991. PALHYDAF (Palaeohydrology in Africa) program: objectives, methods, major results. *Paleogeography Palaeoclimatology, Paleoecology*, **84**, pp. 191–215.

Fornaciari, R., Arru, L., Terenziani, R., and Mercuri, A.M., 2018. The role of morphometry to delineate change in the spikelet shape of wild cereals. The case study of Taborkori (Holocene, Central Sahara, SW Libya. In *Plants and people in the African past*, edited by Mercuri, A.M., d'Adra, A.C., Fornaciari, R., and Höhn, A (Cham: Springer), pp. 100–122.

Fuller, D.O., Barron, A., Champion, M., Dupuy, C., Commelin, D., Raimbault, M., and Denham, T., 2021, Transition from wild to domesticated pearl millet (*Pennisetum glaucum*) revealed in ceramic tempers at three Middle Holocene sites in Northern Mali. *African Archaeological. Review*, **38**, pp. 211–230.

Gabriel, B., 1986, *Die östliche Wüste im Jungquartär.* Berliner Geographische Studien, **19** (Berlin: Institut für Geographie der Technischen Universität Berlin).

Garcea, E.A.A., 2004, An alternative way towards food production: the perspective from the Libyan Sahara. *Journal of World Prehistory*, **18**, 2, pp. 107–154.

Gardi, R., 1978, *Ténéré. Die Wüste, in der man Fische fing.* (Bern: Benteli).

Gartung, W., 1992, *Yallah Tibesti.* Vom Tschadsee zu den Felsenmenschen (Braunschweig: Westermann).

Gast, M., 1968, *Alimentation des populations de l'Ahaggar : Etude ethnographique. Mémoires Centre Recherches, Anthropologique, Préhistorique, Ethnographiques.* **VIII**. (Paris: Arts et Métiers graphiques).

Gast, M., 2000, *Moissons du désert. Utilisations des ressources naturelles du Sahara central.* (Paris: Ibis Press).

Gauthier A., 1984, New data concerning the prehistoric fauna and domestic cattle from Ti-n-Torha (Acacus, Libya). *Paleoecology of Africa*, **16**, pp. 305–309.

Gauthier, A., 1987, Prehistoric men and cattle in North Africa; A dearth of data and a surfeit of models. In: *Arid North Africa*, Essays in honour of Fred Wendorf, edited by Close, A.E. (Dallas: Southern Methodist University Press), pp. 153–187.

Gentilini, M., 1994, *Médecine tropicale.* (Paris: Flamarion).

Giessner, K., 1984, Die Sahara im Überblick. In: *Die Sahara*: Mensch und Natur der größten Wüste der Erde, edited by Göttler, G. (Köln: DuMont Buchverlag), pp. 12–35.

Gifford-Gonzalez, D., 2000, Animal disease challenges to the emergence of pastoralism in Sub-Saharan Africa. *African Archaeological Review*, **17**, pp. 95–139.

Grouzis, M., and Le Floc'h, E., *Un arbre du désert. Acacia raddiana.* (Paris: IRD Editions).

Grunert, J., 1988, Frühholozäne Feuchtzeitböden auf Altdünen der Ténéré und des Tchigai Berglandes – Ost Niger, *Würzburger Geographische Arbeiten*, **69**, pp. 289–304.

Hachid, M., 1998, *Le Tassili des Ajer.* (Paris: Editions Paris-Méditerranée).

Hahn, A., and Kusserow, H., 1998, Spatial and temporal distribution of algae in soil crusts in the Sahel of W-Africa: Preliminary results. *Willdenowia*, **28**, pp. 227–223.

Hahn, K., Schmitt, M., and Thiombino, A., 2019. The potential of wild plant food resources in Burkina Faso/West Africa. *Frankfurter Archäologische Schriften*, **37**, 156–172.

Hiernaux P., Mougin, E., Diarra L., Soumaguel, D., Lavenu F., Tracol, Y., and Diawara, M., 2009, Sahelian rangeland response to change in rainfall over two decades in the Gourma region, Mali. *Journal of Hydrology*, **371**, 1–2, pp. 114–127.

Holl, A.F.C., 2020, Dark side archaeology: climate change and Mid-Holocene Saharan pastoral adaptation. *African Archaeological Review*, **37**, pp. 491–495.

Hugot, H.J., and Bruggmann, M., 1976, *Zehntausend Jahre Sahara.* (Bern: Bucher).

Humann, Cpt., 1933, Notes sur les pratiques médicales des Kel Hoggar. *Archives Institute Pasteur Algerie*, **XI**, 3, pp. 465–512.

Hureiki, J., 2000, *Les médecines touarègues traditionnelles.* (Paris: Edition Karthala).

Huysecom, E., 2012, Un néolithique "très" ancien en Afrique de l'Ouest. *Pour la Science*, **76**, pp. 86–91.

Huysecom, E., Ozainne, S., Raeli, E., Ballouche, A., Rasse, M., and Stokes, S., 2004, Ounjougou, (Mali): A history of Holocene settlement at the Southern edge of the Sahara. *Antiquity*, **78**, 302, pp. 41–53.

Ibrahim, S., and Schulz, E., 2019, At the sources of fear- The Guidimouni record (SE-Niger) and the history of desertification. *Zentralblat für Geologie und Paläontologie*, **I**, 2, pp. 11–25.

Ikiri, K., Saadou, M., and Garba, M., 1984. *Recherches sur la pharmacopée au Niger.* (Niamey: CELTO).

IPCC, 2023, *ARG synthesis report. Climate changes 2023.* (New York. United Nations).

Jacks, B., 1964, Living conditions and nutrition of some tamasheq groups in Mali under the influence of drought. *Ambio*, **23**(7), pp. 438–445.

Jaquadi, W., Hamroudi, L., and Khouija, M.L., 2012, Phénologie de *Acacia tortilis* ssp. *raddiana* dans le Parc National de Bou Hedma en Tunisie, effet du site sur les phénophases de l'espèce. *Bois et Forêts des Tropiques*, **312**, 2, pp. 31–41.

Joseph, A., 1991, L'Air 'Château d'eau' de la bande désertique des Ténérés (Niger). *Revue de Géographie alpine*, **79**, 1, pp. 71–86.

Joseph, A., and Aranyossy, J.F., 1992, Paleohydrologie de la nappe crétacée de Bilma (Niger). *Würzburger Geographische Arbeiten*, **84**, pp. 73–85.

Jousse, H., 2006, What is the impact of Holocene climatic changes on human societies? Analysis of West African neolithic populations dietary customs, *Quaternary International*, **151**, pp. 63–73.

Kaczmarek, T., van der Veen, M., and Gros-Balthazar, M., 2023, Origin and evolution of oasis agriculture in the Sahara: Evidence from morphometric analyses of archaeological date palm seeds. *The Holocene*, 34, 3, pp. 353–365.

Kitz, C., 2004, *Hungersnot in Niger, Bericht über ein Nothilfeprojekt für Caritas internationalis.* (Würzburg: Missionsärztliches Institut).

Klute, G., 1992, *Die schwerste Arbeit der Welt. Alltag von Tuareg-Nomaden.* (München: Trickster).

Krätli, S., 2008, Time to outbreed animal science. A cattle breeding system exploiting structural unpredictability: The WoDaBe herders in Niger. STEPS working paper 7. (Brighton: STEPS Centre).

Krings, T., 2006, *Sahelländer.* Mauretanien, Senegal, Gambia, Mali, Burkina Faso, Niger (Darmstadt: Wissenschaftliche Buchgemeinschaft).

Kröpelin, S.P., 2009, Holozäne Umweltrekonstruktion und Kulturgeschichte der Sahara: Perspektiven aus der sudanesischen Wüste. *Nova Acta Leopoldina*, **108**, 373, pp. 165–191.

Lambrecht, F.L., 1964, Aspects of evolution and ecology of Tsetse flies and trypanosomiasis in prehistoric African environments. *Journal of African History*, **V**, 1, pp. 1–24.

Larrasoane, J.C., Roberts, X.P., and Rolig, E.J., 2013, Dynamics of Green Sahara periods and the role in hominid Evolution, *PLOS One*, **8**/10, e76514.

Lee, R.B., 1970, Ecology of a contemporary San people. In: *The bushmen: San hunters and herders of Southern Africa*, edited by Tobias, P. and Biesele, M. (Toronto: Human and Rousseau).

Leonard, W.Z., 2003, Menschwerdung durch Kraftnahrung. *Spektrum der Wissenschaft*, **5**, pp. 30–38.

Lezine, A.M., and Lemonnier, K., 2021, Changes in the West African landscapes at the end of the African Humid period. *Paleoecology of Africa*, **35**, pp. 63–83.

Lezine, A.M., Zheng, EW., Braconnot, P., and Krinner, G., 2011, Late Holocene plant and climate evolution at Lake Yoa, northern Chad: pollen data and climate simulations *Climate of the Past*, **7**, pp. 1351–1362.

Li, H., Renssen, H., and Roche, D.M., 2022, Comparison of the green – To desert Sahara transition between the Holocene and the last interglacial. *Climate Past*, **18**, pp. 2303–2319.

Linseele, V., van Neer, W., Thys, S. Phillips, R. Cappers, R. Wendrich, W., and Holdaway, S., 2014, New archeozoological data from the Fayum Neolithic with a critical assessment of the evidence for early stock keeping in Egypt. *PLOS One*, **9**, 10, pp. 1–22.

MacDonald, K.C., 1998, Pastoralism and the origin of cultural complexity in the Sahel. In: *Transformations in Africa. Essays on Africa's later past*, edited by Connah, G. (London: Leicester University Press), pp. 92–107.

Magnavita, C., Daugbet, Z., and Bouimen, T., 2019, The Lake Chad region a crossroad. An archaeological and oral historical project on early Kanem-Borno and its intra African connections. *Afrique : Archéologie d'Art*, **15**. https://doi.org/10.4000/000.2654

Maley, J., 1983, Histoire de la végétation et du climat de l'Afrique nord-tropicale au Quaternaire récent. *Bothalia*, **14**, 3, pp. 377–389.

Maley, J., 2004, Le bassin du Tchad au quaternaire récent : formations sédimentaires paléoenvironnements et préhistoire. La question des paléotchads. In*: L'évolution de la végétation depuis deux millions d'années*, edited by Semah, A.M. and Renault-Miskowsky, J. (Paris: Errance Ed. Artcom), pp. 179–217.

Manning, K., 2010, A development history for early West African agriculture. In: *West African Archaeology*, edited by Allsworth-Jones, Ph., *BAR International series* 2104. (Oxford: Archaeopress), pp. 43–53.

Manning, K., and Thimpson, A., 2014, The demographic response to Holocene climate change in the Sahara. *Quaternary Sciences Reviews*, **10**, pp. 28–35.

Manning, K., Breese, P.S., Drake, N., Dunne, J., Casanova, E., and Evershed, R., 2023, Habitat fragmentation and the sporadic spread of pastoralism in the mid-Holocene Sahara. *Quaternary Science Reviews*. https://doi.org/10.1016/j.quascirev.2023.108070

Mauz, B., and Felix-Henningsen, P., 2005, Palaeosols in Saharan and Sahelian dunes of Chad. Archives of Holocene North African climate changes. *The Holocene*, 15, pp. 453–458.

Maydell, H.J., von, 1992, *Trees and shrubs of the Sahel. Their characteristics and uses.* (Weikersheim: Markgraf).

Mercuri, A.-M., and Garcea, E.A.A., 2007, The impact of hunter/gatherers on the vegetation in the Central Sahara. *Groningen Archaeological Studies*, **5**, pp. 87–103.

Mercuri, A.-M., Florenzano, A., Giraudi, C., and Garcea, E.A.A., 2015, Humans and water in desert „refugium" areas: palynological evidence of climate oscillations and cultural developments in Early and Mid-Holocene Saharan edges. *IANSA*, **VI**, 2, pp. 151–160.

Mercuri, A.-M., and Forenziari, A., 2013. Environmental and ethnobotanical data inferred from pollen of Gobero and the dried lakebeds in the surrounding area. *Journal of African archaeology monograph ser.*, **9**, 7, pp. 81–106.

Mercuri, A.-M., Fornaciari, R., Cellinaro, M., Vanin, St., and Di Lernia, S., 2018, Plant behaviour from human imprints, and the cultivation of wild cereals in the Holocene Sahara, *Nature Plants*, **4**, pp. 71–81.

Monod, Th., 1965, *Modes* "contracté et diffus" de végétation saharienne. In: *Biology of deserts*, edited by Thompson, C. (London: Institute of Biology), pp. 35–44.

Monod, Th., 1986, The Sahel zone north of the equator. In: *Ecosystems of the world. Hot deserts and arid shrublands*, edited by Evenary, M., Nov-Meir, J. and Goodall, D.W. (Amsterdam: Elsevier), pp. 203–243.

Munkert, H.C., 2009, Sexual and vegetative regeneration of three leguminous tree species in South African savannas. *South African Journal of Botany*, **75**, pp. 606–610.

Musch, T., 2018, Sorten der Kulturdattel (*Phoenix dactylifera*) im Kaouar-Tal (Niger). *Mitteilungen der Deutschen Dendrologischen Gesellschaft*, **103**, pp. 186–191.

Musch, T., 2020, Am Rande und im Zentrum- die Teda und das Tibesti. *Zentralblatt Geologie Palaeontologie*, 1, pp. 151–181.

Musch, T., 2023, Melipones in the hyper arid desert: Evidence of Hypotrigona ruspolii (Magretti 1898) in the Air Mountains, Central Sahara. *Entomologie heute*, 34, pp. 1–6.

Muzzolini, A., 1991, Le début de la néolithisation du Sahara et les gravures rupestres les plus anciennes. *Préhistoire Ariegeoise*, **XLVI**, pp. 211–233.

Neumann, K., 1988, Die Bedeutung von Holzkohleuntersuchungen für die Vegetationsgeschichte der Sahara- das Beispiel Fachi/Niger. *Würzburger. Geographische Arbeiten*, **69**, 7, pp. 1–86.

Neumann, K., 1989, Zur Vegetationsgeschichte der Ostsahara im Holozän. Holzkohlen aus prähistorischen Fundstellen. *Africa Praehistorica*, **2**, pp. 13–181.

Neumann, K., 1994, Wirtschaftsweisen im Neolithicum der Ostsahara und ihr Einfluß auf die Vegetation. *Colloquium Africanum*, **I**, pp. 47–65.

Neumann, K., 2003, The late emergence of agriculture in Sub Saharan Africa: archaeological evidence and ecological considerations. *Africa Praehistorca*, 15, pp. 71–92.

Nicolaisen, J., 1963. *Ecology and culture of the pastoral Tuareg*. (Copenhagen: National Museum).

Nicolaisen, J., and Nicolaisen, I., 1997, *The Pastoral Tuareg*: Ecology, Culture and Society. (Copenhagen: Rhodos).

Nicoll, K., 2004, Recent environmental change and prehistoric human activity in Egypt and northern Sudan. *Quaternary Science Reviews*, **23**, 561–580.

Nixon-Darcus, L., Meresa, Y., and D'Andrea, A.C., 2024, Models of grinding stones, manufacture use and discard from Tigrai, Ethiopia. Opportunities for cultural comparisons and implications for use ware analysis. *Journal of Field Archaeology*, **49**, 4, pp. 453–456.

Ozaine, S., Major, A., Huysecom, E., and Robion Brunner, C., 2009, Une chronologie pour le peuplement et le climat du pays dogon. La séquence culturelle et environnementale d'Ounjougou (Mali). *Anthropo*, **18**, pp. 37–46.

Pachur, H.-J., and Altmann, N., 2006, *Die Ostsahara im Spätquartär*: Ökosystemwandel im größten hyperariden Raum der Erde (Berlin: Springer).

Pamo, E.T., 1998, Herders and wild game behavior as a strategy against desertification in northern Cameroon. *Journal of Arid Environments*, **39**, pp. 179–190.

Pomel, S., 2008, *La mémoire des sols*. (Bordeaux: Presses Universitaires de Bordeaux).

Pomel, S., Pomel-Rigeaud, F., and Schulz, E., 1994, Indicateurs anthropogènes de la végétation et des sols de quelques savanes de l'Afrique de l'Ouest. In: *Enregistreurs et indicateurs de l'évolution de l'environnement en zone tropicale*, edited by Maire, R., Pomel, S. and Salomon, L.M. (Bordeaux: Presses Universitaires de Bordeaux), pp. 173–200.

Purseglove, J.-W. 1972, *Tropical crops. Monocotyledones*. (London: Longman Group Limited).

Quezel, P., 1965, *La végétation du Sahara du Tchad à la Mauritanie*. (Stuttgart: G. Fischer).

Richerson, P.Y., and Boyd, R., 2001, Institutional evolution in the Holocene: The rise of complex societies. *Proceedings of the British Academy*, **110**, pp. 197–239.

Ritter, H., 1980, *Die letzten Karawanen in der Sahara*. (Luzern: Atlantis).

Ritter, H., 1994, Medizinische Traditionen der Twareg. *Perspektiven afrikanischer Forschung*, pp. 297–322.

Roset, P., 2000, Palaeoclimate and cultural conditions of Neolithic development in the Early Holocene of northern Niger (Air and Ténéré). In: *Prehistory of arid North Africa*, edited by Close, A. (Dallas: Southern Methodists University Press), pp. 211–234.

Sadig, A.M., 2009, Neolithic adaptations and subsistence economy in the Middle Nile Region, Sudan. In: *Water culture and identity, comparing, past and present traditions in the Nile Basis region*, edited by Oestigaard, T (Beren: Bric Press), pp. 237–272.

Scharaika, N., Graef, F., Moset, M., and Becker, K. 2000, Pastoral migration as a method of goal oriented and site specific animal nutrition among the Woodabe in Southeastern Niger. *Die Erde*, **131**, pp. 312–329.

Schultze, E., 1994, Les indicateurs limniques holocènes au nord-est de Niger. In: *Enregistreurs et indicateurs de l'évolution de l'environnement en zone tropicale*, edited by Maire, R., Pomel, S. and Salomon, J.-N. (Bordeaux Presses Universitaires de Bordeaux), pp. 31–54.

Schulz, E., 1990, Zwischen Syrte und Tschad. Der heutige Pollenniederschlag in der Sahara. *Berliner Geographische Studien*, **33**, 193–220.

Schulz, E., 1994a, Indicateurs de l'influence anthropique sur la végétation actuelle *et passée.* In: *Enregistreurs e indicateurs de l'évolution de l'environnement en zone tropicale*, edited by Maire, R., Pomel, S. and Salomon J.-N. (Bordeaux: Presses Universitaires de Bordeaux), pp. 129–144.

Schulz, E., 1994b, The southern limit of Mediterranean vegetation in the Sahara during the Holocene. *Historica Biology*, **9**, pp. 137–156.

Schulz, E., 1999, The southern margin of the Sahara in the Republic of Chad. Vegetation, soil and present pollen rain *Zentralblatt. Geologie. Paläontologie*, **I**, pp. 483–496.

Schulz, E., and Adamou, A., 1992, *Leben in der südlichen Sahara. Die traditionelle Nutzung der Vegetation in Nord-Niger.* (Würzburg: Bericht Institut für Geographie).

Schulz, E., and Adamou, A., 1997, Die Grenzen der "neolithischen Revolution". Gab es einen frühen Ackerbau in der Sahara? *Würzburger Geographische Arbeiten*, **92**, pp. 71–96.

Schulz, E., and Pomel, S., 1994, La bordure Sud du Sahara : du désert a la savane. In: *Enregistreurs et indicateurs de l'évolution de l'environnement en zone tropicale*, edited by Maire, R., Pomel, S. and Salomon, L. M (Bordeaux: Presses Universitaires de Bordeaux), pp. 143–171.

Schulz, E., 2020, Africa in the 19th century. *Zentralblatt. Geologie-Paläontolgie*, **1**, 1, pp. 247–278.

Schulz, E., Abichou, A., Hachicha, T., Pomel, S., Salzmann, U., and Zouari, K., 2002, Sebkhas as ecological archives and the vegetation and landscape history of southeastern Tunisia during the last two millennia. *Journal of African Earth Sciences*, **34**, 3–4, pp. 223–230.

Schulz, E., Adamou, A., Ibrahim, S. Ousseini, I., and Herrmann, L., 2021, Vegetation dynamics. Natural versus cultural and regeneration potential. The example of Sahara-Sahel. In: *Plant Communities and Their Environment* edited by Oliveira, M. T., Candan, F. and Fernandes-Silva, A. (IntechOpen). https://doi.org/10.5772/intechopen.87030.

Schulz, E., Almohamad, H., and Ibrahim, S., 2020, Drugs, flight and migration: A comparison of North and West Africa, Near East and Europe. *Studia UBB Geographia*, **LXV**, 1–2. pp. 9–39.

Schulz, E., Joseph, A., Baumhauer, R., Schultze, E., and Sponholz, B., 1990, Upper Pleistocene and Holocene history of the Bilma region (Kawar, NE-Niger). *Etudes récentes sur la géologie de l'Afrique Nancy: 5th Coll. of African History* (Nancy), pp. 281–284.

Seignobos, Ch., 2017, *Des mondes oubliés.* Carnets d'Afrique (Marseille: Editions Parenthèses).

Sereno, P.C. Garcea, E.A.A., Jousse, H., Stojanowsky, C.M., Saliè, J.F., Maga, A., Ide, O.A., Knudson, K.J., Mercuri. A.M., Stafford, T.W., Kaye, T.G., Giraudi, C., N'siala, I.M., Cocca, E., Moots, H.M., Dutheil, D.B., and Stivers J.P., 2008, Lakeside cemeteries in the Sahara: 5000 years of Holocene population and environmental change *PLoS One*, **3**, pp. 8–22.

Servant, M., 1983, *Séquence continentales et variations climatiques. Evolution du bassin du Lac Chad au cénozoïque supérieur.* (Paris: Travaux, et documents de l'ORSTOM).

Skaf, R., Popov, G.B., and Raffey, J., 1990, The desert locust: an international challenge. *Philosophical Transactions of the Royal Society of London*, ser B, **328**, pp. 525–538.

Smith, A.B., 1974, *Adrar Bous and Karkarichinkat.* (Berkeley: University of California).

Spittler, G., 1989, *Dürren, Krieg und Hungerkrisen bei den Kel Ewey (1900–1985).* Studien zur Kulturkunde, 89 (Wiesbaden: Franz Steiner).

Spittler, G., 1994, Hungerkrisen im Sahel. Wie handeln die Betroffenen? *Geographische Rundschau*, **7–8**, pp 408–413.

Sponholz, B. 1992, Karstverfüllungen in Ostniger. *Würzburger Geographische Arbeiten*, **84**, pp. 131–146.

Sponholz, B., 2009, New discovery of rock fulgurites in the Central Sahara. *Palaeoecology of Africa*, **29**, pp. 127–136

Sponholz, B., Baumhauer, R., and Felix-Henningsen, P. 1993, Fulgurites in the Central Sahara, Republic of Niger and their palaeoenvironmental significance. *The Holocene*, **3**, 2, pp. 97–104.

Stich, A., 2021, Die Ausbreitung vektorübertragener Krankheiten am Beispiel von Malaria und Schlafkrankheit *Petermanns Geographische Mitteilungen*, **201**, 3, pp. 32–38.

Stokker, O., 1962, Steppe-Wüste-Savanne. *Veröffentlichungen des Geobotanischen. Institutes ETH Zürich (Rübel)*, **37**, pp. 234–243.
Stokker, O., 1976, The water-photosynthesis syndrome and the geographical plant distribution in the Saharan desert. In: *Analysis and synthesis*, edited by Lange, O., Kappen, L. and Schultze, E.D. *Ecological Studies*, **11**, pp. 506–521.
Striedter, K.H. 1984, *Felsbilder der Sahara*. (München: Prestel).
Sys, C., 1987, Classification and mapping of desert soils. In. *Eremology*, edited by De Boodt, M. and Hartmann, D. (Gent: Rijksuniversiteit Gent), pp. 95–105.
Tauveron, M. and Striedter, K.H., 2009, Neolithic domestication and pastoralism in Central Sahara. The cattle necropolis of Man Khor (Tadrard, Acacus). *Palaeoecology of Africa*, **29**, pp. 179–189.
Temple, A.H., and Stojakowsky, C.M., 2019, Interrogating the alterity of hunter-gatherers in bioarchaeological context: adaptability, transformation, mobility and resilience of hunter-gatherers in the past. In: *Hunter- gatherers – Adaptation and resilience*, edited by Temple, D.H. and Stojakowsky. (Cambridge: Cambridge University Press), pp. 1–25.
Tews, J., Schurr, F., and Jeltsch, F., 2004, Seed dispersal by cattle may cause shrub encroachment of *Grewia flava* on southern Kalahari rangelands. *Applied Vegetation Science*, **7**, pp. 89–102.
Thiemeyer, H., 2000, From Megachad to Microchad- environmental changes during the Holocene. *Berichte Sonderforschungsbereich*, **268**, 14, pp. 11–19.
Thomas, V. and Nigaru, S., 2018, Twentieth-Century climate change over Africa: seasonal hydroclimate trends and Sahara Desert expansion. *Journal of Climate*, **31**, 9, pp. 3349–3370.
Todd, E., Chauvey, L., Fayes, A., Seguin-Orlando, A., Clavel, P., Khan, U., Perez-Pardarl, L., and Orlando, L. 2020, The genomic history and global expansion of domestic donkeys. *Science*, **377**, 6611, pp. 1172–1180.
Traore, K., Sidibe, d.K., Coulibali, H., and Dayala, J., 2017, Optimizing yield of improving varieties of millet and *Sorghum* under highly variable rainfall conditions using contour ridges in Cinzana Mali. *Agriculture and Food security*, **6**, 11. https://doi.org/10.1186/s40066-016-0086–0
Tubiana, M.J. 1969, La pratique actuelle de la cueillette chez les zasghawa du Tchad, *Journal de l'agriculture traditionnelle et de botanique*, **XVI**, 215, pp. 55–83.
Ullmann, I. 1985, Tagesgänge von Transpiration und stomatärer Leitfähigkeit Sahelischer und Saharischer Akazien in der Trockenzeit. *Flora*, **176**, pp. 389–409.
Vernet, R., 1993. Le Néolithique récent dans le sud-ouest du Sahara. In*: Environmental Changes and human culture in the Nile basin and northern Africa until the second millennium BC*, edited by Krzyzaniak, l., Kobusievicz, M. and Alexander, J. (Poznan), pp. 91–101.
Vernet, R., and Striedter, K.H., 1992. La place du Djado dans le Sahara Central. Hommes et climats à l'Holocene. *Würzburger. Geographische Arbeiten*, **84**, pp. 201–234.
Vogg, R., and Meckelein, W., 1991, Bodengeographische Aspekte zum Südrand der Sahara. *Stuttgarter Geographische Studien*, **116**, pp. 1–44.
Völkl, J., 1988, Zum jungquartären Klimawandel im Saharischen und Sahelischen Ost-Niger aus bodenkundlicher Sicht. *Würzburger Geographische Arbeiten*, **69**, pp. 255–270.
Volpato, G., and Saleh, S.M.L., 2015, Ethnoveterinary of Sahrawi pastoralist of Western Sahara: Camel diseases and remedies. *Journal of Ethnobiology and Ethnomedicine*, **116**, pp. 54–78.
Walter, H., 1964, *Die Vegetation der Erde in ökophysiologischer Betrachtung*. (Jena: VEB Gustav Fischer Verlag).
Walter, H., and Walter, E., 1953, Das Gesetz der relativen Standortskonstanz; das Wesen der Pflanzengemeinschaften. *Berichte. Deutsch. Botanische Gesellschaft*, **66**, pp. 228–236.
Wasylikowa, K., and Dahlberg, J., 1999, Sorghum in the economy of the early neolithic nomadic tribes at Nabta Playa, southern Egypt. In: The *exploitation of plant resources in ancient Africa*, edited by Van der Veen, M. (Dordrecht: Kluver), pp. 11–32.
Watrin, J., Lezine, A.M., and Hely, Ch. 2009, Plant migration and plant communities at the turn of the Green Sahara. *Comptes Rendues. Geosciences*, **341**, pp. 656–670.
Widstrand, C.G., 1975, The rationale of nomad economy. *Ambio*, **4**, pp. 146–153.
Williams, A.j., Buck, B.J., and Beyene, M.A., 2012, Biological soil crusts in the Mojave Desert, USA: Micromorphology and Pedogenesis. *Soil Sciences Society of America Journal*, **76**, pp. 1685–1695.
Wilson, R.T., 1978, The 'gizu' grazing in the South Libyan desert. *Journal of Arid Environments*, **1**, pp. 327–344.
Winchell, T., Stevens, C.J., Murphy, C., Champien, L., and Fuller, D.Q., 2017, Evidence of *Sorghum* domestication in fourth millennium BC Eastern Sudan. Spikelets morphology from ceramics impressions of the Butana group. *Current Anthropology*, **58**, pp. 673–683.
Wright, D.K., 2011, Frontier animal husbandry in the Northeast and East African neolithic. *Journal of Anthropological Research*, **67**, pp. 213–244.

Wright, D.K., 2017, Humans as agents in the termination of the African Humid Period. *Frontiers in Earth Science*. https://doi.org/10.3389/feart.2017.00004

Xue, Y., 2018, A history of male immigration in and out of the Green Sahara. *Genome Biology*, pp. 19–30.

Zabel, G., 1983, *Expérimentation de techniques d'irrigation en zone sahélienne*. (Eschborn: GTZ).

Zehdi-Azouzi, S., Cherif, E., Moussoni, S., Gros-Balthazard, M., *et al.*, 2015, Genetic structure of the date palm (*Phoenix dactylifera*) in the Old World reveals a strong differentiation between eastern and western population. *Annals of Botany*, **116**, 1, pp. 101–112.

Zerboni, A., and Nicoll, K., 2019, Enhanced zoogeomorphological processes in North Africa in the human-impacted landscapes of the Anthropocene, *Geomorphology*, **331**, pp. 22–35.

The Last Three Millennia of Environmental History around Lake Kivu

7

Chantal Kabonyi Nzabandora and Emile Roche†

INTRODUCTION

The environmental evolution of the last three millennia around Lake Kivu has been interpreted from palynological studies carried out on the Congolese ridge within the Kahuzi National Park at high altitude. At low altitude, palynological analyses were carried out around Bukavu city. On the Congo-Nile ridge, the Holocene palaeoenvironmental history of Rwanda was only known through work relating to the main peatlands of the Volcanoes region, to the northwest: Karisoke, Kiguhu and Rugezi. As regards the south of the country, two studies have been carried out on the large Kamiranzovu marsh.

We were careful to consider the two ridges separately in the analysis of the palynological data.

On the eastern ridge of Kivu, subject to rather contrasting seasons, the environmental episodes that followed one another during the Holocene are quite easily discernible through the movements of the vegetation levels on the mountainside.

On the western ridge, however, with more uniform seasons, the indicators of climate change are to be sought within the Afro-mountain vegetation itself, which changes appearance over time according to antagonistic influences. The approach to the problem must therefore be more nuanced here in order to make an appropriate judgment on the evolution of the palaeoenvironment.

In Kivu, the number of months of dry season differs by 1 to 2–3 and annual rainfall of 200–300 mm less and slightly higher temperatures of 1–2 °C in Rwanda (Lebrun, 1956; Sirven *et al.*, 1974).

DOI: 10.1201/9781003727781-7

FIGURE 7.1 Localization of sites

Proportionally, this difference must have existed previously, according to the climatic variations of the past, which explains the different evolutions of the wetlands on the two ridges.

As showed on the Figure 7.1, different sites were prospected on the Congolese ridge and the Congo-Nile ridge during this study.

Objectives

The objectives of this study are two-fold:

1. On the basis of palynological analyses carried out in the Congolese and Congo-Nile ridge, to reconstruct the environmental evolution around Lake Kivu during the last three millennia while distinguishing the climatic and anthropogenic influences that have shaped the landscape.

2. By supplementing the previous work carried out on the study site, draw up a synthesis situating the different evolutionary phases of the forest massif around Lake Kivu in a chronological framework by allowing for evacuation.

GEOGRAPHICAL CONTEXT

Orography

The Congolese ridge is located in the Kivu region in the east of the Democratic Republic of Congo. It is dominated by the Mitumba chain, which culminates at 3310 m altitude at Mount Kahuzi, NW of Bukavu (2°30′50″ S – 28°50′37″ E), capital of South Kivu (Kabonyi Nzabandora *et al.*, 2017).

The Congo-Nile ridge, oriented SW – NE, is affected by a major set of faults; it presents escarpments dissected by small torrential rivers tributary to Lake Kivu (1462 m asl.) located at the bottom of this tectonic gap. Often, the upper course of these rivers presents a senile appearance, forming vast swampy expanses, some of which have evolved into peat bogs (Ilunga, 1991). The Congo-Nile Dorsal, on the other hand, has a width of 20 to 50 km and stretches over a length of 160 km. Its altitude varies between 1,500 and 2,990 m asl. This dominates Lake Kivu and constitutes the watershed zone between the Congo and Nile basins. To the east of the ridge stretches a vast "plateau" whose altitude is between 1,500 m and 2,000 m asl. This Central Plateau or Region of the Hills covers about half of the country over a width of about 80 km. This surface, cut by rivers, has a multitude of elongated hills, with convex slopes, often steep, with a more or less flat summit.

Climate

In intertropical Africa, climatic variations conditioned by the movement of atmospheric masses influence the distribution of large plant communities. In this area, the climate depends on the trade winds.

Submitted to the influence of the trade winds from the southeast, which blow from the Indian Ocean and pass drying over East Africa, Rwanda has a continental climate, but its rigor is tempered by the relief and the nearby lakes. The climate is temperate on the Central Plateau and cool and humid on the ridge.

In Kivu, there is a foehn effect, a wind valley which creates drying conditions at Lake Kivu and its surroundings that favor the development of meso-xerophilic vegetation (Lebrun, 1935).

This zone is also characterized by a regime of alternating winds that determines a local microclimate, lake and land breezes (Leonard, 1962).

Temperatures

The distribution of temperatures in the area of study follows a thermal gradient of altitudinal decrease estimated at 0.7 °C per 100 m of drop (Liben, 1962). Rather small deviations from this theoretical gradient come from local causes, mainly exposure to the wind or topography. Thus, valleys and depressions at high altitude generally have lower than normal temperatures due to the accumulation in the bottom of masses of cold and humid air.

On the Central Plateau, with the elevation of the relief, the average annual temperatures are lower, around 18°–20 °C. The highlands of the ridge have a harsher thermal feature with averages of 15°–17 °C and finally, on the eastern side of Kivu, the annual averages recorded are 21 °C at lake level and 20 °C at 1,600 m asl.

Precipitation

Precipitation in the study region is determined by the combined action of air currents and altitude. The eastern side of the ridge is particularly well watered all year round.

The main engine of the rains that water Rwanda is the trade winds from the southeast. It causes, with altitude, more or less abundant rains up to the level of maximum precipitation located approximately between altitudes 2200 m - 2400 m asl (Liben, 1962). Beyond this zone, there is rather a humid atmosphere maintained by drizzles and fog.

The amount of precipitation is inversely proportional to the temperatures (Ntaganda, 1991). As one moves towards the west, the rainfall increases, passing to 1,200–1,300 mm/year on the Central Plateau and to 1,500 mm/year and more on the dorsal. Going down the western escarpment of the Dorsal, the air masses become drier, and the amount of water released decreases significantly to about 1100 mm/year (Sirven *et al.*, 1974).

In Rwanda, the period from June to August is generally considered to correspond to the dry season. It is three months for the Central Plateau and the edge of the lake. For high altitude regions such as the Congo-Nile Dorsal, the orographic effect is manifested by a reduction of the dry season to just over one month in July-August (Ilunga *et al.*, 2004).

At the lake level, the rainfall is 1,300–1,400 mm/year. On the Congolese ridge, the maximum rainfall occurs at an altitude of between 2,200 m and 2,600 m asl, where it can reach 2,000 mm/year and come in the form of violent showers. At higher altitudes, precipitation decreases to turn into fine rain and fog.

Vegetation

Currently, less than 10% of the area of Rwanda is still covered by natural vegetation. The flora of the eastern reliefs of the Albertine rift is therefore mainly known from works which are already dated by Devred (1958), Lebrun (1956), Liben (1962), Troupin (1966) but which give a more reliable account of the aspect of the ancient landscapes of the country.

In Rwanda, the lower limit of the dense humid mountain forest is around 1700 m asl on the Congo Nile ridge. On the latter, the ombrophile mountain forest characterizes the territories where rainfall is always higher than 1,500 mm per year and where the temperatures (minimum and maximum), between 11 °C and 18 °C, are lowered by the reliefs. This forest is dominated by tall trees whose height can reach 30–40 m.

A peculiarity of the ridge can be seen in the Nyungwe forest, with locally a staggering of the vegetation opposite to that known on the other mountains of intertropical Africa (Ntaganda, 1991; Habiyaremye *et al.*, 1995): stands with *Hagenia abyssinica* are inserted in a discontinuous lower stage of *Philippia johnstonii* surmounted by an upper stage of *Macaranga neomildbraediana* and *Syzygium parvifolium*. This arrangement would correspond to a local microclimate prevailing in the deep valleys where thermal inversions cause harsher conditions in the bottoms and on the slopes, unlike the summits which remain more temperate.

The various depredations of anthropogenic origin have allowed the development, to the detriment of the ombrophilous forest, of fast-growing heliophilous species characteristic of secondary forests; we can mention, in particular: *Bridelia brideliifolia*, *Croton macrostachyus*, *Dombeya goetzenii*, *Hagenia abyssinica*, *Macaranga neomildbraediana*, *Neoboutonia macrocalyx*, *Polyscias fulva*.

However, despite the clearings that reduced the forest area around it from 1955 to 1986 (Peeters, 1989), the area preserved by a prohibition since the 1990s to conserve biodiversity still allows a fairly close representation of what was the original vegetation.

On the central Plateau, between 1,500 m and 1,700 m asl., the relics of wooded formations that once existed in the region have practically disappeared. In the 1950s, one could still find shreds of wooded savannas with various Acacias where other woody species also grew.

The Central Plateau being currently for more than 90% covered with Eucalyptus crops and plantations, it is so difficult to imagine what the ancient vegetation must have been, except by referring to already old publications describing natural environments that were certainly fragmented but still observable in the past (Lebrun, 1956; Liben, 1962). On the Congolese ridge, the classification of vegetation proposed by Liben (1962) appears much more rational. The different stages are defined according to the LANG index, the ratio between precipitation and temperature (P/T). By applying it to the eastern flank of the Congo ridge in the Kivu region, an altitudinal zonation of forest potential can be established as follows:

- P/T = from 50 to 65: from 1500 m to 1600 m – xero-mesophilic formations
- P/T = from 65 to 85: from 1600 m to 1900 m – mesophilic formations
- P/T = from 85 to 150: from 1900 m to 2700 m – mountain rain forest
- P/T = from 150 to 240: from 2700 m to 3200 m – afro-subalpine formations
- P/T > 240: above 3200 m – afro-alpine formations

MATERIALS AND RESEARCH METHODS

Sampling

At the sites studied (fig. 7.1), the soundings were carried out using a "Russian" type core. A brief description of the vegetation of the marshes and their surroundings has been made on the basis of investigations and botanical inventories carried out. Samples were also taken to carry out ^{14}C dating.

Dating

The conventional ^{14}C dating of the sampling carried out on the sequences of the Congolese ridge was carried out at the laboratory of the Centrum voor Isotopen Onderzoek in Groningen (NL). For sites that have already been the subject of previous studies (Runge, 1999, 2001b), the analyses were carried out at the laboratory of the Niedersächsisches Landesamt für Bodenforschung in Hannover (D).

Sample Processing

The fossil material was extracted using the acetolysis process, a technique developed by Erdtman (1960). For samples containing mineral particles (silica, silicates, *etc.*), acetolysis is preceded by HF treatment (40%).

Method of Analysis

The determination of taxa was carried out under a Zeiss Jenaval light microscope by using a collection of reference blades and an iconographic collection from the University of Liège (Belgium). A minimum of 500 grains was counted for each blade.

The proportions of each taxon at all analyzed levels (every 10 cm) were expressed in relative frequencies, i.e., in relation to the pollen grains counted. The relative frequencies of taxa representative of the regional flora were calculated by excluding the local flora, that of the swamp. The latter is expressed as a function of all the pollens counted.

In the pollen diagrams, the identified taxa are classified according to the types of vegetation they represent as described in the botanical works relating to the region concerned (Lebrun, 1935, 1936, 1956,

Robyns 1948; Pecrot and Leonard, 1960; Liben, 1962; Moscol-Olivera and Roche, 1997) and according to field observations made during the study.

Pollen taxa have thus been grouped into palyno-ecological groups characterizing the mountainous areas of the Congolese Kivu ridge, namely:

For the regional flora:

- afro-alpine and subalpine formations
- primary mountain rain forest
- secondary mesophilic mountain forest
- open woody and herbaceous environments of the savanna category

For the local flora:

- marshy environments

On the Rwandan side, the site of Gasare was characteristic of the mountainous areas of the Congo-Nile ridge, and the one of Butare was characteristic of the hills.

RESULTS AND DISCUSSION

The African Holocene is characterized by periods of rain and drought. The latter specializes in opening up the forest environment and lowering the water level of African lakes. The rainy periods are characterized by forest advances and rising water levels. In Africa, the greatest dry Holocene period took place around 4,000 years B.P. and is characterized by a more open stage of forest environments, suggesting a reduction in precipitation and ambient humidity.

In Kivu, the dry period of 4,000 years bp (fig. 7.2) finds its equivalent in a serious drop in the level of Lake Kivu, which, in the long term, will lead to the closure of the basin (Haberyan and Hecky, 1987). In the forest environment, on the other hand, there is an advance of the *Poaceae* which can be assimilated to the OPP (Older Poaceae Period) defined by Runge (1999; 2001b) and which reflects an expansion of altimontane meadows, natural grassy areas whose development is linked to a period of water deficit.

After this regressive period, pollen archiving stops in the various marshes around Lake Kivu, probably under the effect of drought. At the end of this period, we see an advance of *Sinarundinaria* looming which will expand thereafter. This event would result from a wet pulsation located, according to our dating, between ca. 4,000 and ca. 3,000 years bp. It finds its equivalent, in an attenuated way, on the eastern ridge of the Central African Rift, in Rwanda and Burundi (figs. 7.3, 7.4) (Roche *et al.*, 1988; Roche and Bikwemu, 1989; Roche and Ntaganda, 1999).

By combining the data from the palynological analysis of the sedimentary sequences studied on the Congolese ridge and those of the Congo-Nile, the differences in the evolutionary phases of the high and medium altitude environments around Lake Kivu have been highlighted for the last three millennia of the Holocene.

This evolution is as follows (Figure 7.3):

At 4,000 years bp, mountain grasslands occupied large areas within the forest, while at medium and low altitudes, we observe the expansion of savannahs naturally derived from the mesophilic forest environment. After this climatic crisis, we notice that the forest is being reconstituted at all levels of the ridge (Figure 7.2).

Around 3,000 years bp, a meso-ombrophilous forest with intrusive afro-subalpine elements occupies the Congo-Nile ridge, while, on the Central Plateau, a wooded grassy savanna develops. These heterogeneous sets evoke environments subject to cold and dry climatic conditions, with contrasting seasons.

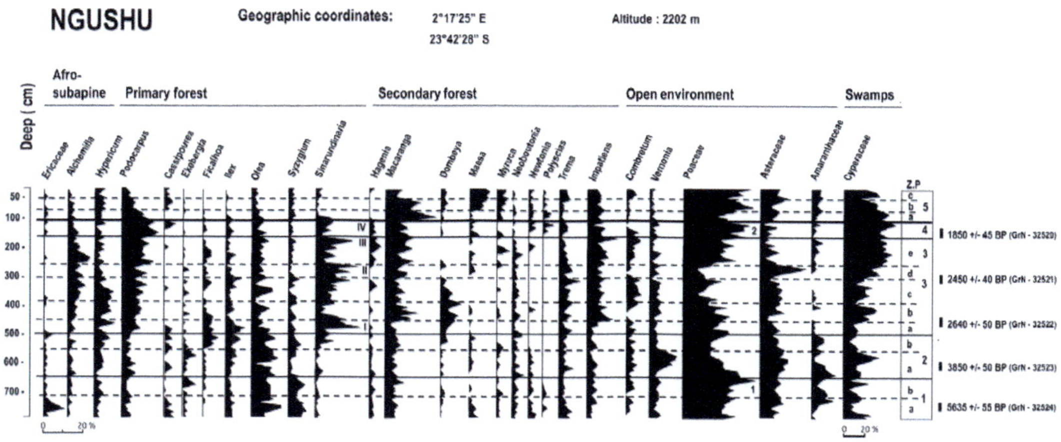

FIGURE 7.2 Detailed pollen diagram of NGUSHU site.

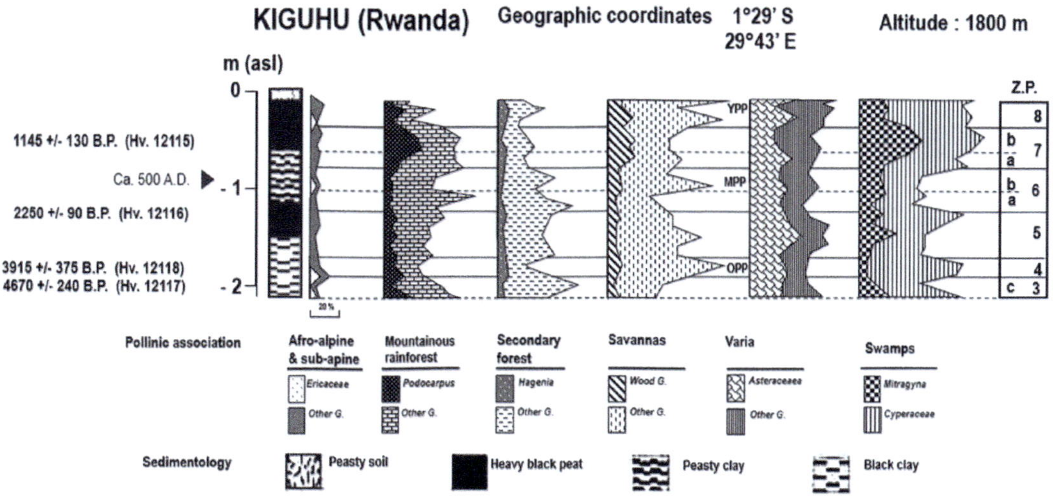

FIGURE 7.3 Detailed pollen diagram of KIGUHU site.

A new period of climatic instability sets in around Lake Kivu around ca 2,600 to ca 2,000 years bp.

Between 2,200 and 1,800 years bp is the expansion of a dense umbro-mesophile forest on the high reliefs, while the cover of the hills is densely wooded. This situation, at its optimum around 2,000 years bp would correspond to an increase in rainfall with a significant lengthening of the rainy season, under milder but still cool temperatures (Figure 7.5).

From the 3rd to the 5th centuries AD, a less dense mountain forest and a savanna environment evolving from a wooded stage to a treed stage testify to a lengthening of the dry season. But the progression of a herbaceous carpet where species of ruderalized environments are reported attests to a growing anthropogenic effect.

In the 6th century, a dry climatic pulsation set in around ca 500 years AD. At the same time, we are seeing the extinction of the so-called "Ancient Iron Age" culture on the hills of Rwanda. In the region of Butare, in Rwanda, the important concentration of Early Iron Age stoves, dated from the 2nd century to the first half of the 6th century AD, located to the east of the locality, mainly on the hills of Gisagara, Kabuye, Gahondo and Muganza, testifies to an intensive metallurgical activity. The study of charcoal

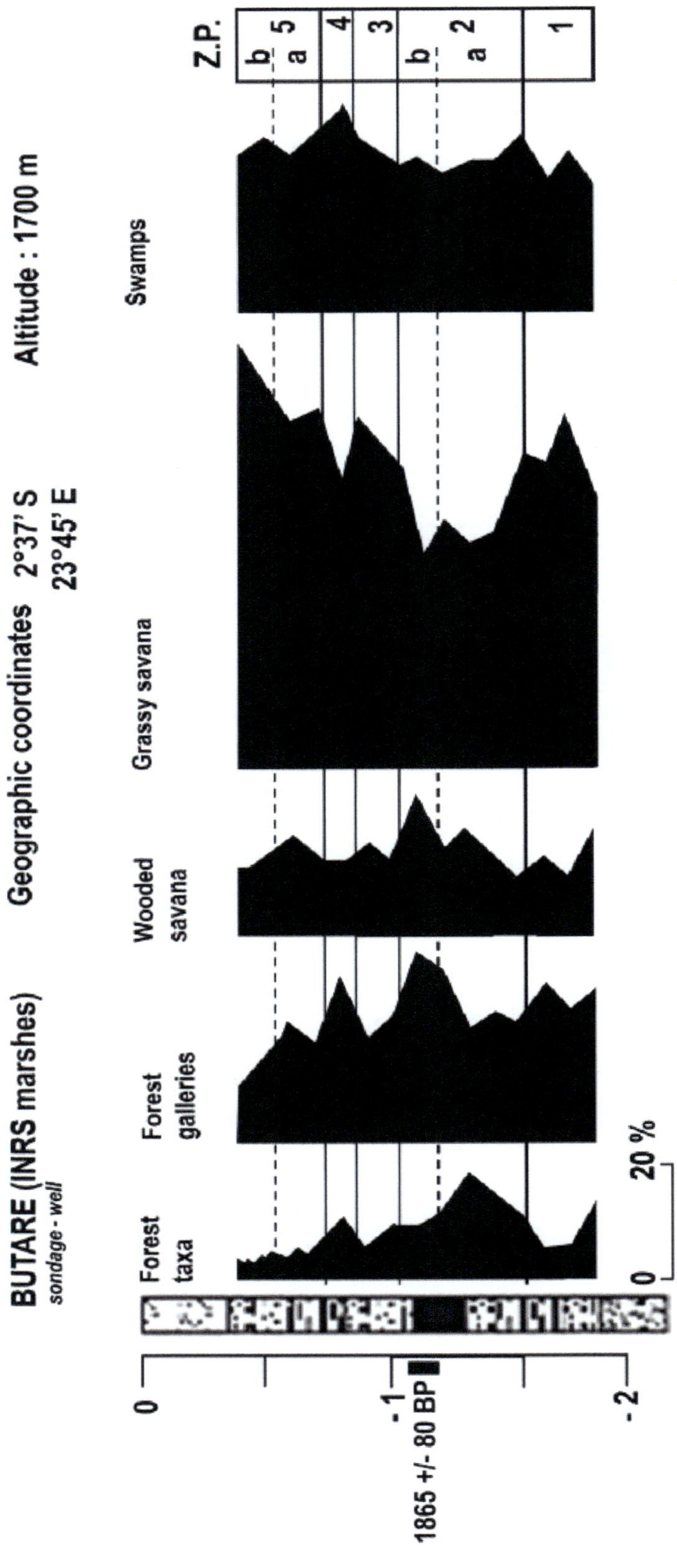

FIGURE 7.4 Detailed pollen diagram of BUTARE site.

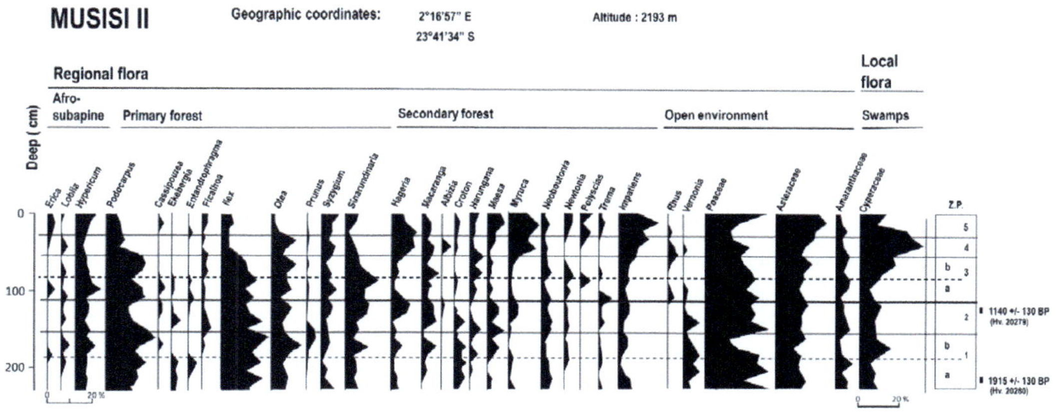

FIGURE 7.5 Detailed pollen diagram of MUSISI II site.

collected in the furnaces provides additional information on the evolution of the regional environment (Van Grunderbeek and Doutrelepont, 1988; Roche and Van Grunderbeek, 2009). A palynological analysis of samples taken from blackened earth at the bottom of the crucibles of three furnaces on the hill of Kabuye (Gisagara) and dated respectively to the beginning of the 3rd century, the end of the 4th century and the beginning of the 6th century AD has, moreover, provided additional information concerning the evolution of the hilly environment during the Early Iron Age (Van Grunderbeek *et al.*, 1982; Roche, 1996; Van Grunderbeek and Roche, 2008). Thus, we notice that we pass from a wooded savannah in the 3rd century via a grassy savannah in the 4th century to a wooded savannah in the 6th century to find a wooded savannah in the 6th century, which follows the evolution evoked by the anthracology.

At this time, known as the Early Iron Age, on the central plateaus of Rwanda and Burundi, populations engaged in metallurgical and agro-pastoral activities at the forest-savannah interface (Van Grunderbeek *et al.*, 1982). The Early Iron Age is characterized by a specific pottery that can be found in all the occupation sites of the period: the "Urewe" pottery with ornamentation by incision. This type of pottery was found in western Rwanda, south of Cyangugu (Boutakoff, 1937; Nenquin, 1967) and in Kivu, south of Bukavu and in Havu and Hunde territories, north of the lake (Hiernaux and Maquet, 1957, 1960; Maquet and Hiernaux, 1969). In Rwanda, many remains of cast iron furnaces have been unearthed, concentrated mainly in the Butare region. In Kivu, the only debris discovered was in Tshamfu, south-west of Bukavu.

But the most interesting element of the analysis is the presence of pollen attributed to finger millet in the three crucibles examined and to *Sorghum* in the two most recent. This suggests that, in parallel with metallurgical activities, a subsistence agriculture based on the cultivation of *Sorghum* and finger millet may have developed, but also that certain varieties of Amaranthaceae, Cucurbitaceae, Fabaceae (Vigna), Musaceae (Ensete) and Solanaceae, from the local vegetation could have been exploited, even cultivated, for consumption. In this type of economy, hunting could constitute an additional nutritional contribution, as well as breeding, attested by the presence of bovine teeth found in certain sites dated from the 3rd century AD (Roche, 1996; Van Grunderbeek and Roche, 2008). This mixed economy being largely dependent on the quality of the environment; the overexploitation of the latter could therefore, in the long term, threaten the survival of the communities of metallurgists of the Early Iron Age in Rwanda. This is what happened in the 6th century AD when a "Global cooling", the result of cataclysmic volcanic eruptions, no longer allowed a weakened biotope to ensure the subsistence of the populations who occupied it. The event is highlighted by the peak of Poaceae archived in the Butare sequence which is similar to the Medium Poaceae Period (MPP) defined by Kabonyi and Roche (2015) in Kivu (Musisi); moreover, we note that the last Old Iron Age furnace inventoried in Rwanda is dated to the first half of the 6th century (Van Grunderbeek, 1992), thus confirming the end of the first Iron Age in the Hills Region. Thus, the pooling of the results of palynological, anthracological, and archaeological research has made it possible to better define the transition from a generally forest landscape to that of a wooded savannah and to follow its evolution during the period extending from the 3rd to the 6th centuries AD.

In the 7th century, we record the return of a more humid but still cold climate, favoring the installation, in the ombro-mesophilic forest, of an inverted mountain stage at Hagenia, on the Central Plateau a wooded savannah is reconstituted. As a result of its regeneration, which would have continued over a fairly long period, probably a century (no occupation site was detected in the 7th century), the natural environment was again likely to welcome new inhabitants. This would have facilitated the installation in Rwanda of a second Iron Age or Late Iron Age, in the 8th century AD.

A second Iron Age or Late Iron Age appears in the Great Lakes region at the end of the 8th century AD. It is characterized by an associated ceramic of a new type, the ceramic decorated with the "braided roulette" introduced by populations South-Nilotic coming from the north (Desmedt, 1991). The successive arrivals in the interlacustrine region (between Lake Victoria and the western branch of the rift) can be seen in the variations in the decoration of pottery, in particular: from ca. 800 AD to ca. 1300 AD the simple braided roulette and, from 1500 AD, the fine braided roulette. The first period of installation of these populations in South Kivu is manifested by a new development of anthropogenic savannahs which, this time at higher altitudes, is in conjunction with the accentuation of the mesophilic character of the Afro-montane forest. This event took place between ca. the 9th–10th centuries AD and ca. the 12th century AD.

Initially attributed to a South Nilotic influx (Desmedt, 1991), this second Iron Age was introduced to Rwanda by a new wave of immigration from the sub-Saharan savannahs of West Africa (Livingstone Smith, 2007). In the Hills Region east of Butare, the first settlement site is dated to 740 AD. It is distinguished by a method of iron casting less elaborate than that of the Early Iron Age and by a ceramic decorated with "twisted roulette".

In the final phase of the Butare sequence, we notice the shaping of the environment by an increasingly incisive human intervention. Like their predecessors, the first occupants of the recent Iron Age practiced an agro-pastoral economy in which pastoralism seems to intensify from the 10th century (Roche, 1996): this is reflected in a significant and gradual regression of wooded areas in favor of secondary grassy savanna areas, the result of the extension of pastures. At the end of the profile, we remark the progression of the *Poaceae* which corresponds to the Younger Poaceae Period (YPP) observed by Runge (2001a, 2001b) in Central Africa from the beginning of the second millennium AD.

The transition from the 7th to the 8th century saw the establishment of an Afro-montane Forest vegetation, in this previously unknown configuration, under climatic conditions heralding those observed today, with less rainfall, in rather cool conditions, maintained by fogs. This installation is similar to that observed in the region of volcanoes, in the north of Rwanda, with the progression of Podocarpus in the Afro-montane upper stage.

In the 8th century, with the persistence of an inverted stage of "forest-park" in Hagenia within the ombro-mesophilic environment, we approach the current climate, humid and cold. On the hills, in an environment of regenerated wooded savannah, once again exploitable, settle the populations of the "Recent Iron Age". The enthronization of the environment was able to continue there under a temperate climate, with moderate rainfall and alternating seasons, similar to the current one. After a period of regeneration of the environment, a second anthropogenic impact, more incisive because it also affected the lower horizon of the primary forest, appeared around the 9th century AD and continued until the end of the 12th century approximately.

Subsequently, the afro-mountain ombro-mesophilic environment sees the perpetuation of enclaves of "forest park" evolving towards an impoverished stage of "forest meadow". A final temporary incursion of Afro-subalpine elements akin to the so-called "Neoglacial" period of Mount Kenya (Hamilton, 1982), recognized in East Africa, then induces vegetation under more aggressive human action than previously: this being probably due to the installation of new populations in the interlacustrine region. This leads, over the following centuries, to a progressive but definitive savanization of the mesophilic formations on the edges of the lake and of the lower horizon of the ombrophilous mountain forest, while the middle horizon of this primary forest is, in turn, secondary. It was probably from the 17th century that the definitive secondarization of the primary forest began. This results in a significant progression of *Hagenia* and the decline of ombrophilous taxa in the middle horizon of the mountain environment.

If the dense forest was able to resist on the ridge, an inhospitable region, until the recent past, we can estimate that the Central Plateau, at least in its southern part, was already ecologically devastated in the first centuries of the second millennium of our era.

CONCLUSION

The first period of forest evolution covering some 3,000 years, from the initial stage of the formation of marshy deposits around Lake Kivu, took place under a relatively lenient, mild, and humid climate, favoring the development of a rather dense, soil-fixing forest and summarizes the following forest history: After the climatic deterioration favorable to the development of savannas and between 2,500 and 2,000 bp., settled around lake Kivu, under a climate close to the present, a forest complex that can be considered as original. At the time, the border slopes of the eastern and western ridges were entirely covered with forests.

From 2,000 years bp, we see the progression of an ombrophilous forest favored by cool climatic conditions. Then occurs, at an estimated date of 500 years AD, an opening of the middle attested by the progression of *Macaranga* and *Hagenia*. On the Congolese ridge of Kivu, ca. 500 years AD, a cold climatic pulsation results in the descent of *Hagenia* from the Afro-subalpine level into the mountain level;

On the Congo-Nile ridge, this episode would be at the beginning of the 6th century AD and would be linked to the Early Iron Age east of Butare. Palynological and anthracological analyses of the latter reveal significant environmental degradation between the end of the 4th century and the beginning of the 6th century AD but, in this case, anthropogenic and climatic effects would be cumulative.

A Second Iron Age, or Late Iron Age, appeared in Rwanda around the 8th century AD, with no apparent contact with the previous culture. The impact of the latter on the environment began to produce obvious effects between the 9th and 12th centuries AD, mainly in the North, where occupation sites have been highlighted. Initially, the effect on the environment is noticed by a degradation of the mesophilic formations occupying the lower slopes of the reliefs, then by a secondarisation affecting the lower horizon of the mountain rain forest. At the same time, we see a significant expansion of open anthropic savannah environments.

From the 2nd millennium AD, the transition forest occupying the lower areas of the reliefs begins to undergo an anthropic influence which will gradually extend in altitude.

The reconstruction of the paleoenvironmental evolution of the border areas of Lake Kivu also shows that, after a climatic crisis, even the most intense, a forest is naturally reconstituted and that, even after anthropogenic degradation, a forest environment contains the potentialities necessary to reconquer the lost ground if it is put in defense. The degradation of many ecosystems in intertropical regions which threatens their balance and their biodiversity is essentially the result of humans, not climatic variations, however unfavorable they may be, which would have existed in the past.

REFERENCES

Boutakoff, I., 1937. First methodical explorations of Stone Age deposits in Ruanda-Urundi. Rock Shelters, Workshops and Outdoor Stations-Preliminary Communication. *Bull. Inst. Royal colonial belge*, VIII, 1: 179–201.

Desmedt, C., 1991. Old pottery decorated with roulette in the Great Lakes region. *Afr. Arch. Rev.*, 9: 161–196.

Devred, R. 1958. La végétation forestière du Congo belge et du Rwanda-Urundi. Bull. Soc. Royale Forest. Belg., 6: 409–468.

Erdtman, G., 1960. The acetolysis method, a revised description. *Swensk Botanisk Tidskrift*, 54: 561–564.

Haberyan, K. A. and Hecky, R.E., 1987. The late Pleistocene and Holocene stratigraphy and paleolimnology of lakes Kivu and Tanganyika. *Palaeogeography, Palaeoclimatology, Palaeoecology*, 61: 169–197.

Habiyaremye, F.X., Roche, E. and Ntaganda, C. 1995. Le Hagenietum abyssinicae sur la dorsale orientale du lac Kivu. Interprétations palynologiques et écologiques. CIFEG- Orléans, P.O.95/31, p.123–144.

Hamilton, A.C., 1982. *Environmental history of East Africa. A study of the Quaternary*. London: 50–90.

Hiernaux, J. and Maquet, E., 1957. Prehistoric cultures of the Metal Age of Ruanda-Urundi and Kivu- First part. *Bull. Acad. Roy. Sc. Coloniales*, n.s., II, 1956–1966: 1126–1149.

Hiernaux, J. and Maquet, E., 1960. Prehistoric cultures of the Metal Age of Ruanda-Urundi and Kivu- Second part. *Same. Acad. Roy. Sc. Outre-Mer-Cl.Sc. Naturelles et Médicales*, n.s., X(2): 5–102.

Ilunga, L. 1991. Morphologie, volcanisme et sédimentation dans le rift du Sud-Kivu. Bull. Soc. Géogr. Liège, 27: 209–228.

Ilunga, L., Muhire, I. and Mbaragijimana, C. 2004. Saisons pluviométriques et origine des pluies au Rwanda, Geo–Eco–Trop. 28, 1–2: 61–68

Kabonyi Nzabandora C. and Roche E., 2015. Six millennia of environmental evolution on the western ridge from Lake Kivu to Mount Kahuzi (R.D.Congo) Palynological analysis of the Ngushu sedimentary sequence. *Geo-Eco-Trop.*, 39, (1): 1–26.

Kabonyi Nzabandora C., Roche E. and Akaegbobi M., 2017. Palaeoenvironments and Palaeoclimates during the Upper-Pleistocene and Holocene in the western Lake Kivu region. *Palaeoecol. Afr.*, 34: 145–159.

Lebrun, J., 1935. The forest species of the Belgian Congo: The forest species of the mountainous regions of Kivu. *INEAC. Ser. Scientist*, 1: 20–58.

Lebrun, J., 1936. Distribution of the equatorial forest and bordering plant formations, Ministry of the Colonies, Publ. General Directorate Agriculture and Livestock, Brussels, 50–89.

Lebrun, J. 1956. The vegetation of the botanical territories of Ruanda-Urundi. *The Belgian Naturalists*, vol. special: 22–48.

Leonard, A. 1962. Les savanes herbeuses du Kivu. Publication INEAC (Bruxelles), Sér. Scientifique, 95, 87p.

Liben, L., 1962. Nature and origin of the plant population (Spermatophytes) of the mountainous regions of eastern Congo. *Same. Acad. Roy. Belgian, Science Class*, 2nd Series, 15, 3: 1–195.

Livingstone Smith, A., 2007. History of roulette decor in sub-Saharan Africa. *Journal of African Archaeology*, 5: 189–216.

Maquet, E. and Hiernaux, J., 1969. Ancient ceramics in Havu and Hunde territories (Kivu, R.D. Congo). Africa Tervuren, XIV- 1968/4: 111–116.

Moscol-Olivera M. and Roche, E., 1997. Palynological analysis of a Holocene sedimentary sequence at Musisi-Karashoma (Kivu, R.D. Congo). Climatic and anthropogenic influences on the environment. *Geo-Eco-Trop.*, 1–4: 1–26.

Nenquin, J., 1967. Contribution to the study of the Prehistoric cultures of Rwanda and Burundi. *Royal Museum for Central Africa, Ann. Humanities*, 7–12.

Ntaganda, C., 1991. Paleoenvironments and paleoclimates of the Upper Quaternary in Rwanda by the palynological analysis of superficial deposits. Doctoral thesis in Botanical Sc., University of Liège, 25–75.

Pecrot, A. and Leonard, A., 1960. The Kivu ridge. Explanatory note of the map of soils and vegetation. Publication INEAC (Brussels), n° 16, 80–100.

Peeters, L. 1989. Cartographie dynamique de la forêt de Nyungwe (Rwanda), Geo – Eco – Trop. 13, 1–4: 81–89.

Robyns, W., 1948. Spermatophyte flora of Albert National Park. Vol. I – Institute of National Parks of Congo, Brussels, 130–160.

Roche, E., 1996. Human influence on the environment during the Iron Age in ancient Rwanda. *Geo- Eco-Trop.*, 20 (1–4): 73–89.

Roche, E. and Bikwemu, G., 1989. Paleoenvironmental change on the Zaire-Nile ridge in Burundi; the last 20,000 years: An interpretation of palynological data from the Kashiru core, Ijenda, Burundi. In: *Quaternary and Environmental Research on East African Mountains*, Edited by W.C. Mahaney (York University, Toronto). A.A. Balkema Publishers, Rotterdam: 231–242.

Roche, E. and Ntaganda, C., 1999. Palynological analysis of the Kiguhu II sedimentary sequence (Birunga region, Rwanda). Evolution of the paleoenvironment and the paleoclimate in the Afro- montane domain of Rwanda during the Holocene. *Geo-Eco-Trop Revue.*, 22: 71–82.

Roche, E. and Van Grunderbeek, M.C., 2009. Environment degradation resulting from human activities. Essay to further recognition through pollen and charcoal analysis. SAFA 2008 Conference: 3–8.

Roche, E., Bikwemu, G. and Ntaganda, C., 1988. Evolution of the Quaternary paleoenvironment in Rwanda and Burundi. Analysis of morphotectonic phenomena and sedimentological and palynological data. *Inst. Fr. Pondicherry, Trav. Dry. Sci. Tech.*, XXV: 105–123.

Roche E., Kabonyi Nzabandora C. and Ntaganda C. (2015). Overview of the Holocene phytodynamics of the mountain environment on the Virunga volcanic chain (Northern Rwanda). *Geo-Eco-Trop.*, 39, (1): 27–54.

Runge, J., 1999. Global Change in Africa: Sedimentanalyse und Kohlenstoffisotope entschlüsseln die Umweltgeschichte tropischer Ökosysteme. *Forschungsforum Paderborn*, 2: 16–19.

Runge, J., 2001a. Landschaftsgenese und Paläoklima in Zentralafrika. Relief, Boden, Paläoklima, Gebr. Borntraeger. Berlin-Stuttgart. 17: 70–90.

Runge, J., 2001b. On the age of stone-lines and hillwash sediments in the Eastern Congo basin- palaeoenvironmental implications. *Palaeoecology of Africa*, 27: 19–36.

Sirven, P., Gotanegre, J.F. and Prioul, C., 1974. *Geography of Rwanda*. Ed. DeBoeck, Brussels, 10–30.

Troupin, G. 1966. Etude phytocénologique du Parc National de l'Akagera et du Rwanda oriental. Recherche d'une méthode d'analyse appropriée à la végétation d'Afrique intertropicale. Inst. Nat. Rech. Sc., Rép. Rwandaise, Publ. 2: 293 p.

Van Grunderbeek, M.C., 1992. Chronological delimitation of the Early Iron Age in Burundi, Rwanda and the Great Lakes region. *Azania (Nairobi)*, XXVII: 53–80.

Van Grunderbeek, M.C. and Doutrelepont, H., 1988. Study of charcoal from metallurgical sites of the Early Iron Age in Rwanda and Burundi. 22, IV.7: 281–295.

Van Grunderbeek, M.C. and Roche, E., 2008. Multidisciplinary Evidence of mixed farming in the early iron age in Rwanda and Burundi. In: *Rethinking Agriculture*; Denham, Iriarte & Vrydaghs, Left Coast Press, Inc.: 299–319.

Van Grunderbeek, M.C., Roche, E. and Doutrelepont, H., 1982. The early iron age in Rwanda and Burundi, archeology and environment. *J. des Africanistes*, 52, (2): 5–58.

The Tree of Life

Biogeographic and Phylogenetic insights into Southwestern Nigerian Tropical Rain Forests

8

Sehinde Akinbiola, Ayobami Salami, Olusegun Awotoye and Ayomide Olubaju

INTRODUCTION

Tropical forests are some of the most biodiverse ecosystems on Earth, hosting an extraordinary variety of plant and animal species (Akinbiola *et al.*, 2025). These forests provide critical ecosystem services such as carbon sequestration, climate regulation, and water cycle maintenance (Edwards *et al.*, 2014; Rajaonarimalala *et al.*, 2024), while also supporting millions of livelihoods globally (Lewis, 2009). The tropical rainforests of Southwestern Nigeria, extending to Ghana and Cameroon, are particularly notable for their rich biodiversity of tree species (Ibrahim *et al.*, 2025). Understanding this diversity is essential for conservation, sustainable management, and addressing the global challenges of climate change and biodiversity loss.

The evolution of angiosperms, commonly known as flowering plants, represents one of the most significant biological events in the history of the evolution of flora, profoundly influencing the structure and function of terrestrial ecosystems (Benton, *et al.* 2022; Morley, 2000). The origin of angiosperms dates back to the Early Cretaceous period, approximately 125–140 million years ago, with their rapid diversification occurring during the mid-Cretaceous to Paleocene epochs (Scutt, 2021; Wang *et al.*, 2022). Angiosperms outcompeted gymnosperms and other plant groups due to their advanced reproductive

DOI: 10.1201/9781003727781-8

strategies, including the development of flowers and fruits, which facilitated more effective seed dispersal and pollination (Wang *et al.*, 2022).

The rise of angiosperms coincided with significant climatic and geological changes (Wang *et al.*, 2022). These changes created conditions conducive to the establishment of dense, multi-layered forests dominated by angiosperms (Benton *et al.*, 2022). By the Eocene epoch, tropical forests resembling modern-day rainforests had formed, characterized by high species richness and complex canopy structures (Benton *et al.*, 2022; Morley, 2000; Scutt, 2021; Wang *et al.*, 2022). The phylogenetic diversity of tropical forest tree species in Southwestern Nigeria reflects this deep evolutionary history, as many lineages present in these forests have ancient origins tied to the evolutionary radiation of angiosperms.

The formation and persistence of tropical forests are intricately linked to specific climatic and bioclimatic conditions (Faith, 2008; Niang *et al.*, 2024). Temperature plays a pivotal role, as tropical forests typically thrive in regions where mean annual temperatures range from 20 °C to 30 °C. Stable and warm temperatures promote continuous growth and reduce seasonal variability, enabling the coexistence of a wide variety of species (Edwards *et al.*, 2014). In addition to temperature, other bioclimatic variables, such as precipitation, humidity, and solar radiation, significantly influence tropical forest ecosystems. High annual rainfall, often exceeding 2,000 mm, is essential for maintaining the lush, evergreen characteristics of these forests. The combination of high temperatures and abundant rainfall creates conditions that support rapid nutrient cycling and high primary productivity (Edwards *et al.*, 2014; Laranjeiras *et al.* 2024; Lewis, 2009).

In Southwestern Nigeria, the interplay between temperature and other bioclimatic variables shapes the region's tropical rainforest (Akinbiola *et al.*, 2023). The region experiences a humid tropical climate with distinct wet and dry seasons, influenced by the Intertropical Convergence Zone (ITCZ). These climatic conditions contribute to the establishment of diverse tree species assemblages adapted to the local environment. However, changes in these climatic factors, driven by global climate change, pose significant threats to the stability and biodiversity of these forests (Akinpelu *et al.*, 2024). Tropical forests are both a driver and a victim of climate change. As carbon sinks, they play a crucial role in mitigating global warming by sequestering large amounts of atmospheric carbon dioxide (Ferreira *et al.*, 2024). However, rising temperatures, shifting precipitation patterns, and increased frequency of extreme weather events threaten the structure and function of these ecosystems (Ferreira *et al.*, 2024; Gota *et al.* 2024).

In Southwestern Nigeria, climate change poses significant risks to the phylogenetic diversity of tropical forest tree species. Increased temperatures may lead to thermal stress, while altered rainfall patterns could result in prolonged dry seasons and reduced water availability. Such changes could disrupt ecological processes, including seed germination, growth, and reproduction, ultimately affecting species composition and diversity. Furthermore, climate change exacerbates other anthropogenic pressures, such as deforestation and habitat fragmentation. The loss of forest cover reduces the resilience of ecosystems to climatic changes and accelerates the extinction of species, particularly those with narrow ecological niches or limited geographic ranges. Addressing these challenges requires a comprehensive understanding of the phylogenetic diversity and ecological dynamics of tropical forest tree species (Agboola *et al.*, 2021; Akinpelu *et al.*, 2024).

According to the International Union for Conservation of Nature IUCN (2024), the tropical forests of Southwestern Nigeria are home to numerous threatened and endemic tree species, e.g., *Pterygota bequaertii*, *Anthocleista scandens*, and *Cola nigerica,* which underscore the region's global conservation significance. Endemic species, those found only within a specific geographic area, contribute disproportionately to the region's phylogenetic diversity and evolutionary heritage. These species are often highly specialized, making them particularly vulnerable to climate change and anthropogenic disturbances (Manes *et al.*, 2021; Prawiradilaga and Irham, 2024). They are listed as vulnerable in the IUCN web, but according to the International Institute of Tropical Agriculture (IITA), they are critically endangered due to their economic importance. Habitat loss due to agricultural expansion, logging, and infrastructure development is a primary driver of species decline. Climate change further compounds these threats by altering habitat conditions and increasing the frequency of extreme weather events.

The loss of phylogenetic diversity associated with the extinction of threatened and endemic species has far-reaching implications. Phylogenetic diversity is a measure of the evolutionary history represented within an ecosystem and is closely linked to ecosystem functioning and resilience (Faith, 2008). Preserving this diversity is essential for maintaining ecosystem services and adaptive capacity in the face of climate change (Faith, 2008; Mahayani *et al.*, 2020; Rajaonarimalala *et al.*, 2024). This study aims to explore the phylogenetic diversity of some selected tropical forest tree species based on their IUCN status in the rainforests of Southwestern Nigeria. By examining the evolutionary relationships among species and their responses to climatic and bioclimatic variables, this study seeks to provide insights into the factors shaping biodiversity patterns in these forests. Additionally, the research highlights the implications of climate change for conservation and management efforts, emphasizing the need to protect threatened and endemic species.

METHODOLOGY

Area of Study

For this study, biophysical forest data were collected from three administratively conserved forests in Southwestern Nigeria. These three conserved forests share common boundaries, allowing the unrestricted flow of biodiversity between the conserved forest reserves, thus forming a contiguous zone of lush forest estate representing the once pristine forest that formed the critical coastal refugium, which is defined by diverse habitats of lowland rainforests, mangrove forests, and mixed deciduous forests, and the combination of varying climates and abundance of genetic diversity. The Omo Conservation Forest (OCR) in Figure 8.1 is the largest forest of the three conserved forests, having several smaller forests such as Omo Biosphere Reserve, which is a United Nations Educational, Scientific and Cultural Organization (UNESCO) World Heritage Site, and the Queen Elizabeth Forest, which is also a world-acclaimed protection site. The remaining two reserves are Oluwa Conservation Forest (OCR2), which has lots of mangrove forests towards the Atlantic coast, and Shahsa Conservation Forest (SCR), which is predominantly evergreen lowland rainforest and mixed deciduous forests

Sampling Design

Plot sampling was conducted using a stratified random sampling technique to ensure representative coverage of all major vegetation strata across the three conservation forests: Omo (OCR), Oluwa (OCR2), and Shasha (SCR). Following the methodology adapted from Akinbiola *et al.* (2023), the study incorporated the concentric plot approach with the central annular plot radius of 17.95 m, subplot radius of 7.32 m, and microplot radius of 2.07 m (Figure 8.2) distributed proportionally across distinct forest zones from the 500 m × 500 m drone gridded plots based on UAV photogrammetry-derived canopy structure and terrain models. These plots were delineated with GPS coordinates and were spaced to capture variability in topography, forest type (evergreen lowland, mangrove, and mixed deciduous), and canopy composition.

Akinbiola *et al.* (2023) utilized unmanned aerial vehicle (UAV) photogrammetry to assess forest structure, identify canopy gaps, and map species network patterns in the tropical forests of Southern Nigeria. The UAV system produced high-resolution orthomosaics and 3D forest models from which stratification zones were identified. This UAV-guided sampling ensured spatially explicit, ecologically informed plot placement that minimized observer bias and maximized detection of forest heterogeneity.

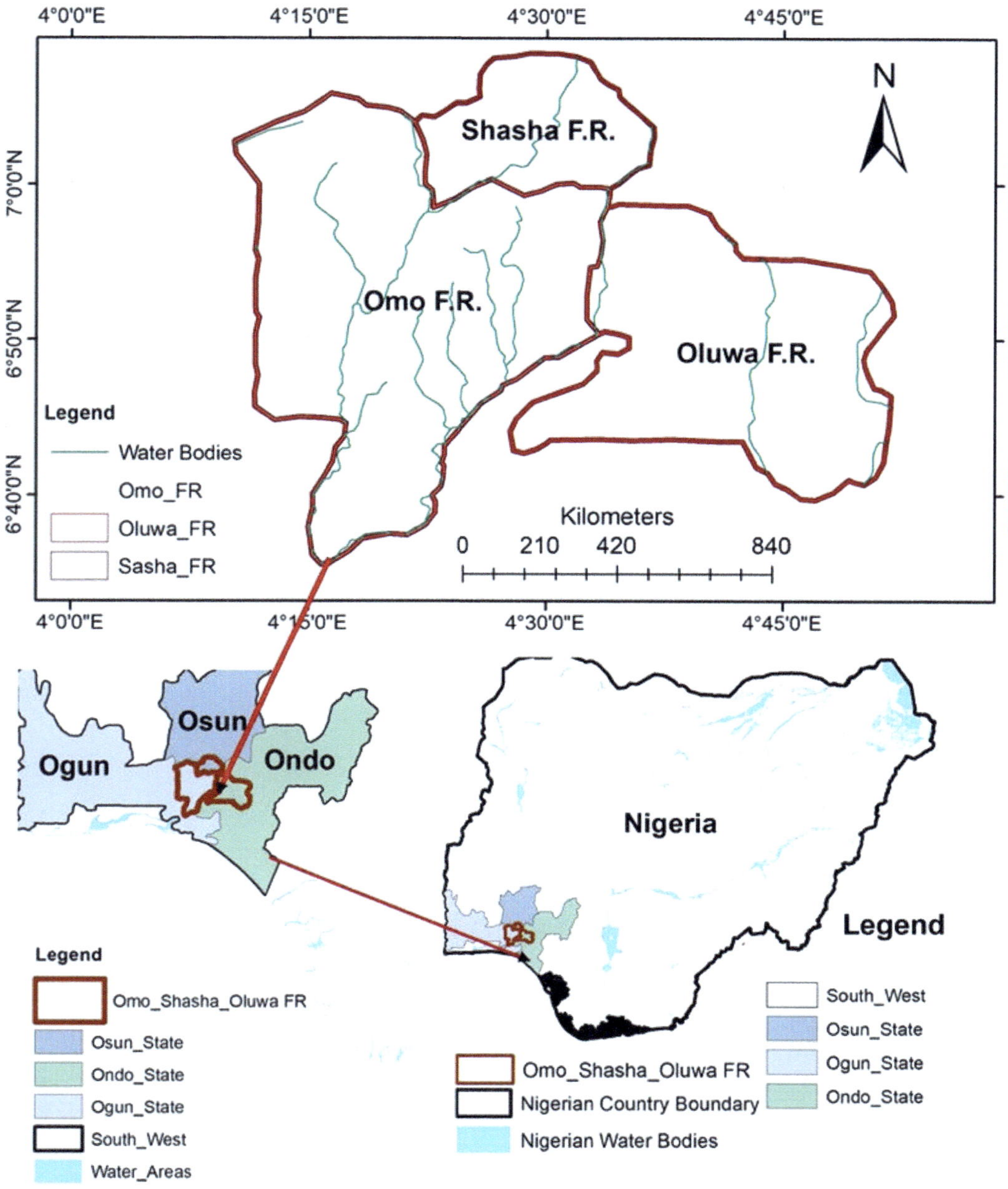

FIGURE 8.1 A study area map showing the three conservation forests: OCR, OCR2, and SCR.

DATA COLLECTION

Within each plot, all tree individuals with a diameter at breast height (DBH) ≥10 cm were recorded in the field. The following biophysical variables were recorded:

- Tree species identity (field ID and verified using Ife Herbarium and GBIF)
- Taxonomic classification (family, genus, and species)

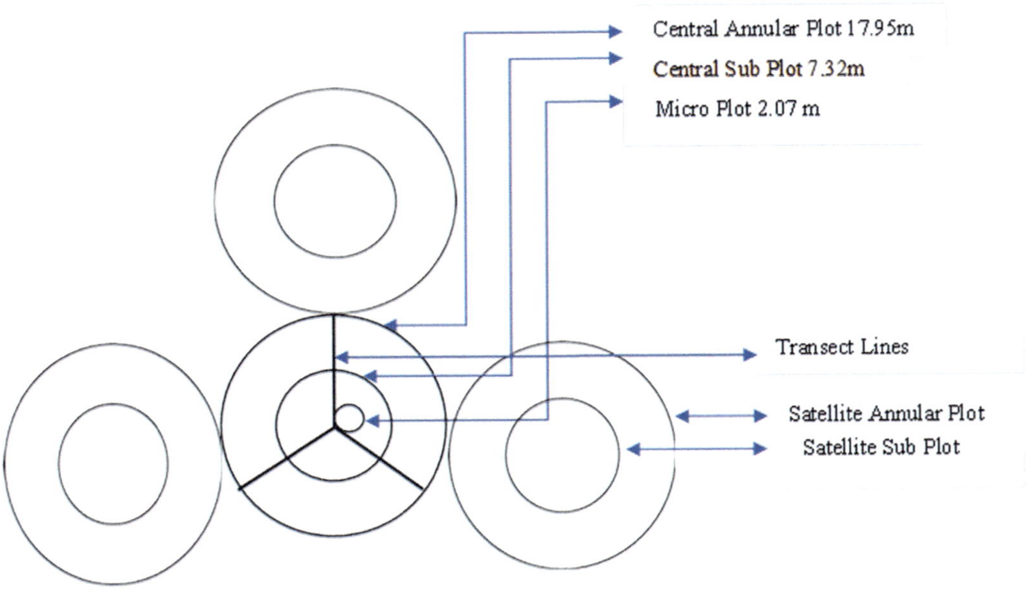

FIGURE 8.2 Field plot, adapted from (Akinbiola *et al.*, 2023), and modified from FIA (USDA Forest Service 2004).

- Diameter at breast height (DBH) at 1.3 m
- Tree height (when accessible)
- Canopy position class (emergent, canopy)
- GPS coordinates (latitude, longitude, and altitude)
- IUCN conservation status (from IUCN Red List, 2024)

Pre-selected focal species were monitored across all plots whenever present. While these tree species are not the only vulnerable tree species, they are the most sought after due to their economic viability e.g., according to IUCN, these tree species are multipurpose trees with a range of uses including timber, charcoal, fodder, and medicinal purposes. The species are *Afzelia africana*, *Khaya ivorensis*, *Brackystegia nigerica*, *Entandrophragma cylindricum*, *Lophira alata*, and *Nauclea diderichii*. Identification was confirmed using local floras and herbarium vouchers. The variables were entered into a digital database, corrected for typographic errors the online Global Biodiversity (GBIF) portal (write off !) used in subsequent phylogenetic and spatial analyses. Table 8.1 shows biophysical variables collected during plot sampling. In addition, six pre-selected tree species (Table 8.2) were recorded whenever they were observed within the sampling plots.

The pre-selected biophysical variables collected *in situ* were recorded on a Microsoft Office Excel spreadsheet, in the order: Plot ID, latitude, longitude, altitude, tree family, tree genus, tree species, and tree dbh (cm). Tree families, genus, and species were corrected for typographical errors using the online GBIF portal and references from Ife Herbarium, Obafemi Awolowo University. The cleaned data sets were then imported to the Conda Environment, after which the required libraries were installed. Furthermore, the gene sequence from the gene bank (www.ncbi.nlm.nih.gov/genbank/) for each of the pre-selected tree species was downloaded and processed in FASTA format. For spatial phylogenetic analysis, randomized trees were created using the PreMock Phylogenetic Tree and the spatial boundary, after which the implementation of the Python scripts for estimating phylogenetic diversity, endemism, and species richness was conducted. Since the data sets used were limited, the two methods for estimating phylogenetic diversity were used. The first method involves obtaining the gene sequence from gene banks for the pre-selected tree species, while the second method uses the tree species boundary as an environmental variable.

TABLE 8.1 Biophysical Variables Collected During Plot Sampling

VARIABLE	*DESCRIPTION*	*MEASUREMENT UNIT*	*INSTRUMENT / SOURCE*
Tree species name	Scientific name of each tree	—	Field ID + Ife Herbarium verification
Family, Genus, and Species	Taxonomic classification	—	GBIF and Ife Herbarium cross-check
DBH	Diameter at Breast Height (measured at 1.3 m)	cm	Forestry tape
Tree height	Estimated or measured vertical tree height	m	Clinometer / visual estimate
Canopy position	Crown class: emergent, canopy	—	Visual estimation during field survey
Latitude, Longitude	4 0 0 E to 4 45 0 E 6 40 0 N to 7 0 0 N	decimal degrees	Garmin GPSMAP 78 Handheld GPS
Altitude	Elevation above sea level	m	Garmin GPSMAP 78
IUCN status	Red List classification (e.g., vulnerable)	—	IUCN Red List (2024): https://www.iucnredlist.org

TABLE 8.2 Selected Tree Species in Southwestern Nigeria Based on the IUCN Red List

S/N	*TREE FAMILY*	*SPECIES*	*NAMES*	*AUTHORITY*	
				IUCN Red List	Class
1	*Lophira alata*	Banks ex Gaertn.	Ochnaceae	vulnerable	Canopy
2	*Khaya ivorensis*	A. Chev.	Meliaceae	vulnerable	Emergent
3	*Entandrophragma cylindricum*	Sprague	Meliaceae	vulnerable	Emergent
4	*Nauclea diderichii*	DeWild. andT. Durand Merrill	Rubiaceae	vulnerable	Emergent
5	*Brackystegia nigerica*	Banks	Fabaceae	vulnerable	Canopy
6	*Afzelia africana*	Sm	Fabaceae	vulnerable	Canopy

RESULTS

Endemism

The analysis revealed the presence of three regionally endemic tree species within the study area: *Afzelia africana*, *Khaya ivorensis*, and *Brachystegia nigerica*. These species exhibit varying degrees of ecological restriction and are largely confined to the West Africa tropical rainforest belt, particularly the Guineo-Congolian phytogeographical zone. This region, which encompasses southern Nigeria and extends westward across Ghana, Cote d'Ivoire, and Liberia, is known for its high rainfall, stratified forest structure, and unique microclimatic conditions that support specialized flora. *Afzelia africana* was predominantly observed along the forest-savanna transition zones, indicating its preference for drier forest margins and open woodlands within the Sudano-Guinean ecotone. *Khaya ivorensis*, in contrast, was limited to mature, moist lowland evergreen forests, reflecting its dependence on humid, undisturbed canopies. *Brachystegia nigerica* exhibited the highest degree of range restriction, being found only within fragmented forest patches in southern Nigeria, confirming its status as a narrowly endemic species with a strong sensitivity to habitat disturbance. The distribution patterns of these species underscore their ecological specificity and highlight the conservation significance of the study area. Their restricted ranges and reliance on intact forest systems suggest that they may serve as key indicators of forest integrity and habitat continuity in the region (Figure 8.3).

Species Richness

The result of the color code for species richness showed that *Afzelia africana* to be the most abundant in the study area; this is followed closely by *Brackystegia nigerica*, while *Entandrophragma cylindricum* and *Lophira alata* have the lowest richness in the study area. Spatial patterns showed an increase in species richness in regions of higher latitudes. The region around the latitudes 6.90 to 7.00 showed the highest biodiversity in the study area of the preselected species.

Phylogenetic Analysis

To evaluate patterns of relatedness among the focal species, two complementary analyses were conducted, one based on ecological and environmental trait similarities and another derived from molecular sequence

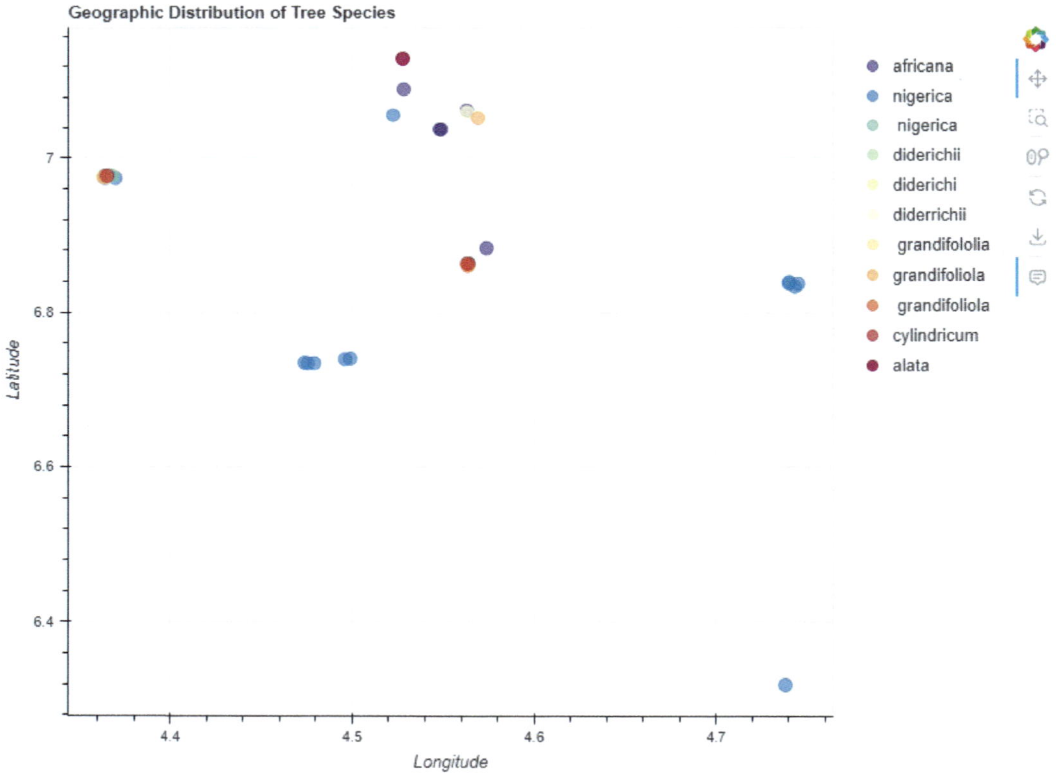

FIGURE 8.3 Species richness of preselected tree species in the study area.

alata. Although both analyses involved the same group of tree species – *Afzelia africana, Brachystegia nigerica, Khaya ivorensis, Khaya grandifoliola, Lophira alata, Entandrophragma cylindricum,* and *Nauclea diderrichii,* the data sources and interpretive frameworks differed substantially.

The first dendrogram (Figure 8.4) was constructed using functional and environmental variables, such as rainfall tolerance, soil preference, canopy structure, and phenological patterns. It revealed clusters of ecological affinity, reflecting species' adaptation to shared environmental conditions. For instance, *Afzelia africana* and *Brachystegia nigerica* grouped closely, indicating similar ecological strategies typical of transitional woodland-savanna systems. In contrast, moist forest species like *Khaya ivorensis* and *Lophira alata* form separate clusters, emphasizing their association with a humid, closed-canopy environment.

Meanwhile, Figure 8.5 depicts a molecular phylogenetic tree based on DNA sequence alignment (e.g., rbcL, matK) and illustrates evolutionary relationships among the species. Here, *Khaya ivorensis* and *Khaya grandifoliola* formed a tight clade, reflecting their close genetic lineage within the Meliaceae family. *Entandrophragma cylindricum,* while taxonomically related, appeared slightly more divergent, and species from other families (*Lophira alata* and *Nauclea diderrichii*) were positioned in more distant branches, consistent with their evolutionary distinctiveness.

Though some topological similarities exist between the two trees, such as the proximity of species sharing habitats, differences in clustering patterns highlight the contrasts between ecological convergence and evolutionary relatedness. This dual approach underscores the importance of integrating functional ecology and molecular phylogenetics to capture both adaptation and ancestry in forest tree communities. Also, placing these diagrams side by side emphasizes the value of integrating both ecological and genetic frameworks in interpreting biodiversity and endemism patterns within the study group.

The results of the phylogenetic trees of the pre-selected tree species in the tropical forest of Southwestern Nigeria (fig. 8.4) showed horizontal lengths representing great distances and four clades. The first clade of *Brackystegia nigerica* formed the most distant clade from other species. The second clade constitutes *Nauclea diderichii,* while the third clade showed a group of tree species of *Lophira alata* and *Afzelia africana.* The last clade for the tree species assessed showed a group of trees for *Khaya ivorensis* and *Entandrophragma cylindricum.* The circular layout showed relationships in a radial format, with species represented at the terminal nodes. The branch points indicated common ancestors, and their position reflected evolutionary divergence.

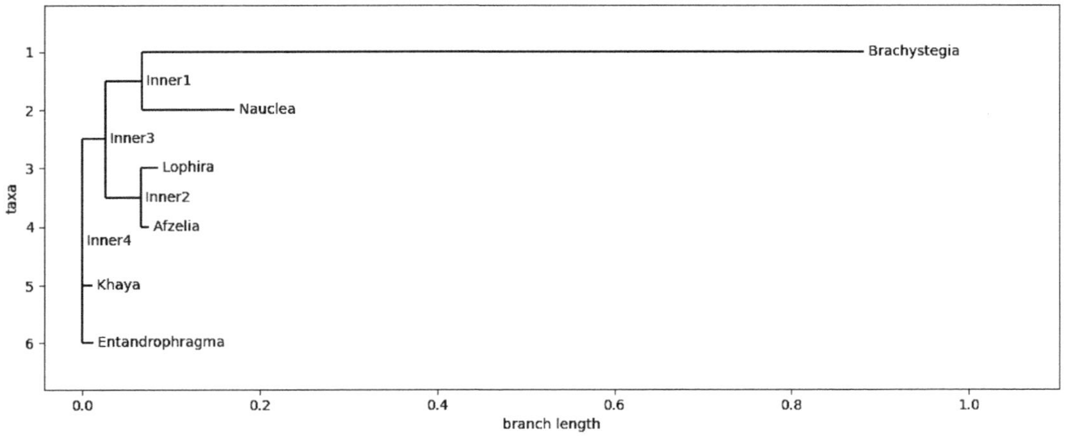

FIGURE 8.4 Functional clustering dendrogram of selected species based on environmental and ecological trait similarity.

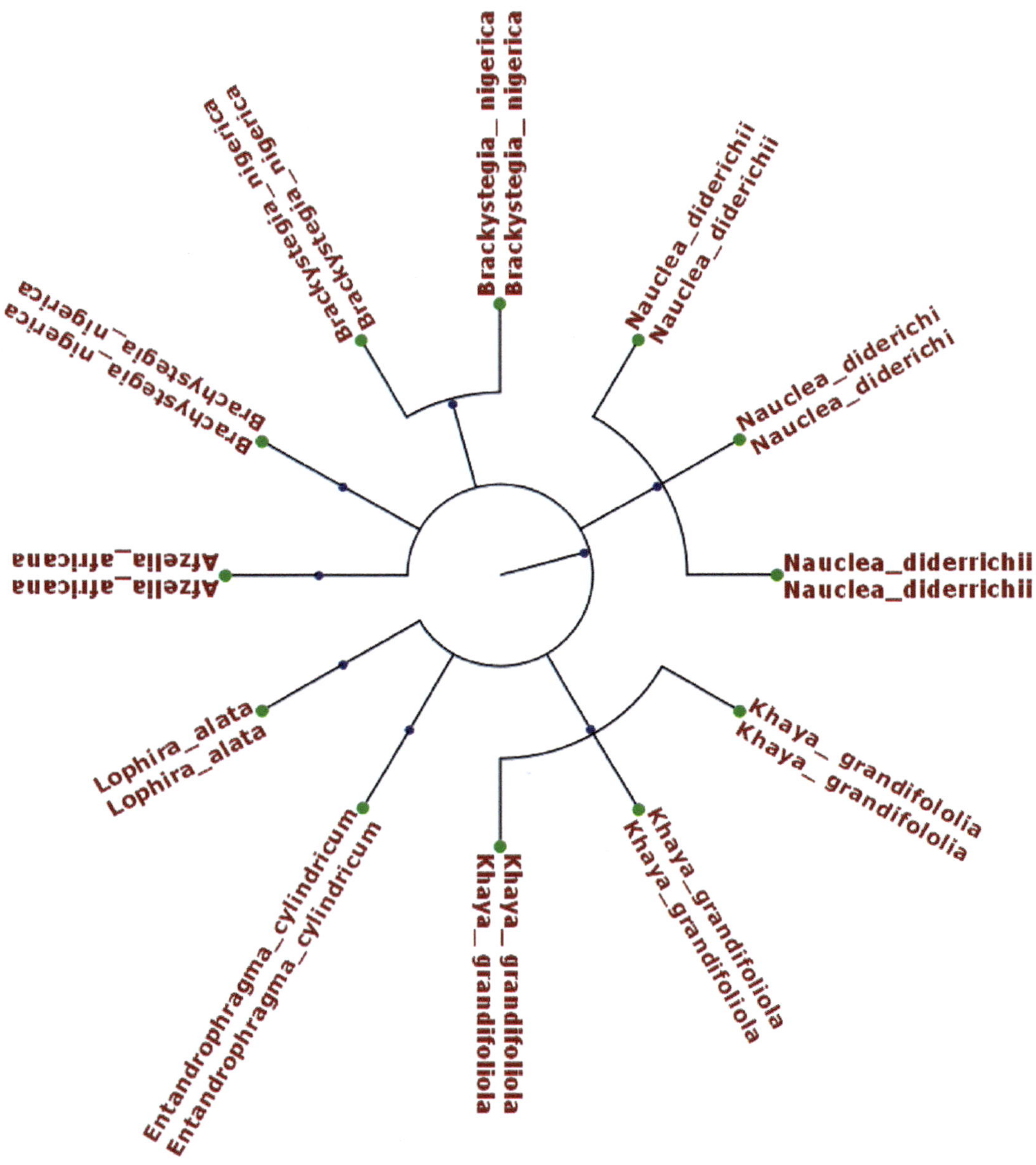

FIGURE 8.5 Molecular phylogenetic tree based on gene sequence data (ITS, rbcL). Evolutionary relationships were inferred from maximum likelihood analysis and Nosaea was used as an outgroup to root the tree.

DISCUSSION

Southwestern Nigeria serves as a critical refugium for numerous endemic tropical forest tree species, including *Afzelia africana*, *Khaya ivorensis*, and *Brachystegia nigerica*. These species are not only integral to the ecological integrity of the region's forests but also represent unique evolutionary lineages. *Afzelia africana* is a leguminous tree species that holds significant ecological and economic importance in tropical Africa. Its restricted distribution makes it particularly susceptible to habitat loss and climate change (Assunção-Silva, *et al.* 2021; Barreto *et al.*, 2023; Gota *et al.*, 2024). The ecological role of *Afzelia africana* includes nitrogen fixation, which enhances soil fertility and supports the growth of other plant species. However, overexploitation for timber and traditional medicine has led to a decline in their population and current studies have shown that they may already be threatened (IUCN, 2024). *Khaya ivorensis*, commonly known as African mahogany, is also a valuable hardwood species endemic to the tropical forests of West Africa, including Southwestern Nigeria. This species is a cornerstone of the region's forest ecosystems, contributing to carbon sequestration and habitat provision for numerous animal species. Despite its ecological significance, *Khaya ivorensis* faces severe threats from illegal logging and habitat degradation. The high demand for its timber, prized for its durability and aesthetic qualities, has significantly reduced its population, necessitating urgent conservation *measures. Brachystegia nigerica* is a lesser-known but ecologically vital tree species endemic to Nigeria's tropical rainforests. It belongs to the Fabaceae family and plays a crucial role in maintaining forest structure and function. The species' endemism highlights its evolutionary distinctiveness and adaptation to local environmental conditions. However, habitat fragmentation and agricultural encroachment threaten its survival. Conservation efforts focusing on habitat restoration and protection are essential to ensure the persistence of *Brachystegia nigerica* and the biodiversity it supports (Akinpelu *et al.*, 2024; Arroyo-Rodríguez *et al.*, 2012; Cerasoni *et al.*, 2023). Species richness is a crucial metric for assessing biodiversity in the study area. The data points for *Afzelia africana* appeared clustered in areas of moderate latitude, reflecting the species' preference for specific microhabitats within the tropical forest mosaic. This distribution suggested that *Afzelia africana* thrives in regions with relatively stable climatic conditions that provide adequate water availability and nutrient-rich soils (Akinbiola *et al.*, 2025). The occurrences of *Khaya ivorensis* are concentrated in regions with higher latitudes compared to *Afzelia africana*. This species' distribution may be indicative of its reliance on dense, humid tropical forest conditions. Its habitat preference aligns with areas that receive consistent rainfall and exhibit minimal seasonal variation in temperature (Rajaonarimalala *et al.*, 2024; Santos *et al.* 2023; Wikramanayake *et al.*, 2022; Zhou *et al.*, 2023). The data for *Brachystegia nigerica* demonstrated a scattered pattern, possibly reflecting its adaptability to various microclimates or the fragmented nature of its remaining habitats due to anthropogenic pressures. Its occurrence in disparate locations highlights the urgent need to conserve its habitats to prevent further fragmentation (Akinbiola *et al.*, 2025). The distribution patterns revealed a key biodiversity hotspot along latitudes 6.9 to 7.0 and longitudes 4.45 to 4.69 within Southwestern Nigeria, where multiple species coexist. These areas are likely characterized by favorable bioclimatic conditions such as high annual rainfall, stable temperatures, and fertile soils. Protecting these hotspots is crucial for conserving the region's overall phylogenetic diversity (Lewis, 2009; Maure *et al.*, 2023).

For genetic diversity, *Brackystegia nigerica* had the longest branch indicating greater genetic and evolutionary divergence. According to Faith (2008), *Brackystegia nigerica*, being the most divergent, might have unique ecological roles or genetic traits requiring specialized conservation efforts. The tree suggested an early divergence for *Brackystegia nigerica*, possibly due to unique ecological pressures or geographic isolation. This is in line with previous studies which indicated that *Brackystegia nigerica* is geographically limited to Nigeria and some parts of Cameroon (IUCN, 2024). From the results, *Lophira alata* and *Afzelia africana* are closely related. This implied that the two species shared more common ancestors and traits compared with other species from different clades; for example, *Khaya ivorensis* and *Entandrophragma ivorensis* are from the same clades and showed distinct evolutionary history to *Lophira alata* and *Afzelia africana* or Brackystegia nigerica. The implication of this finding is that *Lophira alata* and *Afzelia africana* may share vulnerabilities to climate change and may benefit from similar conservation strategies, while *Brackystegia nigerica* showed greater genetic diversity within the group, which is crucial for resilience to climate change. Furthermore, the clustering of species like *Nauclea diderichii* and *Khaya ivorensis* highlighted their shared evolutionary history, suggesting possible overlap in their historical geographic range or selective pressures (Arroyo-Rodríguez *et al.*, 2012; Carvalho *et al.*, 2021; Faith, 2008; Laranjeiras *et al.*, 2024; Mahayani *et al.*, 2020).

CONCLUSION

The findings of this study have important implications for biodiversity conservation and climate change mitigation. Understanding the phylogenetic diversity of tropical forest tree species can inform the development of conservation strategies that prioritize evolutionary distinctiveness and ecosystem resilience. Furthermore, this research contributes to global efforts to achieve biodiversity targets, such as those outlined in the Convention on Biological Diversity (CBD) and the United Nations Sustainable Development Goals (SDGs).

In conclusion, the tropical rainforests of Southwestern Nigeria represent a critical component of the global biodiversity hotspot. Their phylogenetic diversity, shaped by millions of years of evolutionary history, underscores the importance of preserving these ecosystems in the face of mounting environmental challenges. Through a multidisciplinary approach that integrates phylogenetics, ecology, and climate science, this study aims to advance our understanding of tropical forest biodiversity and contribute to the sustainable management of these invaluable ecosystems. Conservation efforts in the study area should prioritize *Brackystegia nigerica* because of its high genetic diversity to maintain ecosystem stability.

REFERENCES

Agboola, O. O., Mayowa, F., Adeonipekun, P. A., Akintuyi, A., Gbenga, O., Ogundipe, O. T., . . . Alabi, S. (2021). A rapid exploratory assessment of vegetation structure and carbon pools of the remaining tropical lowland forests of Southwestern Nigeria. *Trees, Forests and People*, *6*, 100158. https://doi.org/10.1016/j.tfp.2021.100158

Akinbiola, S., Akinsola, J. E. T., Ajagbe, S. A., Salami, A., Olubaju, A., Awotoye, O., and Awoleye, O. M. (2025) Artificial intelligence techniques for prediction of carbon stocks and uncertainty estimates in tropical forests. *SN Computer Science*, *6*, 342. https://doi.org/10.1007/s42979-025-03888-4

Akinbiola, S., Salami, A. T., Awotoye, O. O., Popoola, S. O., and Olusola, J. A. (2023). Application of UAV photogrammetry for the assessment of forest structure and species network in the tropical forests of Southern Nigeria. *Geocarto International*, *38*(1), 2190621. doi:10.1080/10106049.2023.2190621

Akinpelu, O. T., Arimoro, F. O., Ayanwale, A. V., Chukwuemeka, V. I., Adedapo, A. M., and Akindele, E. O. (2024). Comparative investigation provides further insights on how riparian deforestation and different land uses impact the distribution of freshwater macroinvertebrates in Nigerian streams. *Ecohydrology & Hydrobiology*. https://doi.org/10.1016/j.ecohyd.2024.01.006

Arroyo-Rodríguez, V., Cavender-Bares, J., Escobar, F., Melo, F. P., Tabarelli, M., and Santos, B. A. (2012). Maintenance of tree phylogenetic diversity in a highly fragmented rain forest. *Journal of Ecology*, *100*(3), 702–711.

Assunção-Silva, C. C., Bueno, M. L., and Assis, L. C. d. S. (2021). Centers of richness and endemism of Lauraceae in the Espinhaço Range, Brazil. *Flora*, *285*, 151943. https://doi.org/10.1016/j.flora.2021.151943

Barreto, J. R., Pardini, R., Metzger, J. P., Silva, F. A. B., and Nichols, E. S. (2023). When forest loss leads to biodiversity gain: Insights from the Brazilian Atlantic Forest. *Biological Conservation*, *279*, 109957. https://doi.org/10.1016/j.biocon.2023.109957

Benton, M. J., Wilf, P., and Sauquet, H. (2022). The Angiosperm Terrestrial Revolution and the origins of modern biodiversity. *New Phytologist*, *233*(5), 2017–2035. https://doi.org/10.1111/nph.17822

Carvalho, C. d. S., Martello, F., Galetti, M., Pinto, F., Francisco, M. R., Silveira, L. F., and Galetti Jr, P. M. (2021). Environmental heterogeneity and sampling relevance areas in an Atlantic forest endemism region. *Perspectives in Ecology and Conservation*, *19*(3), 311–318. https://doi.org/10.1016/j.pecon.2021.05.001

Cerasoni, J. N., Hallett, E. Y., Orijemie, E. A., Ashastina, K., Lucas, M., Farr, L., . . . Scerri, E. M. L. (2023). Human interactions with tropical environments over the last 14,000 years at Iho Eleru, Nigeria. *iScience*, *26*(3), 106153. https://doi.org/10.1016/j.isci.2023.106153

Edwards, D. P., Tobias, J. A., Sheil, D., Meijaard, E., and Laurance, W. F. (2014). Maintaining ecosystem function and services in logged tropical forests. *Trends in Ecology & Evolution, 29*(9), 511–520. https://doi.org/10.1016/j.tree.2014.07.003

Faith, D. P. (2008). Threatened species and the potential loss of phylogenetic diversity: Conservation scenarios based on estimated extinction probabilities and phylogenetic risk analysis. *Conservation Biology, 22*(6), 1461–1470.

Ferreira, G. M. D., Leme, E. M. C., Aona, L. Y. S., Souza, F. V. D., Moura, H. C. P., and Souza, E. H. d. (2024). Microendemism, high diversity and taxonomic novelties in Bromeliaceae: Studies from an Atlantic Forest hotspot in Bahia, Brazil. *Journal for Nature Conservation, 79*, 126599. https://doi.org/10.1016/j.jnc.2024.126599

Gota, P., Massingue, A., and Lencastro, C. (2024). Vegetation structure, composition and biodiversity in locally protected forests and woodlands in Inhambane Province, southern Mozambique. *South African Journal of Botany, 172*, 180–191. https://doi.org/10.1016/j.sajb.2024.07.028

Ibrahim, F., Adu-Bredu, S., Buechling, A., Duah-Gyamfi, A., Addo-Danso, S., Malhi, Y., . . . Matula, R. (2025). Disentangling the impacts of soil moisture and vapour pressure deficit on biodiversity patterns and conservation value of tropical forests in Ghana, West Africa. *Biological Conservation, 302*, 110952. https://doi.org/10.1016/j.biocon.2024.110952

IUCN. (2024). The IUCN Red List of Threatened Species. Retrieved from https://www.iucnredlist.org

Laranjeiras, T. O., Ribas, C. C., and Cohn-Haft, M. (2024). Patterns of endemism in Amazonian floodplain birds. *Perspectives in Ecology and Conservation, 22*(3), 306–314. https://doi.org/10.1016/j.pecon.2024.08.001

Lewis, O. T. (2009). Biodiversity change and ecosystem function in tropical forests. *Basic and Applied Ecology, 10*(2), 97–102. https://doi.org/10.1016/j.baae.2008.08.010

Mahayani, N. P. D., Slik, F. J., Savini, T., Webb, E. L., and Gale, G. A. (2020). Rapid recovery of phylogenetic diversity, community structure and composition of Bornean tropical forest a decade after logging and post-logging silvicultural interventions. *Forest Ecology and Management, 476*, 118467.

Manes, S., Costello, M. J., Beckett, H., Debnath, A., Devenish-Nelson, E., Grey, K.-A., . . . Vale, M. M. (2021). Endemism increases species' climate change risk in areas of global biodiversity importance. *Biological Conservation, 257*, 109070. https://doi.org/10.1016/j.biocon.2021.109070

Maure, L. A., Diniz, M. F., Pacheco Coelho, M. T., Molin, P. G., Rodrigues da Silva, F., and Hasui, E. (2023). Biodiversity and carbon conservation under the ecosystem stability of tropical forests. *Journal of Environmental Management, 345*, 118929. https://doi.org/10.1016/j.jenvman.2023.118929

Morley, R. (2000). *Origin and Evolution of Tropical Rain Forests* (Vol. 15).

Niang, F., Marchand, P., Sambou, B., and Fenton, N. (2024). Exploring the effects of forest management on tree diversity, community composition, population structure and carbon stocks in sudanian domain of Senegal, West Africa. *Forest Ecology and Management, 559*, 121821. https://doi.org/10.1016/j.foreco.2024.121821

Prawiradilaga, D. M., and Irham, M. (2024). Endemism. In S. M. Scheiner (Ed.), *Encyclopedia of Biodiversity* (Third Edition) (pp. 72–83). Oxford: Academic Press.

Rajaonarimalala, R., Korol, Y., Andrianarimisa, A., Dröge, S., Fulgence, T. R., Grass, I., . . . Martin, D. A. (2024). Complex stands in forested tropical landscapes harbor more endemic biodiversity and ecosystem functions. *Global Ecology and Conservation, 54*, e03154. https://doi.org/10.1016/j.gecco.2024.e03154

Santos, C. V. d., Silva, F. M. e., and Faria, L. I. L. d. (2023). The Brazilian Atlantic Forest genetic resources in patents and the challenges to control the economic use of biodiversity. *World Patent Information, 74*, 102218. https://doi.org/10.1016/j.wpi.2023.102218

Scutt, C. P. (2021). The origin of angiosperms. In *Evolutionary Developmental Biology: A Reference Guide* (pp. 663–682): Springer.

USDA Forest Service. (2004). Forest Inventory and Analysis fiscal year 2004 business report. U.S. Department of Agriculture, Forest Service. usda.gov

Wang, Q., Huang, J., Zang, R., Li, Z., and El-Kassaby, Y. A. (2022). Centres of neo- and paleo-endemism for Chinese woody flora and their environmental features. *Biological Conservation, 276*, 109817. https://doi.org/10.1016/j.biocon.2022.109817

Wikramanayake, E., Fernando, S., Wickramaratne, C., Akbarally, Z., and Wikramanayake, S. (2022). Sri Lankan moist forests ecoregion: An imperiled island rainforest. In D. A. DellaSala and M. I. Goldstein (Eds.), *Imperiled: The Encyclopedia of Conservation* (pp. 37–52). Oxford: Elsevier.

Zhou, R., Ci, X., Hu, J., Zhang, X., Cao, G., Xiao, J., . . . Li, J. (2023). Transitional areas of vegetation as biodiversity hotspots evidenced by multifaceted biodiversity analysis of a dominant group in Chinese evergreen broad-leaved forests. *Ecological Indicators, 147*, 110001. https://doi.org/10.1016/j.ecolind.2023.110001.

Sedimentary Charcoal Analysis from African Grassy Ecosystems with Contrasting Local Fire Regimes and Biomass Stability

9

Abraham Nqabutho Dabengwa and Marion Bamford

INTRODUCTION

Approximately 40% of Earth's terrestrial ecosystems are covered by grasslands, where fire is an important and common disturbance that permits the maintenance of tree canopy openness (Bond, 2021). In C4 grass-dominated rangelands, fires facilitate grazing, land clearing, nutrient cycling, biodiversity enhancement, and carbon capture (Archibald and Hempson, 2016; Conant *et al.*, 2017). Local extinction of vulnerable species as well as ecosystem degradation is an example of adverse effects from increases or decreases in fire activity (Fuhlendorf *et al.*, 2009). Many vegetation communities in open ecosystems that are prone to fires show positive vegetation-fire feedback mechanisms that enhance community stability (Mutch, 1970). As a result, plant community compositions tend to shift towards greater levels of flammability, or fire tolerance (Keeley *et al.*, 2011; Keeley and Rundel, 2005). These functional trait aspects of vegetation biomass influence fire parameters, including frequency, intensity, size, return, and seasonality, which define a property known as the fire regime (Bond and Keeley, 2005). Although there is knowledge of the important fire regime for healthy grassy ecosystems functioning (Archibald *et al.*, 2013; Archibald and Hempson, 2016; Ekblom and Gillson, 2010b; Leys *et al.*, 2018), very little is known about regimes over long periods of time (Sayedi *et al.*, 2024).

DOI: 10.1201/9781003727781-9

Fire regimes across space and time can be understood as phase states along a vegetation or climate continuum (Figure 9.1). The responses of these phase states to environmental drivers range from linear to nonlinear (Holling, 1973; Petraitis, 2013). From an evolutionary standpoint, fire-related traits such as flammability are expected to increase along fire gradients (Beckage and Ellingwood, 2009; Colombaroli *et al.*, 2014; Pausas and Schwilk, 2012). This knowledge can aid in inferring and comparing fire regimes using sedimentary charcoal and pollen records (Bremond *et al.*, 2024; Power *et al.*, 2008; Whitlock *et al.*, 2010). Still, huge uncertainties remain over charcoal production rates. Additionally, vegetation biomass can exist in alternative stable states (ASS) maintained by positive feedback mechanisms: short-grass states with high grazer biomass, tall grass states with frequent fire, and open tree-grass states under intermediate fire and grazing conditions (Hempson *et al.*, 2019; Noy-Meir, 1975; Perrings and Walker, 1997). Common ways of dealing with these challenges include treating forest and grassy ecosystem charcoal separately (Leys *et al.*, 2018) and rescaling charcoal of any size using Z-scores for comparison (Marlon *et al.*, 2008). While these approaches have been effective, they lack a functional eco-evolutionary foundation.

This study compares Late Holocene charcoal records from mesic grasslands and semi-arid savannas in South Africa's fire-prone biomes to infer past fire regimes. These sites differ in climate, vegetation, and large herbivore biomass, leading to expected variations in fire regimes. For example, they contribute to different growing season lengths, resistance to disturbances, and rates of grass curing (Archibald *et al.*, 2019; Ellery *et al.*, 1995; Simpson *et al.*, 2022). Flammability is defined as the ability of vegetation to ignite and sustain fire, which is controlled by grass curing (Pausas *et al.*, 2017; Simpson *et al.*, 2016), which affects charcoal production. Since growing seasons and temperature optima may differ between grasses

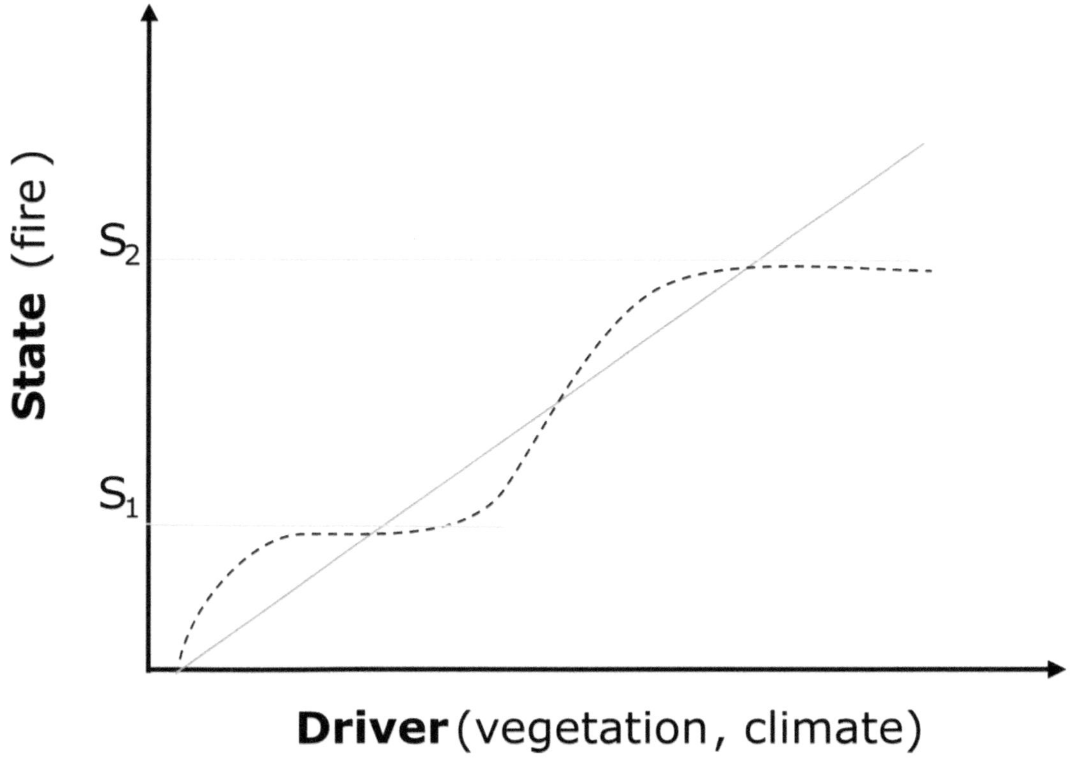

FIGURE 9.1 A state-driven phase diagram for assessing fire regimes driven by climate or vegetation. The solid diagonal line represents a linear relationship, while the dotted line is nonlinear. Generally, steady state phases S1 and S2 may indicate short versus tall grass dominance (e.g., Hempson *et al.*, 2019) or closed forest versus open canopy grassy ecosystems (e.g., Pausas and Bond, 2020). This suggests that in paleofire studies, vegetation-fire dynamics from proxies can indicate flammability or vegetation states along fire gradients (cf. Pausas and Schwilk, 2012).

using C3 and C4 photosynthetic pathways (Ehleringer and Monson, 1993), flammability is expected to be higher in systems dominated by the latter (Colombaroli *et al.*, 2014; Ripley *et al.*, 2010). The objectives of this study were to (1) determine whether flammability regimes are influenced by biome-controlled herbaceous biomass and (2) compare the stability nature of flammability dynamics.

METHODS

Study Site

The grassland and savanna biome study sites uMngeni Vlei (29° 29.304' S, 29° 49.554' E, 1,840 m above sea level) and uMfolozi (28° 20.46' S, 31° 57.12' E, 50 m above sea level) are situated in uMngeni Vlei Nature Reserve and Hluhluwe-iMfolozi Park (HiP), KwaZulu-Natal (KZN) Province, South Africa (Figure 9.2). uMngeni receives higher rainfall, averaging approximately 980 mm/year, compared to the semi-arid uMfolozi, which receives around 635 mm/year (Balfour and Howison, 2001; Kotze and O'Connor, 2000). Additionally, uMngeni is cooler, with a mean annual temperature of about 12°C, while uMfolozi has a warmer mean annual temperature of approximately 21°C. Sedimentary cores were collected at both sites using the vibracorer method in December 2011 (Meadows and Baxter, 1999).

From a rangeland classification narrative, uMngeni falls within the highland 'sourveld' category, characterised by nutrient-poor dystrophic soils, while uMfolozi is situated in the fertile floodplain of the 'mixed to sweetveld' lowveld (Tainton *et al.*, 1989). Grasslands generally have nutrient-poor dystrophic soils, while savanna is situated in a fertile floodplain (e.g., Ellery *et al.*, 1995). The uMngeni grassland is dominated by *Themeda triandra* and *Hyparrhenia* C4 grasses, forming extensive open grasslands with flammable tall grass tussocks interspersed with patches of C3 *Festuca* species. In contrast, uMfolozi's

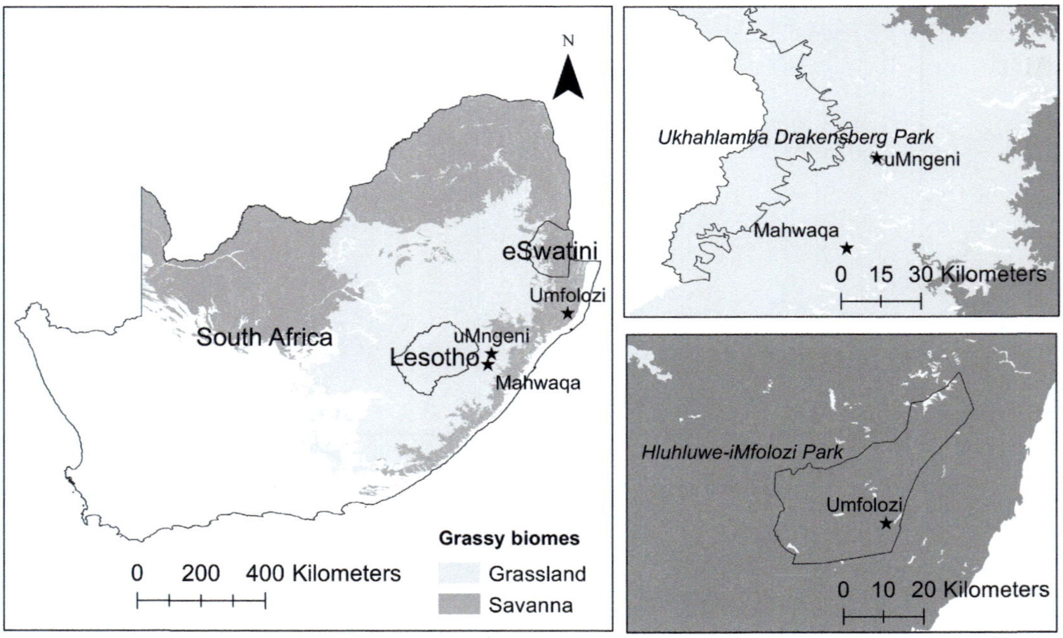

FIGURE 9.2 Location of study sites within major grassy biomes of South Africa according to Mucina and Rutherford (2006).

landscape features tall grass patches interspersed with tree and shrub cover. Notably, O'Connor *et al.* (2011) found that wetlands in KZN experienced fewer fire events compared with surrounding landscapes.

Herbivore biomass at uMfolozi is high and dominated by white rhinoceros (*Ceratotherium simum*) at ~ 12,500 kg km-2. Although uMngeni is fenced, local cattle are permitted seasonal grazing at low density. Consequently, the uMfolozi wetland has short C4 lawn grasses, including *Urochloa mossambicensis*, *Panicum coloratum*, and *Cynodon dactylon*. C3 herbaceous biomass at wetland fringes is dominated by *Phragmites australis* at uMfolozi compared with *Typha capensis* and *Carex* spp. at the cored section of uMngeni Vlei. These differences in climate, vegetation, and herbivore dynamics highlight the contrasting ecological conditions between the two sites.

Chronology

The chronological control of the uMngeni and uMfolozi Late Holocene sedimentary sequences was established using accelerator mass spectrometry (AMS) radiocarbon dates. These dates were analysed using the Bacon package in the R Statistical programming environment (Blaauw and Christeny, 2011). Radiocarbon ages were calibrated to years before present (cal yr bp) using the SHCal13 curve (Hogg *et al.*, 2020) to get models for interpolating mean ages at each depth.

Palaeoecological Analyses

Herbaceous Biomass Change

Local herbaceous biomass was inferred indirectly from accumulated soil/sediment organic carbon (e.g., Ingram *et al.*, 2008; Schaeffer *et al.*, 2025) as herbaceous biomass thrived in these seasonally flooded wetlands. Our expectation is that organic carbon accumulation is related to the above herbaceous biomass's potential via litter production (e.g. Elschot *et al.*, 2013; Ingram *et al.*, 2008; Schaeffer *et al.*, 2025). Thus, the organic carbon loss-on-ignition (LOI) method (Wang *et al.*, 2011) was used to estimate sediment organic carbon (SOC) in the sedimentary sequences with sampling every 2 cm for uMngeni and 2–4 cm for uMfolozi.

Fire Activity and Flammability

The role of local fire activity and flammability as drivers of charcoal production and biomass burning was investigated by analysing microcharcoal (< 150 μm) and macrocharcoal (> 150 μm) particle concentrations (Pitkänen *et al.*, 1999). Charcoal samples were collected at 2–4 cm resolution. Microcharcoal was divided into five size-classes (i.e. 10–15 μm, 16–30 μm, 31–50 μm, 51–70 μm, and 71–149 μm). Microcharcoal was enumerated at x400 magnification from pollen slides spiked with a known quantity of *Lycopodium* spores (Finsinger and Tinner, 2005; Stockmarr, 1971). Sieved macrocharcoal was counted using a stereomicroscope at X10 to X150 magnification (Ekblom and Gillson, 2010a). Microcharcoal and macrocharcoal concentrations were added together to derive Total Charcoal because these are strongly correlated in small basins (Pitkänen *et al.*, 1999). Relationships and variations between charcoal concentrations in size classes and study sites were analysed using principal component analysis (PCA) with log10 transformed charcoal (Legendre and Birks, 2012). Charcoal accumulation rates (CHAR) were estimated by the product of accumulation rates from the age-depth model and Total Charcoal (Aleman *et al.*, 2013). For paleofire site comparisons, CHAR values from uMfolozi, uMngeni and nearby Mahwaqa (Neumann *et al.*, 2014) were converted, box-cox transformed and rescaled with Z-scores (Brucher *et al.*, 2014). Flammability of local herbaceous biomass was explored by interpreting the main charcoal principal component axis (PC1) scores (Dabengwa, 2019), varying with climate and or vegetation type (Colombaroli *et al.*, 2014).

Herbaceous Biomass Responses to Fire Activity or Flammability

Relationships between responses of local herbaceous biomass to flammability gradients were assessed with Z-scores of Charcoal PC1 scores and SOC values using generalised additive models (GAMs; Hastie and Tibshirani, 1986). GAMs allow for flexible linear and nonlinear local regression models. We used the generalised cross-validation (GCV) method to estimate smoothing parameters (s) using the 'mgcv' package in R (Wood, 2017), with smoothers with p-values < 0.05 considered statistically significant. The Charcoal PC1 axis values associated with Total charcoal concentrations were transformed by raising them to e(PC1 +0) and Z-score transformed for GAMs. Similarly, combined LOI values were rescaled between 0–1. Statistical analyses were done using the R programme (R Core Team, 2023). Data plots were made in R and the C2 program (Juggins, 2011).

RESULTS

Chronology

Chronological control for the samples was provided by three AMS dates on uMngeni and four on uMfolozi (Figure 9.3; Table 9.1). The basal dates were 2,311–2,018 cal yr bp (2,200 ± 30 bp) for uMfolozi at 201 cm and 2,746–2,438 cal yr bp (2,560 ± 35 bp) at uMfolozi. The age-depth model for uMfolozi excluded the age reversal date at 105 cm which indicated bioturbation. Sediment accumulation rates for uMfolozi were constant compared with an increasing trend at uMngeni between 23–30 cm.

Biomass and Fire Dynamics

Sedimentary records of local herbaceous biomass and fire dynamics represented by sediment organic carbon and charcoal concentrations at uMngeni (n = 19) and uMfolozi (n = 24) showed contrasting patterns (Figures 9.4 and 9.5). The longer uMngeni record had higher LOI (13.1 ± 4.9% g-1) and lower Total Charcoal (141,760 ± 62,170 particles cm-2) compared with uMfolozi at 6.1 ± 4.9% g-1 and 207,595 ±

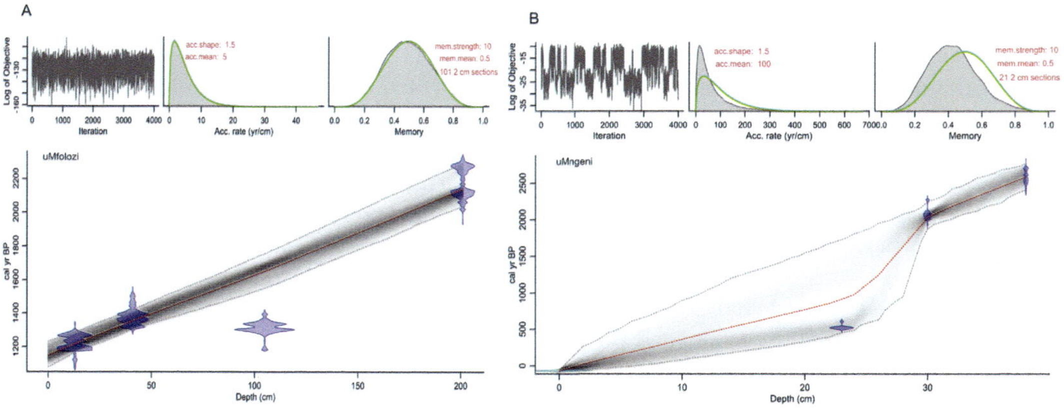

FIGURE 9.3 Estimated Bacon age-depth models for (A) uMfolozi and (B) uMngeni sedimentary records. Shaded regions represent the 95% confidence intervals of age-depth models from calibrated radiocarbon dates shown by blue violin plots.

TABLE 9.1 Radiocarbon Ages and Calibration Information for uMfolozi and uMngeni Records.

SITE	*DEPTH (CM)*	*LAB*	*LAB CODE*	*RADIOCARBON AGE (BP)*	*CALIBRATED DATES (CAL YR BP)*	*PROBABILITY*
uMfolozi	13	Beta-Analytic	Beta-410179	1360 ± 30	1292-1258	31.6
					1252-1178	63.3
	41	Beta-Analytic	UBA-26949	1538 ± 32	1471-1461	1.3
					1453-1441	1.7
					1427-1309	92
	105	Direct AMS	D-AMS 004503	1457 ± 29	1359-1283	94.9
	201	Beta-Analytic	Beta-410176	2200 ± 30	2311-2228	36.3
					2177-2171	0.7
					2154-2044	56.9
					2027-2018	1.1
uMngeni	23	CHRONOS	26946	558 ± 28	563-505	94.7
	30	Beta-Analytic	410178	2150 ± 30	2289-2274	3.8
					2136-2005	91.2
	38	Poznan	Poz-61759	2560 ± 35	2746-2485	92.5
					2482-2465	1.8
					2446-2438	0.7

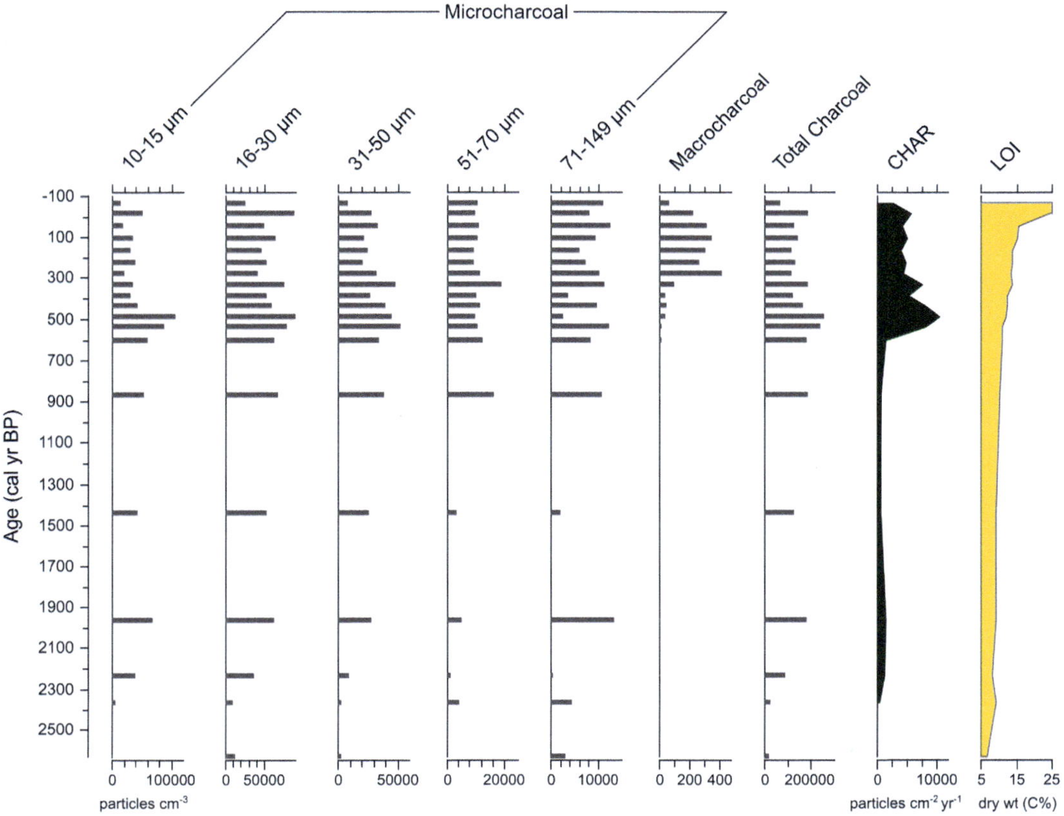

FIGURE 9.4 Summary diagram of charcoal concentrations and sediment organic carbon from loss-on-ignition (LOI) for the uMngeni core, uMngeni Vlei Nature Reserve, South Africa.

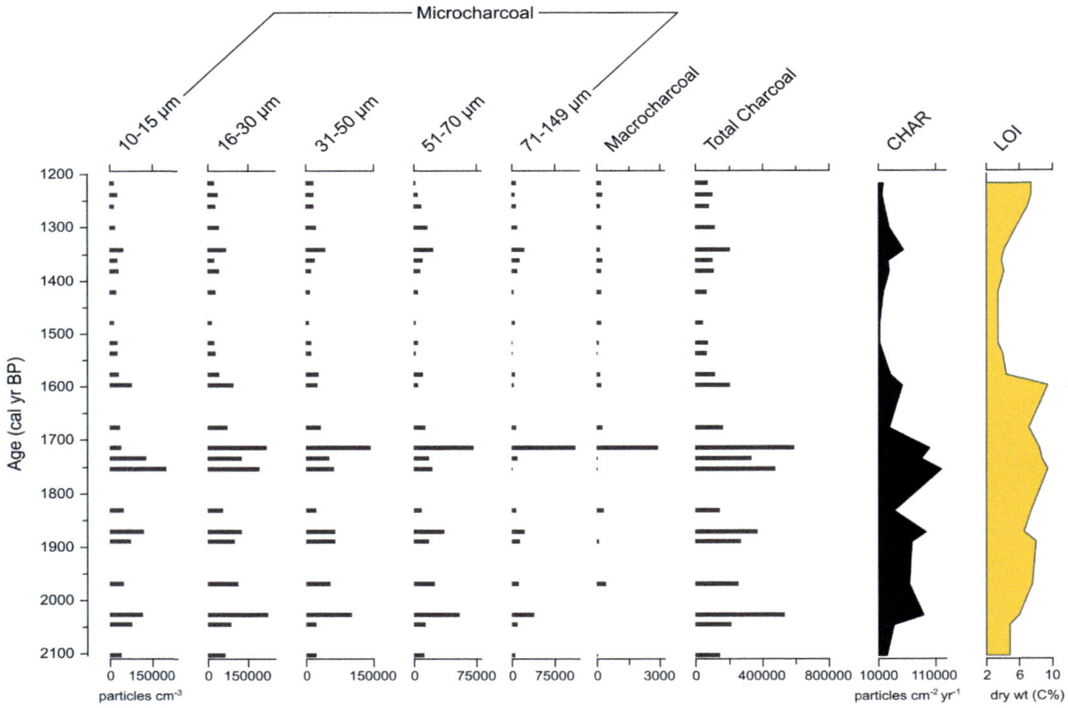

FIGURE 9.5 Summary diagram of charcoal concentrations and sediment organic carbon from loss-on-ignition (LOI) for the uMfolozi core, Hluhluwe-iMfolozi Park, South Africa.

155,600 particles cm-2. Charcoal accumulation rates at uMfolozi were higher at 45,408 ± 2,970 particles cm-2 y-1 compared with 4,230 ± 3,180 particles cm-2 y-1 from uMngeni. Local herbaceous vegetation and fire indicated by LOI and charcoal at uMfolozi showed more fluctuation whereas charcoal variability at uMngeni only increased from ~620 cal yr bp years as LOI steadily increased. Curiously, at uMfolozi, charcoal declined from ~1,700 cal yr bp followed by LOI from ~1,600 cal yr bp.

The first two principal components of the ordination of charcoal concentrations at uMngeni and uMfolozi accounted for 84.1% variation in the dataset (fig. 9.6). There was an overlap of samples from sites in ordination space. Principal component one (PC1) scores were linked with charcoal concentrations and PC2 scores with charcoal size classes. The log10 of Total Charcoal was linearly related to PC1 (r2 = 0.97, df = 22, p-value < 0.0001; Supplementary material 2). There was a stronger positive correlation between the first principal component of the charcoal ordination and Z-scores of charcoal accumulation rates of uMfolozi (r = 0.91, df = 22, p-value < 0.0001) compared with uMngeni (r = 0.57, df = 16, p-value = 0.012).

Fire Regimes and Biomass Stability

Positive herbaceous biomass to flammability relationships were found at both sites using LOI and charcoal PC1 scores (Figure 9.7A). The relationship was stronger at uMfolozi (r = 0.59, df = 22, p-value = 0.002) compared with uMngeni (r = 0.31, df = 13, p-value = 0.2671) and four outliers were found from the bottom and top two samples for uMngeni. Modelled response curves of flammability versus local herbaceous biomass had significant positive relationships for individual sites and a weak relationship in the combined dataset (figs. 9.7 B-D; Table 9.2). uMfolozi patterns indicated multimodal responses compared to uMngeni's unimodal response.

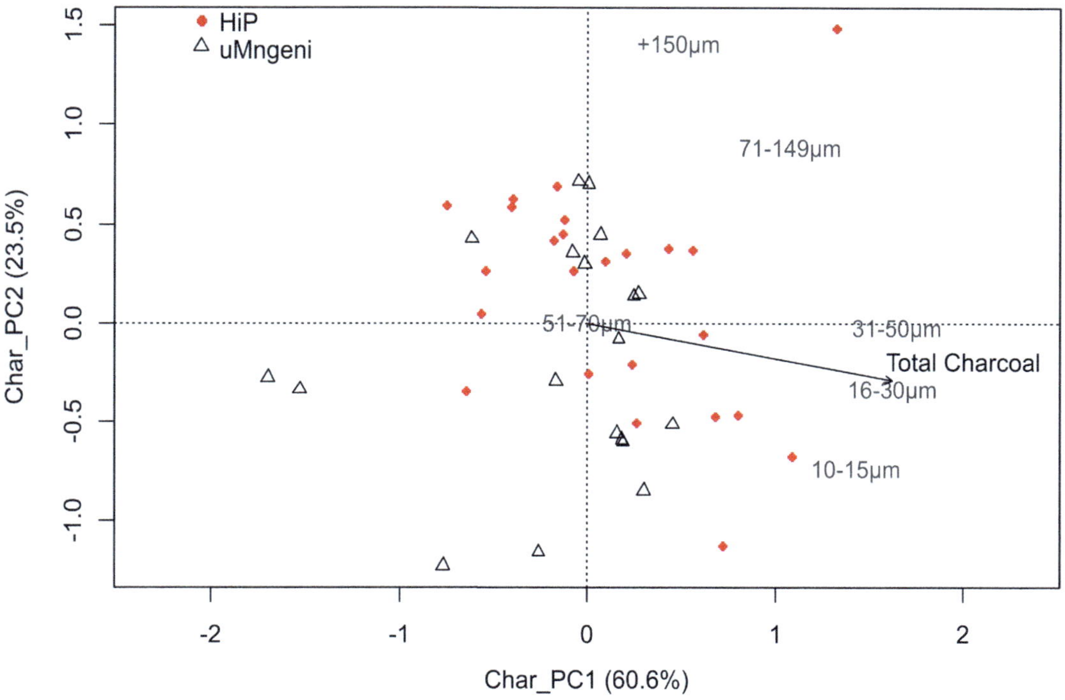

FIGURE 9.6 Principal component analysis ordination of uMngeni and uMfolozi charcoal concentrations of fragment size classes. Total charcoal concentration from all size classes is included as a covariate.

DISCUSSION

The relationship between plants and fire plays a critical role in shaping vegetation patterns across various scales, from small patches to biomes (Bond, 2005). In C4 grassy ecosystems, herbaceous plants – especially grasses – have adapted to coexist with fire and seasonal droughts over evolutionary timescales (Keeley and Rundel, 2005). Studying the long-term impacts of fire on these ecosystems has often relied on sedimentary charcoal and vegetation indicators such as fossil pollen. While these methods have proven effective for analysing climate-vegetation-fire dynamics, the conceptual foundation for assessing fire regimes is weak. Therefore, we examined Late Holocene fire dynamics at two fire-prone South African grassy ecosystems differing in climate, vegetation, and herbivore biomass to understand factors influencing fire regimes

Local Herbaceous Biomass Dynamics

Despite covering only a few overlapping periods during the Late Holocene, the changes in local herbaceous biomass potential – represented by soil organic carbon – differed significantly between the wetter uMngeni highland and the drier uMfolozi lowland (Figure 9.8A). This indicates that rainfall acts as a key constraint on the upper limit of plant biomass in grasslands (Bond and Keeley, 2005), particularly in wetland areas. Findings from this study align with modern ecological research in the region, showing that grass biomass increases with rainfall (Archibald and Hempson, 2016; Fynn and O'Connor, 2005; Pausas and Ribeiro, 2013). For instance, declines in herbaceous biomass at uMfolozi during periods of reduced regional rainfall (~1,750–1,680 cal yr bp) and (1,600–1,580 cal yr bp) from Chevalier and Chase (2015).

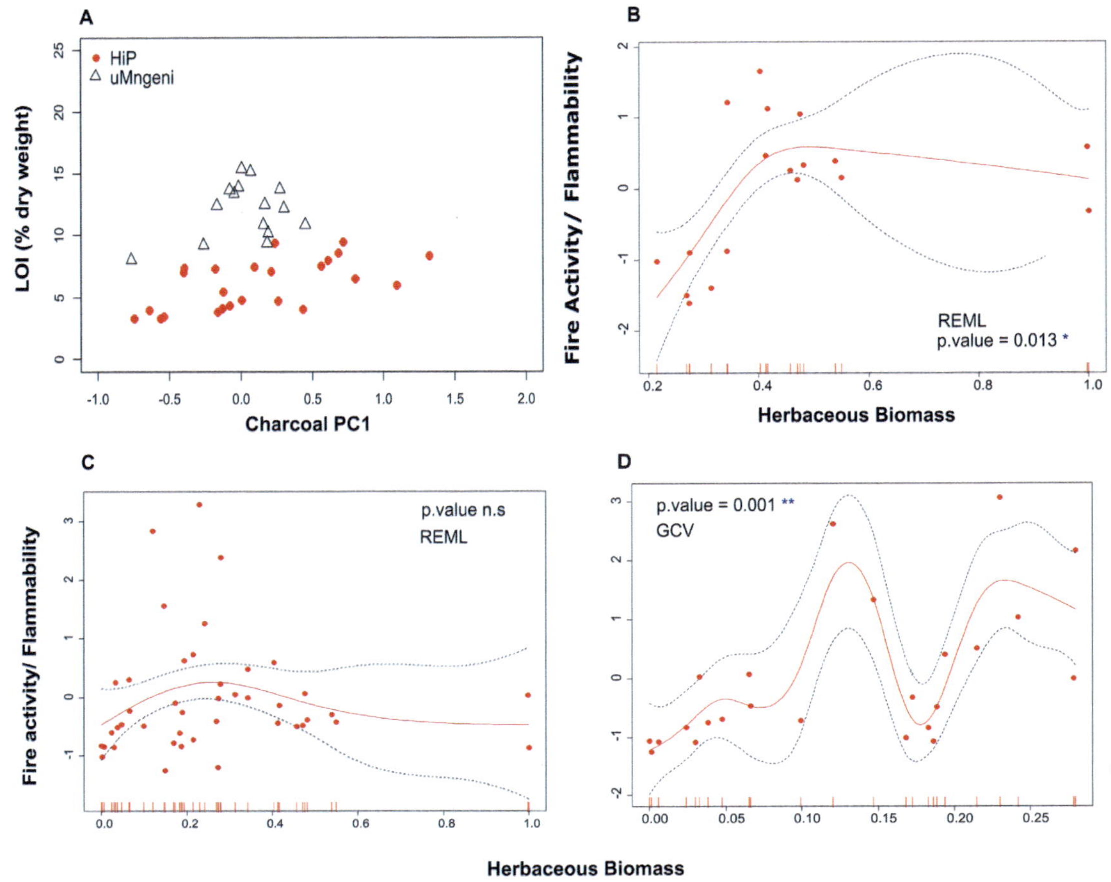

FIGURE 9.7 Summary of local herbaceous biomass responses to fire activity (or flammability) represented by sedimentary organic carbon (LOI) and charcoal metrics at uMngeni grassland and uMfolozi savanna, South Africa. (A) Combined charcoal principal component one scores (CHAR_PC1) relationship with LOI; (B) LOI GAM model representing the relationship between combined local herbaceous biomass and fire activity (Total charcoal) at uMngeni and uMfolozi; (C and D) Responses of LOI linked to local herbaceous biomasses with fire activity (charcoal) at uMngeni; and (uMfolozi. Black dashes represent 95% confidence intervals around means in the models. Letters REML and GCV indicate models used for selecting smoothing parameters, i.e., restricted maximum likelihood and generalised cross-validation. Significant models are represented by blue stars following p-values.

TABLE 9.2 Results of Three Generalised Additive Models Comparing Responses of Fire Activity to Herbaceous Biomass Represented by Charcoal PC1 Scores and LOI at uMngeni and uMfolozi.

			F	*DEVIANCE*
SITES	*RESPONSE VARIABLE*	*ADJUSTED R^2*	*LOI*	
All sites	Charcoal_PC1	0.002	0.37*ns*	3.83%
uMngeni	Charcoal_PC1	0.612	37.15**	70.5%
uMfolozi	Charcoal_PC1	0.641	5.336**	75.9%

***p-value* < 0.01
ns = not statistically significant

In contrast, local herbaceous biomass at uMngeni showed no clear relationship with rainfall, as SOC levels increased over time, with a particularly notable rise in the last 600 years, which was more frequently sampled.

Our results also suggest that the broad rangeland classification of 'sweet' versus 'sourveld' provides a useful framework for comparing long-term grass biomass dynamics in palaeoecological studies. In mixed and sweet veld, frequent changes in herbaceous biomass can be linked to repeated grazing, which reduces grass biomass or productivity, creating gaps for forbs and tree seedlings to establish (e.g. Hempson *et al.*, 2015; Veen *et al.*, 2008). On the opposite end of the spectrum, infrequently grazed highland sourveld accumulates more biomass and fine fuel for fires (e.g. Archibald and Hempson, 2016; Ellery *et al.*, 1995). Grazing in sourveld is often facilitated by regular burning, which appears to have little impact on species diversity or herbaceous biomass (Everson and Everson, 2016; Morris *et al.*, 2020).

Fire Dynamics

Sedimentary charcoal records often struggle to represent fire signals of fire-prone grassy ecosystems whose peak concentrations after fire events smooth out over time (e.g. Ekblom and Gillson, 2010b; Leys *et al.*, 2017). An alternative approach is to focus on charcoal production and concentration patterns to detect flammability states. Compared with previous studies arguing microcharcoal (< 150 μm) and macrocharcoal (> 150 μm) have different source areas (Ekblom and Gillson, 2010b; Vachula, 2021), these different sizes were considered together for our small study basins as an indicator of flammability. Our results suggest that considering a range of charcoal sizes may be a better way of tracking charcoal production, fire activity, and local vegetation flammability compared with Z-scores (figs. 9.4; 9.5. and 9.7. B and C).

The Z-scores of charcoal accumulation rates were still useful for comparing fire patterns from uMfolozi, uMngeni and nearby Mahwaqa. The Z-score patterns reflected expected patterns in flammability which depend on grass curing rates, with higher charcoal concentration at uMfolozi because it likely dried out faster than uMngeni. Also, charcoal patterns at uMfolozi and Mahwaqa tracked each other well between ~2,010–1,200 cal yr bp, which implicated a regional climatic driver. Peaks in charcoal at uMfolozi correspond with periods of human settlement indicated by recovered artefacts (Hall, 1981). In comparison, the nearby sites uMngeni and Mahwaqa lagged each other in increased fire activity peaks from ~600–490 cal yr bp in the former and ~470–330 cal yr bp in the latter. These lags may be due to different occupation periods by Late Iron Age farmers in the region (e.g. Hall, 1981; Maggs, 1976).

However, Z-scores performed poorly for tracking the initial charcoal accumulation phase of uMngeni sediment from ~2,600–600 cal yr bp and the last 600 years. From a flammability perspective (Figure 9.8B), after the initial increase in flammability from ~2600–1970 cal yr bp, flammability remained high and stable, only dropping sharply in the last 20 years, corresponding with an updated conservation status. Because fire activity is still generally high because of lightning and land use (O'Connor *et al.*, 2011), the aggregated flammability signal was a more robust measure of fire dynamics.

Synthesis of Grassy Ecosystem Paleofire Regimes

Paleofire approaches for inferring fire regimes using charcoal analysis in African grassy ecosystems need revision. From a classical ecological modelling perspective, it is essential to identify relationships between vegetation biomass and fire as a disturbance driver (Beckage *et al.*, 2009; Perrings and Walker, 1997) and to investigate the nature of variability or stability dynamics (Clarke and Lawes, 2011; Perrings and Walker, 1997; Petraitis, 2013). Generally, positive vegetation biomass and fire interactions arise when fire acts as an ecosystem state attractor in frequently burned grassy ecosystems (Bond, 2005; Hempson *et al.*, 2019; Keeley and Rundel, 2005; Pausas and Ribeiro, 2013), likely leading to increased vegetation openness and charcoal production. In this study, we found evidence of strong and distinctive positive correlations between charcoal concentrations and sedimentary organic carbon, which served as proxies for local

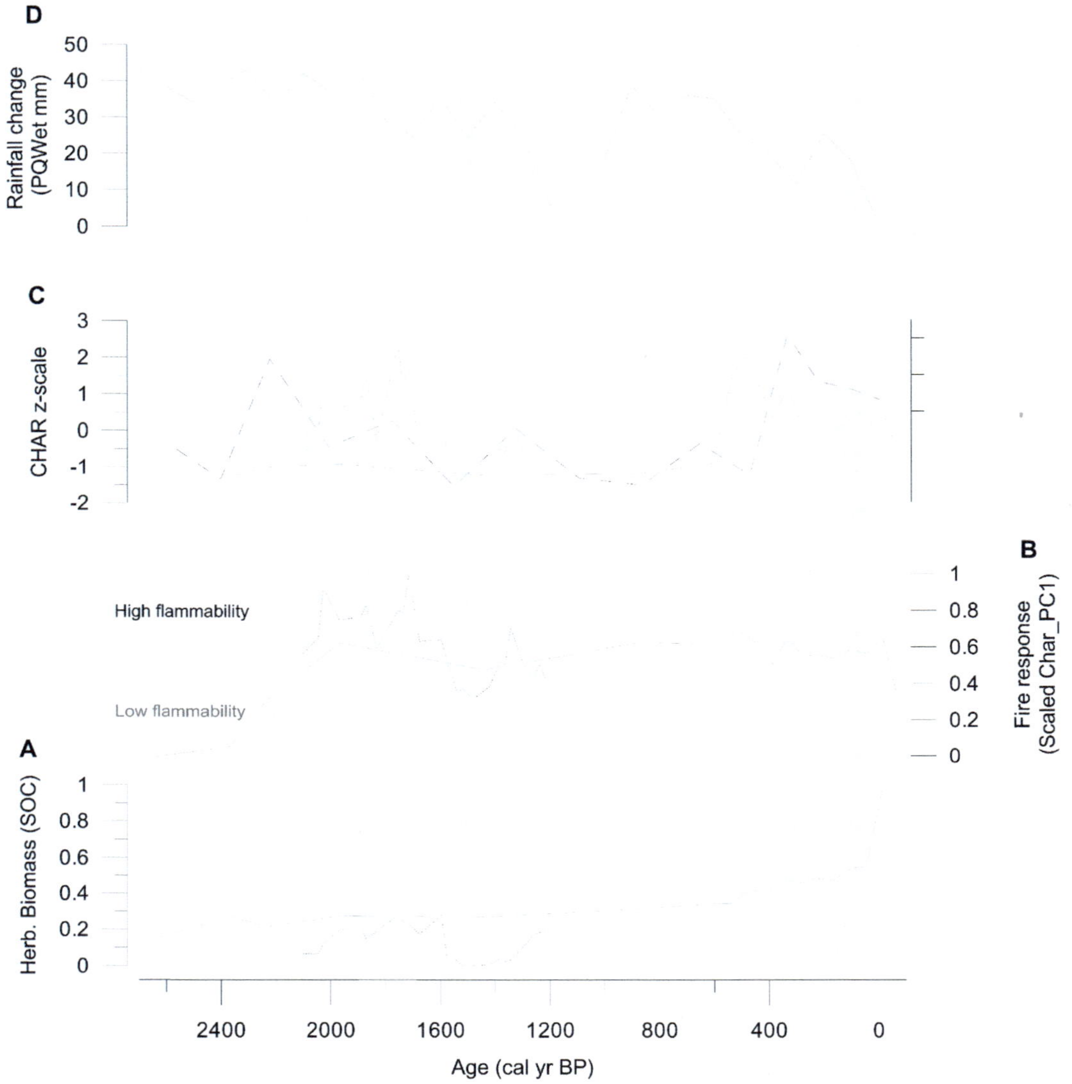

FIGURE 9.8 Summarised changes in local herbaceous biomass, fire, and climate at uMngeni (solid lines) and uMfolozi (dotted lines), South Africa, over the last 2,500 years. (A) Standardised sediment organic carbon amounts representing herbaceous biomass potential at uMngeni and uMfolozi. (B) Standardised charcoal principal component one (Char_PC1) scores indicating the gradient of flammability. (C) Changes in Z-score transformed Total Charcoal accumulation rates (CHAR), including Mahwaqa in bold dash lines (Neumann *et al.*, 2014). (D) Pollen-derived changes in regional rainfall from Chevalier and Chase (2015). Dry climatic periods or phases are indicated by shaded bands. Neumann *et al.*, 2014. (D) Pollen-derived changes in regional rainfall from Chevalier and Chase (2015). Dry climatic periods or phases are indicated by shaded bands.

fire activity/flammability and herbaceous biomass at uMngeni and uMfolozi biome contrasts (Figures 9.7A and 9.9A).

At the broader vegetation classification level, the observed patterns suggest that differences in paleofire regimes between grassland and savanna biomes were likely driven by climate (Colombaroli *et al.*, 2014; Ekblom and Gillson, 2010c), maximum potential herbaceous biomass (Bond and Keeley, 2005), and local C_3/C_4 vegetation flammability (Colombaroli *et al.*, 2014; Ripley *et al.*, 2010). Our findings therefore support the rationale for biome-specific comparisons of paleofire records (e.g. Leys *et al.*, 2018) by further demonstrating separation within grassy ecosystems. At the local level, while a unimodal vegetation-fire relationship was observed at uMngeni grassland (figs: 9.7 B and C), the uMfolozi savanna site's behaviour suggested the possibility of multiple stable states and multimodal fire regimes (Figure 9.7D). Consumer control over grass biomass by large herbivores, particularly megaherbivores historically present (McCracken, 2008), may have prevented herbaceous biomass from reaching its full climatic potential. Thus, this study provides compelling palaeoecological evidence for theoretically predicted multimodal fire states and regimes (e.g. Hempson *et al.*, 2019; Noy-Meir, 1982; Perrings and Walker, 1997; Ratajczak *et al.*, 2014).

The grassland and savanna sites exhibited contrasting flammability dynamics. At uMngeni, flammability showed stable or damped fluctuations over time, oscillating around an equilibrium but consistently trending towards high vegetation flammability (figs. 9.8 B and 9.9 B). These findings reinforce existing evidence of long-term biomass stability in frequently burned highland sourveld grasslands, which has been used to justify more intensive fire management practices (e.g. Everson and Everson, 2016). Notably, while charcoal concentrations at the grassland site increased over the last 600 years, flammability remained stable, exhibiting damped oscillations. This suggests heightened local burning activity without significant changes in vegetation structure (e.g. Finch *et al.*, 2021; Neumann *et al.*, 2014). The results imply that flammability and vegetation may have biophysical limits (ceiling values), with charcoal concentrations exceeding these limits likely reflecting repeated wetland burning for purposes such as agriculture, hunting, and livestock management.

In contrast, the savanna site displayed irregular fluctuations in flammability, indicating instability or disequilibrium (Figures 9.8B and 9.9B). Modern savannas are often hypothesised to experience disequilibrium due to factors such as high rainfall variability, episodic droughts, grazing pressure, and fire (Beckage *et al.*, 2009; Bond, 2005; Ellis and Swift, 1988; Gillson, 2004). Therefore, from a flammability perspective, uMngeni and uMfolozi represent steady-state (equilibrium) versus unstable (disequilibrium) fire regimes, respectively.

Implications for Past Fire Regimes in African Grassy Ecosystems

Our analysis of two Late Holocene sequences reveals that interpreting fire patterns in African grassy ecosystems from a palaeoecological perspective is highly complex. Charcoal concentrations, often assumed to directly reflect biomass burning (c.f. Daniau *et al.*, 2013; Hantson *et al.*, 2016; Marlon *et al.*, 2015), may instead be more closely linked to flammability states or regimes (Figure 9.9B). This suggests that changes in fire activity over space and time are better understood as shifts between alternative low and high flammability states (e.g. Pausas *et al.*, 2017). For instance, charcoal records from African forest ecosystems indicate stable low flammability over the last 4,500 years, followed by a significant increase in flammability in the last ~1,500 years, marking a transition from closed-canopy forests to open savanna (Bremond *et al.*, 2024).

In contrast, Southern Africa's grassy biomes exhibited fluctuations between low and high flammability states throughout the same period, driven primarily by climatic variability. We further propose that stability in fire regimes is associated with fuel or vegetation homogeneity, which was higher in forests compared to the heterogeneity typical of open tree-grass ecosystems. These findings underscore the importance of biome-specific syntheses of charcoal-fire patterns (Leys *et al.*, 2018) and the need to carefully interpret the nature of charcoal fluctuations.

We also suggest that charcoal concentrations should not be directly equated with biomass burning, but rather with the areas under flammability and vegetation biomass graphs for defined intervals (Figure 9.9A). This approach allows vegetation structure (growth-form) summaries derived from REVEALS pollen (Tabares *et al.*, 2021) or phytolith datasets (Dabengwa *et al.*, 2022) can be used to discover spatial scales at which flammability is indicated by charcoal data. A key challenge for vegetation reconstructions is the difficulty of linking pollen estimates to vegetation biomass states in dynamic global vegetation models to be used to identify the spatial scales at which flammability, as indicated by charcoal, operates.

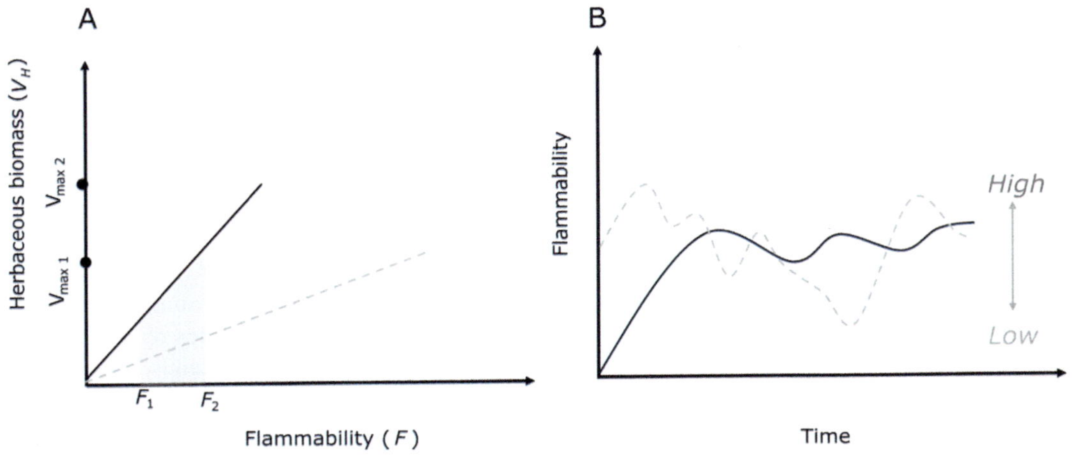

FIGURE 9.9 Synthesis of observed fire regimes from relationships between herbaceous biomass and flammability over time at uMngeni grassland (solid line) and uMfolozi savanna (dotted line) Late Holocene sites. (A) Relationships between flammability and herbaceous biomass suggested potential maximum herbaceous biomasses (Vmax). We inferred that between flammability points F1 and F2, uMngeni had a higher shaded area compared to uMfolozi, corresponding to more potential burned biomass (BV). Thus, we hypothesise gradients (ΔVH / ΔF) of herbaceous biomass (VH) and flammability (F) are related to properties affecting fire resistance modified by rainfall, humans, and herbivores. (B) Summary of flammability regimes over time representing steady-state fluctuations* at uMngeni versus uMfolozi's irregular rhythm. Note: *Fluctuations are superimposed only for illustration.

A significant challenge in vegetation reconstructions lies in linking pollen estimates to vegetation biomass states in dynamic global vegetation models (Dallmeyer *et al.*, 2023). However, we circumvented this limitation by assuming that SOC was directly linked to aboveground herbaceous biomass or productivity. Focusing on the herbaceous layer was particularly relevant because our study sites were in seasonally flooded wetlands dominated by grasses with varying flooding tolerances (Kotze and O'Connor, 2000).

Advancing our understanding of paleofire regimes will require integrating modelling, vegetation proxies, and calibration studies. While we have proposed models to summarise flammability regimes (Figure 9.9), these conceptual models require mathematical elaboration to explore their dynamical interactions (e.g. McCormick *et al.*, 2023; Perrings and Walker, 1997). This could also involve investigating flammability state thresholds or tipping points, and resilience (Archibald and Hempson, 2016; Dabengwa *et al.*, 2022; Mckenzie and Kennedy, 2012). Pollen and phytolith analyses will be critical for resolving taxonomic uncertainties, as sediment organic carbon alone does not provide sufficient detail. Calibration studies are also essential to test our conceptual models of herbaceous biomass versus flammability using contemporary remotely sensed data alongside charcoal, phytolith, and pollen datasets. Additionally, charcoal production processes need to be better understood and compared across different grassland systems. Successful calibration will facilitate the integration and refinement of vegetation-fire relationships in dynamic global vegetation models (e.g. Brucher *et al.*, 2014; Dallmeyer *et al.*, 2023). Furthermore, analysing dung fungal spores from savannas with a long eco-evolutionary history of megaherbivore grazing will be crucial for exploring alternative stable herbaceous biomass states influenced by grazing pressure.

CONCLUSION

The examination of Late Holocene long-term fire and herbaceous biomass changes from fire-prone grassland and savanna sites in South Africa provided evidence of multiple fire states or regimes. By linking charcoal concentrations from several fragment size classes to vegetation flammability, we showed differences in vegetation-fire patterns and flammability states (Supplementary material 3). At biome level, herbaceous biomass and flammability potential between the two sites were different. At local level, fluctuations of flammability in grassland were stable compared with savanna irregularity. Hypothesised causes of differences in fire fluxes included vegetation homogeneity, climate stability, presence of humans and large herbivores. Interestingly, flammability in savanna was multimodal, providing crucial evidence for multiple stable states and alternative stable states. We also found that burnt biomass depended on flammability and vegetation biomass potential which vary spatiotemporally. To further examine alternative stable states, we need to analyse multimodality caused among contrasts between short *versus* tall grass states or wooded savanna *versus* open grassland states. Our study shows the value of developing better insights for discovering ancient fire regimes in African grassy ecosystems which are useful for shaping future conservation and management.

ACKNOWLEDGMENTS

I would like to express my gratitude to Professors William Bond and Lindsey Gillson who oversaw the early development of these ideas. The work would not have been possible without the field teams (especially Mathew Britton), dedicated game guards, and permission from Ezemvelo KwaZulu-Natal Wildlife Authority. Ms Sue van Rensburg from the South African Earth Observation Network (SAEON) helped me navigate around the KwaZulu-Natal Province and its bureaucracy. Sponsorship for this work was provided by the African Climate and Development Initiative (ACDI), National Research Foundation (NRF) and SAEON.

SUPPLEMENTARY MATERIAL

Data used in this manuscript can be accessed at 10.5281/zenodo.14843172

REFERENCES

Aleman, J. C., Blarquez, O., Bentaleb, I., Bonté, P., Brossier, B., Carcaillet, C., Gond, V., Gourlet-fleury, S., Kpolita, A., Lefèvre, I., Oslisly, R., Power, M. J., Yongo, O., Bremond, L., Aleman, J. C., Blarquez, O., Bentaleb, I., Bonté, P., Brossier, B., … Favier, C. (2013). *The Holocene sedimentary charcoal in the Afrotropics*. https://doi.org/10.1177/0959683613508159

Archibald, S., and Hempson, G. P. (2016). Competing consumers: contrasting the patterns and impacts of fire and mammalian herbivory in Africa. *Philosophical Transactions of the Royal Society B: Biological Sciences*, *371*(1703), 20150309. https://doi.org/10.1098/rstb.2015.0309

Archibald, S., Hempson, G. P., and Lehmann, C. (2019). A unified framework for plant life-history strategies shaped by fire and herbivory. *New Phytologist*, *224*(4), 1490–1503. https://doi.org/10.1111/nph.15986

Archibald, S., Lehmann, C. E. R., Gómez-dans, J. L., and Bradstock, R. A. (2013). Defining pyromes and global syndromes of fire regimes. *Proceedings of the National Academy of Sciences*, *110*(16), 6442–6447. https://doi.org/10.1073/pnas.1211466110/-/DCSupplemental.www.pnas.org/cgi/doi/10.1073/pnas.1211466110

Balfour, D. A., and Howison, O. E. (2001). Spatial and temporal variation in a mesic savanna fire regime: responses to variation in annual rainfall. *African Journal of Range & Forage Science*, *19*(1), 45–53. https://doi.org/10.2989/10220110209485773

Beckage, B., and Ellingwood, C. (2009). Fire feedbacks with vegetation and alternative stable states. *Complex Systems*, *18*(1), 159–173. https://doi.org/10.25088/complexsystems.18.1.159

Beckage, B., Platt, W. J., and Gross, L. J. (2009). Vegetation, fire, and feedbacks: a disturbance-mediated model of savannas. *American Naturalist*, *174*(6), 805–818. https://doi.org/10.1086/648458

Blaauw, M., and Christeny, J. A. (2011). Flexible paleoclimate age-depth models using an autoregressive gamma process. *Bayesian Analysis*, *6*(3), 457–474. https://doi.org/10.1214/11-BA618

Bond, W. J. (2005). Large parts of the world are brown or black: a different view on the 'Green World' hypothesis. *Journal of Vegetation Science*, *16*(3), 261–266. https://doi.org/10.1658/1100-9233(2005)016

Bond, W. J. (2021). Out of the shadows: ecology of open ecosystems. In *Plant Ecology and Diversity* (Vol. 14, Issues 5–6, pp. 205–222). Taylor & Francis. https://doi.org/10.1080/17550874.2022.2034065

Bond, W. J., and Keeley, J. E. (2005). Fire as a global "herbivore": the ecology and evolution of flammable ecosystems. *Trends in Ecology and Evolution*, *20*(7), 387–394. https://doi.org/10.1016/j.tree.2005.04.025

Bremond, L., Alemanby\, J. C., Favier, C., Blarquez, O., Colombaroli, D., Connor, S. E., Cordova, C. E., Courtney-Mustaphi, C., Dabengwa, A. N., Gil-Romera, G., Gosling, W. D., Hamilton, T., Montade, V., Razafimanantsoa, A. H. I., Power, M. J., Razanatsoa, E., Yabi, I., and Vannière, B. (2024). Past fire dynamics in sub-Saharan Africa during the last 25,000 years: climate change and increasing human impacts. *Quaternary International, July*. https://doi.org/10.1016/j.quaint.2024.07.012

Brucher, T., Brovkin, V., Kloster, S., Marlon, J. R., and Power, M. J. (2014). Comparing modelled fire dynamics with charcoal records for the Holocene. *Climate of the Past*, *10*(2), 811–824. https://doi.org/10.5194/cp-10-811-2014

Chevalier, M., and Chase, B. M. (2015). Southeast African records reveal a coherent shift from high- to low-latitude forcing mechanisms along the east African margin across last glacial–interglacial transition. *Quaternary Science Reviews*, *125*, 117–130. https://doi.org/10.1016/j.quascirev.2015.07.009

Clarke, P. J., and Lawes, M. J. (2011). A burning issue: community stability and alternative stable states in relation to fire. In *The Balance of Nature and Human Impact* (Vol. 9781107019614). https://doi.org/10.1017/CBO9781139095075.009

Colombaroli, D., Ssemmanda, I., Gelorini, V., and Verschuren, D. (2014). Contrasting long-term records of biomass burning in wet and dry savannas of equatorial East Africa. *Global Change Biology*, *20*(9), 2903–2914. https://doi.org/10.1111/gcb.12583

Conant, R. T., Cerri, C. E. P., Osborne, B. B., and Paustian, K. (2017). Grassland management impacts on soil carbon stocks: a new synthesis. *Ecological Applications*, *27*(2), 662–668. https://doi.org/10.1002/eap.1473

Dabengwa, A. N. (2019). *Long-term ecosystem dynamics of contrasting grasslands in South Africa*. University of Cape Town.

Dabengwa, A. N., Gillson, L., and Bond, W. J. (2022). Resilience modes of an ancient valley grassland in South Africa indicated by palaeoenvironmental methods. *Environmental Research Letters*, *16*(5). https://doi.org/https://doi.org/10.1088/1748-9326/abdf87

Dallmeyer, A., Poska, A., Marquer, L., Seim, A., and Gaillard, M. J. (2023). The challenge of comparing pollen-based quantitative vegetation reconstructions with outputs from vegetation models – a European perspective. *Climate of the Past*, *19*(7), 1531–1557. https://doi.org/10.5194/cp-19-1531-2023

Daniau, A.-L., Fernanda, M., Goñi, S., Martinez, P., Urrego, D. H., Goñi, M. F. S., Martinez, P., Urrego, D. H., Bout-Roumazeilles, V., Desprat, S., and Marlon, J. R. (2013). Orbital-scale climate forcing of grassland burning in southern Africa. *Proceedings of the National Academy of Sciences of the United States of America*, *110*(13), 5069–5073. https://doi.org/10.1073/pnas.1214292110

Ehleringer, J. R., and Monson, R. K. (1993). Evolutionary and ecological aspects of photosynthetic pathway variation. *Annual Review of Ecology and Systematics*, *24*, 411–439. http://www.jstor.org/stable/2097185

Ekblom, A., and Gillson, L. (2010a). Dung fungi as indicators of past herbivore abundance, Kruger and Limpopo National Park. *Palaeogeography, Palaeoclimatology, Palaeoecology*, *296*(1–2), 14–27. https://doi.org/10.1016/j.palaeo.2010.06.009

Ekblom, A., and Gillson, L. (2010b). Fire history and fire ecology of Northern Kruger (KNP) and Limpopo National Park (PNL), southern Africa. *The Holocene*, *20*(7), 1063–1077. https://doi.org/10.1177/0959683610369499

Ekblom, A., and Gillson, L. (2010c). Hierarchy and scale: testing the long term role of water, grazing and nitrogen in the savanna landscape of Limpopo National Park (Mozambique). *Landscape Ecology*, *25*, 1529–1546. https://doi.org/10.1007/s10980-010-9522-x

Ellery, W. N., Scholes, R. J., and Scholes, M. C. (1995). The distribution of sweetveld and sourveld in south africa's grassland biome in relation to environmental factors. *African Journal of Range and Forage Science*, *12*(1), 38–45. https://doi.org/10.1080/10220119.1995.9647860

Ellis, J. E., and Swift, D. M. (1988). Stability of African pastoral ecosystems: alternate paradigms and implications for development. *Journal of Range Management*, *41*(6), 450–459. https://doi.org/10.2307/3899515

Elschot, K., Bouma, T. J., Temmerman, S., and Bakker, J. P. (2013). Effects of long-term grazing on sediment deposition and salt-marsh accretion rates. *Estuarine, Coastal and Shelf Science*, *133*, 109–115. https://doi.org/10.1016/j.ecss.2013.08.021

Everson, C. S., and Everson, T. (2016). The long-term effects of fire regime on primary production of montane grasslands in South Africa. *African Journal of Range & Forage Science*, *33*(1), 1–9. https://doi.org/10.2989/10220119.2015.1124922

Finch, J. M., Hill, T. R., Meadows, M. E., Lodder, J., and Bodmann, L. (2021). Fire and montane vegetation dynamics through successive phases of human occupation in the northern Drakensberg, South Africa. *Quaternary International*. https://doi.org/10.1016/j.quaint.2021.01.026

Finsinger, W., and Tinner, W. (2005). Minimum count sums for charcoal concentration estimates in pollen slides: accuracy and potential errors. *The Holocene*, *2*, 293–297.

Fuhlendorf, S. D., Engle, D. M., Kerby, J. A. Y., and Hamilton, R. (2009). Pyric herbivory: rewilding landscapes through the recoupling of fire and grazing. *Conservation Biology*, *23*(3), 588–598. https://doi.org/10.1111/j.1523-1739.2008.01139.x

Fynn, R. W. S., and O'Connor, T. G. (2005). Determinants of community organisation of a South African mesic grassland. *Journal of Vegetation Science*, *16*(1), 93–102. https://doi.org/10.1658/1100-9233(2005)016[0093:DOCOOA]2.0.CO;2

Gillson, L. (2004). Testing non-equilibrium theories in savannas: 1400 years of vegetation change in Tsavo National Park, Kenya. *Ecological Complexity*, *1*, 281–298. https://doi.org/10.1016/j.ecocom.2004.06.001

Hall, M. (1981). Settlement patterns in the Iron Age of Zululand: an ecological interpretation. *British Archaeological Research International Series*, *119*, 1–191.

Hantson, S., Kloster, S., Coughlan, M., Daniau, A. L., Vannière, B., Brücher, T., Kehrwald, N., and Magi, B. I. (2016). Fire in the earth system: bridging data and modeling research. *Bulletin of the American Meteorological Society*, *97*(6), 1069–1072. https://doi.org/10.1175/BAMS-D-15-00319.1

Hastie, T., and Tibshirani, R. (1986). Generalized additive models. *Statistical Science*, *1*(3). https://doi.org/10.1214/ss/1177013604

Hempson, G. P., Archibald, S., Bond, W. J., Ellis, R. P., Grant, C. C., Kruger, F. J., Kruger, L. M., Moxley, C., Owen-Smith, N., Peel, M. J. S., Smit, I. P. J., and Vickers, K. J. (2015). Ecology of grazing lawns in Africa. *Biological Reviews*, *90*(3), 979–994. https://doi.org/10.1111/brv.12145

Hempson, G. P., Archibald, S., Donaldson, E., and Lehmann, C. E. R. (2019). Alternate grassy ecosystem states are determined by palatability – flammability trade-offs. *Trends in Ecology & Evolution*, *xx*(Box 1), 1–4. https://doi.org/10.1016/j.tree.2019.01.007

Hogg, A. G., Heaton, T. J., Hua, Q., Palmer, J. G., Turney, C. S. M., Southon, J., Bayliss, A., Blackwell, P. G., Boswijk, G., Bronk Ramsey, C., Pearson, C., Petchey, F., Reimer, P., Reimer, R., and Wacker, L. (2020). SHCal20 Southern Hemisphere Calibration, 0–55,000 Years cal BP. *Radiocarbon*, *62*(4), 759–778. https://doi.org/10.1017/RDC.2020.59

Holling, C. S. (1973). Resilience and stability of ecological systems. *Annual Review of Ecology and Systematics*, *4*(1), 1–23. https://doi.org/10.1146/annurev.es.04.110173.000245

Ingram, L. J., Stahl, P. D., Schuman, G. E., Buyer, J. S., Vance, G. F., Ganjegunte, G. K., Welker, J. M., and Derner, J. D. (2008). Grazing impacts on soil carbon and microbial communities in a mixed-grass ecosystem. *Soil Science Society of America Journal*, *72*(4), 939. https://doi.org/10.2136/sssaj2007.0038

Juggins, S. (2011). *C2 data analysis*. Newcastle University. https://www.staff.ncl.ac.uk/stephen.juggins/software/C2Home.htm

Keeley, J. E., Pausas, J. G., Rundel, P. W., Bond, W. J., and Bradstock, R. A. (2011). Fire as an evolutionary pressure shaping plant traits. *Trends in Plant Science*, *16*(8), 406–411. https://doi.org/10.1016/j.tplants.2011.04.002

Keeley, J. E., and Rundel, P. W. (2005). Fire and the Miocene expansion of C4 grasslands. *Ecology Letters*, *8*(7), 683–690. https://doi.org/10.1111/j.1461-0248.2005.00767.x

Kotze, D. C., and O'Connor, T. G. (2000). Vegetation variation within and among palustrine wetlands along an altitudinal gradient in KwaZulu-Natal, South Africa. *Plant Ecology*, *146*, 77–96.

Legendre, P., and Birks, H. J. B. (2012). From Classical to Canonical Ordination. In H. J. B. Birks, A. F. Lotter, S. Juggins, and J. P. Smol (Eds.), *Tracking Environmental Change Using Lake Sediments, Volume 5: Data Handling and Numberical Techniques* (Vol. 5, pp. 201–248). Springer. http://dx.doi.org/10.1007/978-94-007-2745-8

Leys, B. A., Commerford, J. L., and McLauchlan, K. K. (2017). Reconstructing grassland fire history using sedimentary charcoal: considering count, size and shape. *PLoS ONE*, *12*(4), 1–15. https://doi.org/10.1371/journal.pone.0176445

Leys, B. A., Marlon, J. R., Umbanhowar, C., and Vannière, B. (2018). Global fire history of grassland biomes. *Ecology and Evolution*, *8*(17), 8831–8852. https://doi.org/10.1002/ece3.4394

Maggs, T. (1976). Iron age patterns and Sotho history on the southern Highveld: South Africa. *World Archaeology*, *7*(3), 318–332. https://doi.org/10.1080/00438243.1976.9979644

Marlon, J. R., Bartlein, P. J., Carcaillet, C., Gavin, D. G., Harrison, S. P., Higuera, P. E., Joos, F., Power, M. J., and Prentice, I. C. (2008). Climate and human influences on global biomass burning over the past two millennia. *Nature Geoscience*, *1*(10), 697–702. https://doi.org/10.1038/ngeo313

Marlon, J. R., Kelly, R., Daniau, A.-L., Vannière, B., Power, M. J., Bartlein, P., Higuera, P., Blarquez, O., Brewer, S., Brücher, T., Feurdean, A., Gil-Romera, G., Iglesias, V., Maezumi, S. Y., Magi, B., Mustaphi, C. J. C., and Zhihai, T. (2015). Reconstructions of biomass burning from sediment charcoal records to improve data-model comparisons. *Biogeosciences Discussions*, *12*(22), 18571–18623. https://doi.org/10.5194/bgd-12-18571-2015

McCormick, A. R., Phillips, J. S., Botsch, J. C., Einarsson, Á., Gardarsson, A., and Ives, A. R. (2023). Reconstructing midge consumer–resource dynamics using carbon stable isotope signatures of archived specimens. *Ecology*, *104*(2), 1–11. https://doi.org/10.1002/ecy.3901

McCracken, D. P. (2008). *Saving the Zululand Wilderness: An Early Struggle for Nature Conservation*. Jacana Media (Pty) Ltd.

Mckenzie, D., and Kennedy, M. C. (2012). Power laws reveal phase transitions in landscape controls of fire regimes. *Nature Communications*, *3*, 726. https://doi.org/10.1038/ncomms1731

Meadows, M. E., and Baxter, A. J. (1999). Late Quaternary palaeoenvironments of the southwestern Cape, South Africa: a regional synthesis. *Quaternary International*, *57–58*, 193–206. http://www.scopus.com/inward/record.url?eid=2-s2.0-0033369626&partnerID=40&md5=1b4445dc55d8e30313bb72c6b7560843

Morris, C. D., Everson, C. S., Everson, T. M., and Gordijn, P. J. (2020). Frequent burning maintained a stable grassland over four decades in the Drakensberg, South Africa. *African Journal of Range and Forage Science*. https://doi.org/10.2989/10220119.2020.1825120

Mucina, L., and Rutherford, M. C. (2006). The vegetation of South Africa, Lesotho and Swaziland. In *Strelitzia* (Vol. 19). South African National Biodiversity Institute.

Mutch, R. W. (1970). Wildland fires and ecosystems - A hypothesis. *Ecology*, *51*(6), 1046–1051. https://doi.org/http://www.jstor.org/stable/1933631 .

Neumann, F. H., Botha, G. A., and Scott, L. (2014). 18,000 years of grassland evolution in the summer rainfall region of South Africa: evidence from Mahwaqa Mountain, KwaZulu-Natal. *Vegetation History and Archaeobotany*, *23*, 665–681. https://doi.org/10.1007/s00334-014-0445-3

Noy-Meir, I. (1975). Stability of grazing systems: an application of predator-prey graphs. *Journal of Ecology*, *63*(2), 459–481. https://doi.org/10.2307/2258730

Noy-Meir, I. (1982). Stability of plant herbivore models and possible applications to savanna. In B. J. Huntley and B. H. Walker (Eds.), *The Ecology of Tropical Savannas* (pp. 591–609). Springer-Verlag.

O'Connor, T. G., Mulqueeny, C. M., and Goodman, P. S. (2011). Determinants of spatial variation in fire return period in a semiarid African savanna. *International Journal of Wildland Fire*, *20*, 540–549.

Pausas, J. G., and Bond, W. J. (2020). Alternative biome states in terrestrial ecosystems. *Trends in Plant Science*, *25*(3), 250–263. https://doi.org/10.1016/j.tplants.2019.11.003

Pausas, J. G., Keeley, J. E., and Schwilk, D. W. (2017). Flammability as an ecological and evolutionary driver. *Journal of Ecology*, *105*(2), 289–297. https://doi.org/10.1111/1365-2745.12691

Pausas, J. G., and Ribeiro, E. (2013). The global fire-productivity relationship. *Global Ecology and Biogeography*, *22*(6), 728–736. https://doi.org/10.1111/geb.12043

Pausas, J. G., and Schwilk, D. (2012). Fire and plant evolution. *New Phytologist*, *193*(2), 301–303. https://doi.org/10.1111/j.1469-8137.2011.04010.x

Perrings, C., and Walker, B. H. (1997). Biodiversity, resilience and the control of ecological-economic systems: the case of fire-driven rangelands. *Ecological Economics*, *22*(1), 73–83. https://doi.org/10.1016/S0921-8009(97)00565-X

Petraitis, P. (2013). Multiple stable states in natural ecosystems. In *Multiple Stable States in Natural Ecosystems*. 1–200. https://doi.org/10.1093/acprof:osobl/9780199569342.001.0001

Pitkänen, A., Lehtonen, H., and Huttunen, P. (1999). Comparison of sedimentary microscopic charcoal particle records in a small lake with dendrochronological data: evidence for the local origin of microscopic charcoal produced by forest fires of low intensity in eastern Finland Aki Pitka. *The Holocene*, *9*(5), 559–567. https://doi.org/10.1191/095968399670319510

Power, M. J., Marlon, Ortiz, N., Bartlein, P. J., Harrison, S. P., Mayle, F. E., Ballouche, A., Bradshaw, R. H. W., Carcaillet, C., Cordova, C., Mooney, S., Moreno, P. I., Prentice, I. C., Thonicke, K., Tinner, W., Whitlock, C., Zhang, Y., Zhao, Y., Ali, A. A., … Wilmshurst, Æ. N. W. Æ. J. (2008). Changes in fire regimes since the last glacial maximum: an assessment based on a global synthesis and analysis of charcoal data. *Climate Dynamics*, *30*(7–8), 887–907. https://doi.org/10.1007/s00382-007-0334-x

R Core Team. (2023). *R: A language and environment for statistical computing.* R Foundation for Statistical Computing. http://www.r-project.org/

Ratajczak, Z., Nippert, J. B., Briggs, J. M., and Blair, J. M. (2014). Fire dynamics distinguish grasslands, shrublands and woodlands as alternative attractors in the Central Great Plains of North America. *Journal of Ecology, 102,* 1374–1385. https://doi.org/10.1111/1365-2745.12311

Ripley, B., Donald, G., Osborne, C. P., Abraham, T., Ripley, B., Donald, G., Osborne, C. P., Abraham, T., and Martin, T. (2010). Experimental investigation of fire ecology in the C3 and C4 subspecies of Alloteropsis semialata. *Journal of Ecology, 98*(5), 1196–1203. http://www.jstor.org/stable/40929069

Sayedi, S. S., Abbott, B. W., Vannière, B., Leys, B., Colombaroli, D., Romera, G. G., Słowiński, M., Aleman, J. C., Blarquez, O., Feurdean, A., Brown, K., Aakala, T., Alenius, T., Allen, K., Andric, M., Bergeron, Y., Biagioni, S., Bradshaw, R., Bremond, L., … Daniau, A. L. (2024). Assessing changes in global fire regimes. *Fire Ecology, 20*(1). https://doi.org/10.1186/s42408-023-00237-9

Schaeffer, K. E., Bestelmeyer, B. T., Burkett, L. M., and McLaren, J. R. (2025). The potential for using soil carbon, soil texture, and elevation as indicators of grass-cover response in Chihuahuan Desert grassland restoration practices. *Journal of Arid Environments, 227*(May 2024), 105326. https://doi.org/10.1016/j.jaridenv.2025.105326

Simpson, K. J., Archibald, S., and Osborne, C. P. (2022). Savanna fire regimes depend on grass trait diversity. *Trends in Ecology and Evolution, 37*(9), 749–758. https://doi.org/10.1016/j.tree.2022.04.010

Simpson, K. J., Ripley, B. S., Christin, P. A., Belcher, C. M., Lehmann, C. E. R., Thomas, G. H., and Osborne, C. P. (2016). Determinants of flammability in savanna grass species. *Journal of Ecology, 104*(1), 138–148. https://doi.org/10.1111/1365-2745.12503

Stockmarr, J. (1971). Tablets with spores used in absolute pollen analysis. *Pollen et Spores, 13,* 615–621.

Tabares, X., Ratzmann, G., Kruse, S., Theuerkauf, M., Mapani, B., and Herzschuh, U. (2021). Relative pollen productivity estimates of savanna taxa from southern Africa and their application to reconstruct shrub encroachment during the last century. *Holocene, 31*(7), 1100–1111. https://doi.org/10.1177/09596836211003193

Tainton, N. M., Zacharias, P. J. K., and Hardy, M. B. (1989). The contribution of veld diversity to the agricultural economy. In B. J. Huntley (Ed.), *Biotic Diversity in Southern Africa: Concepts and Conservation.* Oxford University Press.

Vachula, R. S. (2021). A meta-analytical approach to understanding the charcoal source area problem. *Palaeogeography, Palaeoclimatology, Palaeoecology, 562*(August 2020), 110111. https://doi.org/10.1016/j.palaeo.2020.110111

Veen, G. F. C., Blair, J. M., Smith, M. D., and Collins, S. L. (2008). Influence of grazing and fire frequency on small-scale plant community structure and resource variability in native tallgrass prairie. *Oikos, 117*(6), 859–866. https://doi.org/10.1111/j.2008.0030-1299.16515.x

Wang, Q., Li, Y., and Wang, Y. (2011). Optimizing the weight loss-on-ignition methodology to quantify organic and carbonate carbon of sediments from diverse sources. *Environmental Monitoring and Assessment, 174*(1–4), 241–257. https://doi.org/10.1007/s10661-010-1454-z

Whitlock, C., Higuera, P. E., McWethy, D. B., and Briles, C. E. (2010). Paleoecological perspectives on fire ecology: revisiting the fire-regime concept. *The Open Ecology Journal, 3*(2), 6–23. https://doi.org/10.2174/1874213001003020006

Wood, S. N. (2017). Generalized additive models: an introduction with R, second edition. In *Generalized Additive Models: An Introduction with R, Second Edition.* https://doi.org/10.1201/9781315370279

10

Palaeoenvironmental Radiocarbon Dates for Environmental Change and Land Use Research Across Eastern Africa

An Updated Open Access Data Set

Colin J. Courtney-Mustaphi, Esther N. Githumbi, Anna Shoemaker, Rebecca Kariuki, Linus Munishi, Julie Aleman, Makarius Peter Itambu, Alexes Mes, Johnny Panga Brushworth, Kate Bailey-Neale, Pierre Lapellegerie, Robert Marchant and Paul J. Lane

INTRODUCTION

Radiocarbon dating is one of the most widely used radiometric geochronological age determination methods in palaeoenvironmental, palaeontological, and archaeological research of the Late Quaternary (Walker and Lowe, 2007; Lowe and Walker, 2014; Wood, 2015; Hajdas *et al.*, 2021). Radiocarbon age determinations, calendar year calibrations, and age-depth models are crucial to our understanding of geochronologies, uncertainties on the timing of past events (Grimm *et al.*, 2009; Parnell *et al.*, 2008;

DOI: 10.1201/9781003727781-10

Buckland *et al.*, 2020; Rey *et al.*, 2023), rates of change (Urrego *et al.*, 2009; Mottl *et al.*, 2021; King *et al.*, 2024), and spatiotemporal patterns of palaeoenvironmental and palaeoanthropological variability (Blois *et al.*, 2011; Robinson *et al.*, 2019).

Compilations of radiocarbon dates are used for assessments of research effort and remaining spatiotemporal knowledge gaps (Morlan, 1999; Riede, 2009; Marchant *et al.*, 2018) at multiple scales from local (Prendergast *et al.*, 2014) to regional and global (Gajewski *et al.*, 2011; Chaput *et al.*, 2015; Gayo *et al.*, 2015; Crann *et al.*, 2015; Chaput and Gajewski, 2016; Goldstein *et al.*, 2022; Boisard and Ben Arous, 2024). Theoretical and statistical techniques continue to be refined for both stratigraphic age-depth modelling (Blaauw and Heegaard, 2012) and the application of radiocarbon to archaeological and palaeodemographic research (Attenbrow and Hiscock, 2015; Torfing, 2015; Wright, 2017). Digital open access repositories extend the lifecycle of such datasets and improve access and findability to facilitate independent, collaborative, and interdisciplinary research efforts, and future expansion (Brewer *et al.*, 2012; Grimm *et al.*, 2013; Goring *et al.*, 2015; Williams *et al.*, 2018; Courtney Mustaphi *et al.*, 2019a). Further development of open access software tools promotes data reuse in palaeoenvironmental research as well as interdisciplinary studies (Vidaña and Goring, 2023; Nelson *et al.*, 2024; Gillson *et al.*, 2025). A longitudinal literature search of published and new publications has been ongoing since 2015 as part of environmental change data collation and spatiotemporal research gap identification during the Resilience in East African Landscapes project (REAL; Courtney Mustaphi *et al.*, 2014; Marchant *et al.*, 2018; EU CORDIS, 2024) and the Adaptation and Resilience to Climate Change project (ARCC; Courtney Mustaphi *et al.*, 2019b; This study presents an update to an open access dataset of radiocarbon dates collated from study sites in eastern Africa (Courtney Mustaphi and Marchant, 2016), with additional data, improved citation and referencing of data, and stratigraphic information.

Context

Spatial Scope

Scientific literature and data that used and reported radiocarbon dates determined from study sites in eastern Africa were targeted for inclusion in the data set. The geographic focus was the East African countries of Burundi, Kenya, Tanzania, Rwanda, and Uganda and included occasional data from neighbouring countries and marine areas.

Temporal Coverage and Units

The coverage range was from the analytical and theoretical limit of approximately 55,000–60,000 ^{14}C yr bp to "modern" radiocarbon ages, delimited as 1950 Common Era (*Anno Domini*), 0 calibrated years bp (bp, before present), which also included post-bomb and ^{14}C enriched determinations. Radiocarbon dating methods include both the conventional method radiocarbon (beta counting) and accelerator mass spectrometry (AMS; Hajdas, 2008; Dunbar *et al.*, 2016; Ascough *et al.*, 2024). Reported radiocarbon ages with 95% uncertainties are the most reported values in the literature and are the result of δ^{13}C correction after measurement by the radiocarbon dating facility. Reported radiocarbon determinations are not calibrated. Calibrated results produced by the original authors were not collated owing to variable use of calibration tools and frequent updating of the OxCal calibration programme (Bronk Ramsey, 1995; Bronk Ramsey and Lee, 2013; Van der Plicht *et al.*, 2020), and periodic revisions to the available calibration curves, such as the Northern Hemisphere (IntCal20) and Southern Hemisphere (SHCal20) products (Hogg *et al.*, 2020; Reimer *et al.*, 2020). Occasionally, measured radiocarbon ages with uncertainties that were not δ^{13}C corrected are also reported in the literature and have been collated when available, for example, see Casanova and Hillaire-Marcel (1992). When available from the original publications, modern and post-bomb radiocarbon determinations have been included as per cent Modern Carbon, *p*MC, with uncertainties when available, which may then be calibrated by data analysts with post-bomb and other radiocarbon calibration curves (Hua and Barbetti, 2004; Reimer *et al.*, 2004).

Data Sources

The primary targets for collating information were peer-reviewed scientific literature in academic journals, as well as monographs, scientific reports, PhD theses, and associated data available from online data repositories and archives, journal supplementary information, and thesis data. Upon discovery and identification of radiocarbon determinations, effort was undertaken to reference and cite the earliest publication, but this was not followed exhaustively in cases where radiocarbon determinations have been reused in two or more publications. The references provided should not be assumed to be the first instance of publication of the radiocarbon determination. No effort has been made to document cross-referencing of radiocarbon data with the previous version (Courtney Mustaphi, 2016) or other data repositories. All data are cited and referenced for findability and in some cases, additional detail was sought through direct communication with the authors or staff at radiocarbon dating facilities.

METHODS

Literature Review

A longitudinal literature search of previously published studies and alerts for new publications that presented radiocarbon data has been ongoing since 2015. Published radiocarbon dates were entered from palaeoenvironmental studies, including palaeoclimatology, palaeoecology, environmental archaeology, and palaeontology resources. The literature review was inherently non-exhaustive. Literature was collated through searches and monitoring of new publications through Google Scholar, ResearchGate, and Web of Science and monitoring of some academic journals and the known most active research groups with an interest in palaeoenvironmental change during the past approximately 60,000 years across eastern Africa. English language search terms included the following keywords with spelling and Boolean variants: age-depth models, AMS, archaeology, bones, chronologies, climate change, coral, coastal, deltaic, depositional environments, geochronologies, lacustrine, lakes, Late Pleistocene, marine, microfossils, palustrine, palaeoclimatology, palaeontology, palaeoenvironments, Quaternary, radiocarbon dating, radiometric, sediment cores, sediments, and sedimentology; among others. Author and known institutional searches were conducted for researchers and research groups that are well established. Terms were searched throughout the titles, abstracts, and main text of publications. Manual searches and re-digitisation of radiocarbon data were undertaken for print books and digital sources that were not machine readable. The temporal scope of the literature review spanned the early phase of application of radiocarbon measurements to environmental and archaeological research, during the mid 1950s until February 2025. The literature review undertaken was not exhaustive and there are likely more resources available and also in other languages, potentially French, Italian, Swahili, and Japanese, among others.

In the identified published studies, radiocarbon-dated material was collected from geologic deposits and detritus (organic sediments, plant macroremains, pollen concentrates, and stromatolites); anthropogenic deposits (hearths, fire pits, and pottery); plant (charcoal, wood, and seeds); and faunal remains (eggshell, bone apatite, bone collagen, and dung). Other sources of radiocarbon-dated material were noted in the accompanying data files (Courtney Mustaphi, 2016). Metadata were also collated on the associated study site information, dated material, and radiocarbon determinations, such as published geographic coordinates, laboratory code identifiers, and radiocarbon ages and errors (Stuiver and Polach, 1977; Van der Plicht and Hogg, 2006; Millard, 2014). Bibliographic metadata were also collated, and digital online identifiers (DOIs) or permalinks (PURLs) were included to facilitate referencing.

Data and Format

The data were entered into a modified spreadsheet template used for the Canadian Archaeological Radiocarbon Database (Morlan, 1999) with additional columns to include new data—such as sample depth in a stratigraphy, location uncertainty rankings, *p*MC measurements, and DOIs (described below). Data and metadata were collated from the original publication text, tables, supplementary materials, and digital supplements, such as data repositories. Radiocarbon date determinations were entered primarily with the reported radiocarbon age with associated 95% uncertainty (in ^{14}C years, not calibrated). The collated data maintained the original authors' choice with regard to rounding of reported ages, often to the nearest decade (Stuiver and Polach, 1977), while in other cases, the author(s) reported ages directly from the laboratory results without rounding. Modern dates vary in presentation styles when reported and are often omitted from publications by the original authors. Some modern carbon dates were categorically listed as "modern" or as a negative bp value, and these dates were collated, especially if the fraction (or percent) of modern carbon was reported. Laboratory code identifiers for individual dates have been listed when available and appear as "UNK-0000" if the code was not found. A list of radiocarbon dating facility acronyms curated by the team of the journal title *Radiocarbon* is available online from the journal website (University of Arizona, 2025).

Spatiality

Efforts were made to collect spatially explicit radiocarbon dates and to document the relative certainty of the location. Geographic coordinates were recorded when published by the authors. In a few cases, the coordinates were estimated from maps provided in the publication, or were requested from the original authors by email, or were corrected or improved from those provided in the original publication. Geographic coordinates are presented as decimal degrees (negative values for the Southern Hemisphere) with the World Geodetic System of 1984 (WGS84) datum in Google Earth Pro (version 7.3.6.10201, 64-bit). Some coordinate entries that could not be estimated were left with no location data and can be filtered out of the data set, or locations may be established in the future. Each entry has a matching rank as a semi-quantitative descriptor of geographic uncertainty from 0, no geographic coordinate estimate, 1 = relatively low spatial uncertainty (10s of metres), 2 = moderate uncertainty (approximately 100 metres), and 3 = relatively uncertain location (> a few 100s of metres). Elevation was occasionally collated when conspicuously reported by authors. Data re-users should improve upon this field with other data products in cases when elevation is deemed important.

Completeness

In a few cases, radiocarbon dates that lacked some minimal metadata were still included as placeholders for when and if the full data are made available or recovered. These cases are relevant for scientific consideration of past research effort, planning sampling strategies, data syntheses, and literature reviews. Units have been provided in the header row for each column in the data set. Users of this radiocarbon determination compilation should preprocess the spreadsheet under minimal data criteria before further analysis is undertaken.

RESULTS

Availability of the Results

An open access data set of radiocarbon dates, metadata, and references associated with this study is available online (Courtney Mustaphi, 2016 and 2025), appended to a previously published version and description (Courtney Mustaphi and Marchant, 2016). The updated data set is available alongside the previous version, which has received over 2,000 downloads as of February 2025, and will remain available for posterity and reuse (n = 793 dates; Courtney Mustaphi, 2016). A data availability statement is presented below.

Summary of Collated Data

A total of 2,193 published radiocarbon date determinations were identified in the literature review (Figures 10.1 and 10.2), of which 283 dates were archaeological, 1,789 were geological, and 121 were paleontological. Some (n = 48) date determinations were presented by the original authors as "modern"; some of these dates had an associated percent or fraction of modern carbon data, and some did not. Unknown age determinations have been included as placeholders (n = 49), and the data may be resolved in the future. Only a small number of reported failed measurements are included (n = 7; Temoltzin-Loranca *et al.*, 2023). Data sources spanned publication years from 1956 to 2025 and predominantly referenced peer-reviewed journal articles, PhD theses (i.e. Mehlman, 1989; Manyanga, 2006; Higgins, 2017; radiocarbon lists (Rubin and Suess, 1956; Barendsen *et al.*, 1957; Vogel, 1970; Barker and Mackey, 1960), and books (Hamilton, 1982; Andama and Lejju, 2012). Palaeoenvironmental studies of sediment cores collected from Lake Victoria (Kendall, 1969; Stager *et al.*, 2000, 2003; Temoltzin-Loranca, 2023; Wienhues *et al.*, 2024) and Lake Chala (Blaauw *et al.*, 2011), and archaeological sites such as Engaruka, Tanzania (Sassoon, 1966; Sassoon, 1967; Robertshaw, 1986; Thornton-Barnett, 2018), provided many radiocarbon dates.

The primary target locations included the East African countries of Kenya, Tanzania, Burundi, Uganda, Rwanda, and neighbouring Ethiopia, Somalia, Chad, Egypt, Sudan, and the Democratic Republic of Congo. Occasional entries were from sites in Madagascar, Malawi, Mauritius, Mozambique, Zimbabwe, South Africa, Cameroon, and two marine areas—the Gulf of Aden (Tierney *et al.*, 2015) and western Indian Ocean (Kuhnert *et al.*, 2014; Rippert *et al.*, 2015). Reported radiocarbon dates were commonly collected from organic lacustrine and palustrine sediments (Colombaroli *et al.*, 2018), plant and animal fossil remains in sediments (Rijsdijk *et al.*, 2009), soils (Cooremans and Mahaney, 1990; Montade *et al.*, 2018; Courtney Mustaphi *et al.*, 2025), and carbonaceous deposits (Casanova and Hillaire-Marcel, 1992). Some remains were taxonomically identified, namely plant or fauna remains (Bower *et al.*, 2012). The youngest reported dates were 10 ± 20 ^{14}C years bp (Beta-323991; Shilling, 2013) and 11 ± 30 ^{14}C years bp (Wk-25720; Courtney Mustaphi *et al.*, 2021). The oldest determination was 57804 ± 9260 (Erl-4301; Zech, 2006); these are the same youngest and oldest determinations identified in Courtney Mustaphi and Marchant (2016). The gamut of reported *p*MC measurements ranged from 0.4 ± 0.3 (UQ-1456; Bonnefille and Riollet, 1988) to 135.5 ± 0.6 (Beta-172859; Ashley *et al.*, 2004; 2013).

Stylistic Variations in Data Presentation in the Data Set

The data are predominantly presented as reported radiocarbon date determinations with (uncalibrated) age and 95% uncertainties (^{14}C years bp) as listed from the dating laboratories. A total of 48 dates were presented as "modern". The radiocarbon determination Wk-23591 has a date listed as "modern" carbon,

FIGURE 10.1 Map of the spatially explicit radiocarbon determinations identified in eastern Africa, surrounding islands, and marine areas, and some neighbouring countries (n = 2016, "x" symbols). Inset abbreviations: KE, Kenya; TZ, Tanzania; and UG, Uganda. A KML file is also available for use with Google Earth Pro or other spatial data visualisation tools (Courtney Mustaphi, 2016 and 2025).

and the measurement pMC = 110.8 ± 0.3 is listed in a separate column (Finch *et al.*, 2014) and represents a reported determination enriched in modern carbon. Dates reported categorically as "modern" may have associated per cent modern carbon values that were reported to the original authors but not presented in the associated publication. In a few of the identified records (n = 4), modern ages were presented as negative years bp with an uncertainty, for example, an archaeological date GX-10615 presented as modern −65 ± 100 (Bower and Chadderdon, 1986; and see Schüler, 2012). In cases where an insufficient carbon mass was used for an isotopic measurement, the field is listed as "MEASUREMENT FAILED" (Temoltzin-Loranca *et al.*, 2023). Two dates were not legible due to the resolution of diagrams and may be updated in the future (Mahaney *et al.*, 1991). In some instances, radiocarbon dates near the limits of resolving a measurement may not have reported uncertainties, for example, Wk-23584 has an uncalibrated age of 46,664 ^{14}C years bp (Finch *et al.*, 2014). A new column was created to designate data entries where individual radiocarbon determinations were not included in the age-depth model provided by the authors

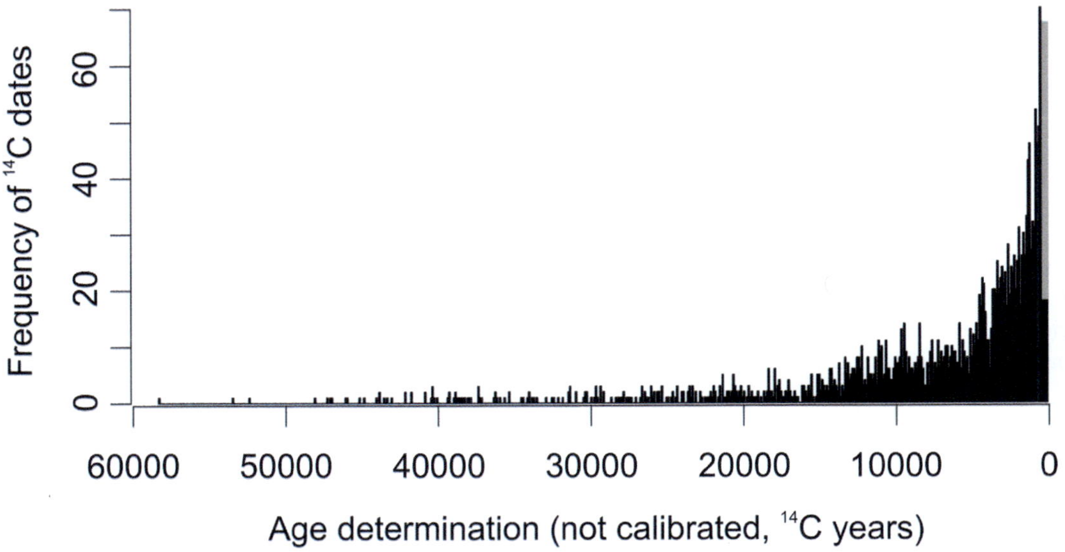

FIGURE 10.2 Histogram of reported radiocarbon ages (n = 2,137) in 100-year bins. Date determinations presented by the original authors qualitatively as "modern" have nominally been assigned 1 ^{14}C years for visualisation (grey bar, far right;n = 48). Unknown age determinations (n = 49) and failed measurements (n = 7) have been removed for visualisation purposes.

of the original publications (n = 99 dates; see examples in Beuning *et al.*, 1997, and Temoltzin-Loranca *et al.*, 2023). In the future, new techniques may have more objective methods of including or excluding age determinations from models.

DISCUSSION

Data Use

The dataset is available in a CSV spreadsheet format (Courtney Mustaphi, 2016 and 2025) that can be manually downloaded, linked or accessed through custom API connections (Schmid *et al.*, 2019). Only a limited effort was applied to assess data entries that are duplicated or cross-listed with other databases and open access data. Users of this dataset should create and implement protocols to check for duplicates when merging or using this dataset with other available data.

Geochronologies in Eastern Africa

Bulk sediments (n = 944) and charcoal or charred plant remains (n = 349) were the predominant material used for radiocarbon dating. More effort has been applied to age determinations for refining age-depth modelling of sediment cores globally through a stepwise cascade of lab procedures by fine sieving sediments to target terrestrial carbon in plant macroremains (Grimm *et al.*, 2009; Rey *et al.*, 2023; Temoltzin-Loranca *et al.*, 2023), then targeting fine organic concentrates (n = 60), dissolved organic fractions (Perrott, 1982), and bulk sediment dates. Lake Victoria sediment cores have been subject to various approaches to develop age-depth models and investigate radiocarbon age offset (Temoltzin-Loranca *et al.*, 2023). The region hosts two very exemplary studies of age-depth models of lacustrine sediments that may serve as demonstrations for other regions and for new studies because of the high number of radiocarbon determinations and the quality of the source carbon in the sediments (plant macroremains; Blaauw *et al.*, 2011; Temoltzin-Loranca *et al.*, 2023). Further development for palaeoenvironmental records that make use of many radiocarbon age determinations and meticulous sieving of sediment core material for short-lived terrestrial plant remains will produce new insights on spatial patterns, timing and rates of change, and inter-site synchroneity. Few palaeoenvironmental examples of radiocarbon dating of soil stratigraphies were identified (Tomlinson, 1974; Heckmann *et al.*, 2014; Montade *et al.*, 2018; Courtney Mustaphi *et al.*, 2025) and will benefit from further taphonomic investigations in sites that are repeatedly wet and dried, experience subterranean bioturbation, and post-depositional transport of plant remains, like charcoal (Courtney Mustaphi *et al.*, 2022).

The spatial distribution of radiocarbon dates in eastern Africa reflects the research efforts of the past century, beginning with the earliest published radiocarbon dates from archaeological and palaeoenvironmental sites in the region (Rubin and Suess, 1956; Barendsen *et al.*, 1957; Brachi, 1960; Stuiver *et al.*, 1960; van Zinderen Bakker, 1962). Spatial gaps identified in this data collation include Somalia, northern and northeastern Kenya, northern Uganda, central and western Tanzania, and southern Tanzania extending into northern Mozambique (Figure 10.1). Radiocarbon dates predominantly span the Holocene to recent geological time, with few dates focusing on the oldest possible range for radiocarbon measurements (60–50,000 ^{14}C years bp; Figure 10.2). Bulk dates predominate the radiocarbon determinations from geological sources, and future research effort should focus on terrestrial plant remains with rapid growth before deposition, such as plant seeds (Temoltzin-Loranca *et al.*, 2023) (Figure 10.3). These and other biases in the collated data offer guidance for future research efforts, both spatially and temporally, as well as data quality considerations that can improve precision (calibrate uncertainties) for individual radiocarbon dates.

Reporting Radiocarbon Determinations and Metadata

Surveying the radiocarbon determinations literature beginning in the 1950s to the present has identified reporting pattern norms within academic communities and the potential for improvements in reporting and data reuse, especially with regard to palaeoenvironmental studies that have different considerations than archaeological and palaeontological research. Radiocarbon dates tend to be reported by laboratory identification code, reported age and uncertainty, dated material, and sample depth. It is critical for study reproducibility and replication that descriptors of the material dated are included, and stratigraphic depth is optimally presented as an interval. We have introduced an uncertainty ranking for the geographic location of samples that has been applied retrospectively to the study site presentation by the authors, based on the resolution of reported geographic coordinates and their correctness as well as qualitative interpretation of maps or site drawings. The purpose was to offer data users a guide for future fieldwork logistics and for data analysts to justify data processing decisions. Generally, older studies presented radiocarbon data within stratigraphic diagrams, and more recent publications include a data table or supplementary, machine-readable versions of data reflecting improved guidance from publishing outlets on how to report radiocarbon dates and the associated metadata (Millard, 2014). Similar efforts for reporting guidelines come from professional institutions, such as those by the African Chronometric Dating Fund (ACDF), an initiative of the Society for Africanist Archaeologists (SAfA), the PanAfrican Archaeological Association (PAA), and the British Institute in Eastern Africa (BIEA).[1] Decisions to round radiocarbon dating results are less clear for AMS dates relative to conventional dates (Stuiver and Polach, 1977; Van der Plicht and Hogg, 2006; Millard, 2014) and clarity of reporting may become increasingly important as research effort targets carbon sources from plant remains to reduce the reported uncertainties of dates (Rey *et al.*, 2023; Grudzinska *et al.*, 2024).

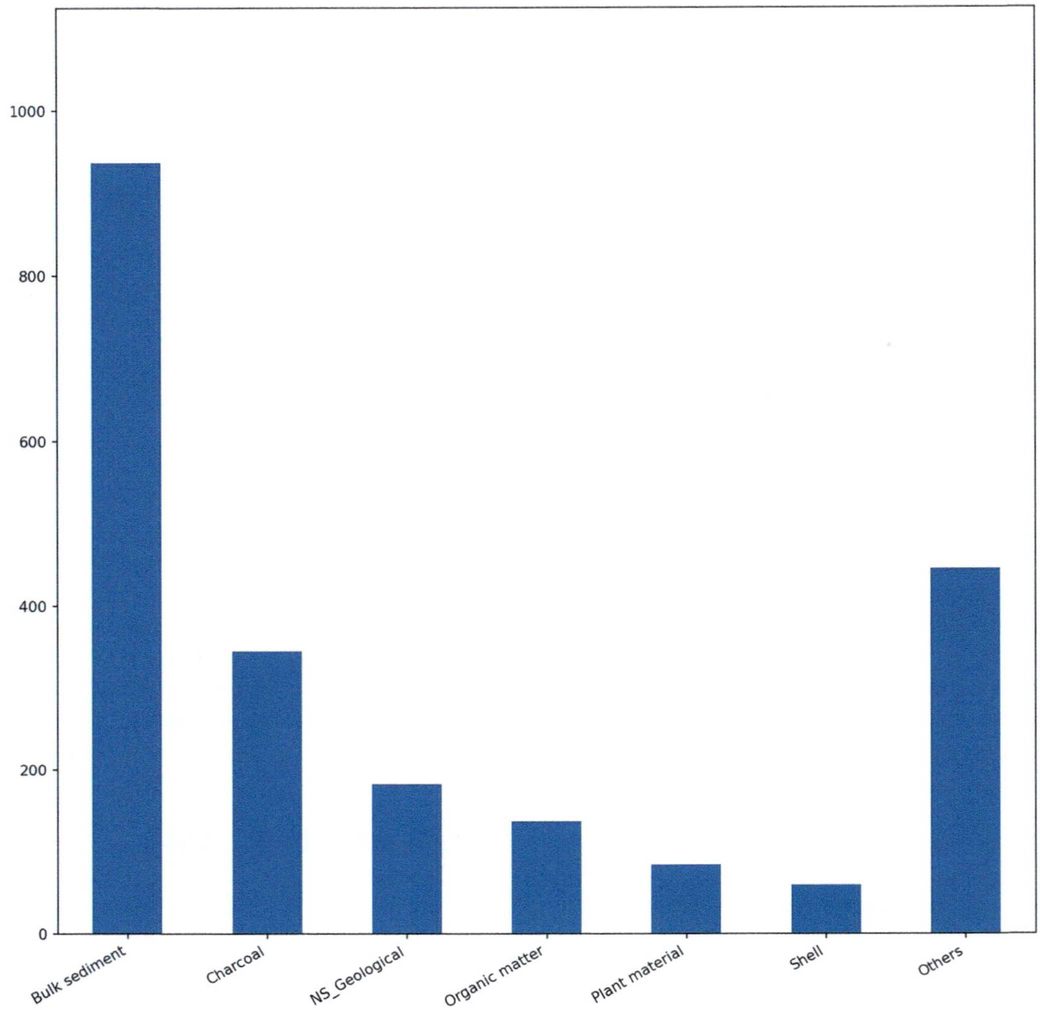

FIGURE 10.3 The most predominant sources of radiocarbon age determinations in the new data set (NS_ geological = not specified geological).

Other Radiocarbon Resources

Several open-access data sets and compendia of radiocarbon determinations that included eastern Africa have been identified, and this list is not exhaustive. Recent years have seen an uptick in the availability of open access radiocarbon databases from archaeological contexts spanning sub-Saharan Africa, including the *Southern African Radiocarbon Database* (SARD) that includes 2,728 dates from 635 archaeological sites in Southern Africa (Loftus *et al.*, 2019) and *Archive des datations radiocarbones d'Afrique centrale* (aDRAC) that covers 664 archaeological sites with 2,197 associated radiocarbon dates from Central Africa. In addition, there are databases aimed at combining different types of archaeological information, such as the AfriArch isotopic database (Goldstein *et al.*, 2022), a database of 5,568 entries that integrates bioarchaeological stable isotope and archaeological radiocarbon measurements from the African Holocene. Global radiocarbon databases, such as pk314c (Bird *et al.*, 2022), provide a collation of regional databases, such as SARD and aDRAC, and offer a helpful overview of datasets available in other regions of interest. Global repositories and archives of paleoenvironmental data often include reported radiocarbon and other geochronological determinations, such as the Neotoma (Grimm, 2008; Jacobson, 2021), and Paleobiology Databases (Varela *et al.*, 2015; Uhen *et al.*, 2018), and global charcoal databases (Hawthorne *et al.*, 2018). Other regional data sets have also been compiled and published from sub-Saharan Africa (Mes, 2024), including West and Central Africa (De Maret *et al.*, 1977; Russell and Steele, 2009; Oslisly *et al.*, 2013; Clist *et al.*, 2023).

Data Availability Statement

The collated radiocarbon dates and metadata are available in both a CSV spreadsheet and MS Excel file in a modified CARD2.0 template (Morlan, 1999) with a persistent DOI from the Harvard Dataverse (Courtney Mustaphi, 2016 and updated 2025) under a CC-BY 4.0 open access Creative Commons licence.

CONCLUSION

Previous compendia of radiocarbon dates have been used to analyse spatial patterns of fine or very broad (up to global) patterns in archaeological (Prendergast *et al.*, 2014), palaeobiogeography, and palaeoenvironmental studies. The results may connect to other palaeoenvironmental databases and repositories and serve as an entry point to the literature on Late Quaternary environmental change and related research efforts in eastern Africa. Our dataset augments those already in the public domain and is intended to facilitate comparison of trends reflected in palaeoenvironmental records with those evident from archaeological and other lines of evidence from across East Africa. It has also been designed to enable researchers to share new dating evidence as it is generated and to input any additional existing radiocarbon dates that our survey has not captured, thereby making this a dynamic rather than a static archive and research tool.

ACKNOWLEDGEMENTS

We thank the palaeoenvironmental research community and authors who have published and shared data, and Dave Hay for technical support. We especially thank Gail Ashley, Jessica Blois, Enrico Crema, Gijs de Cort, Anneli Ekblom, Jemma Finch, Konrad Gajewski, Lindsey Gillson, Karin Holmgren, Katie Loakes, Andrew Martindale, Stefania Merlo, Veronica Muiruri, Cassian Mumbi, Rebecca Muthoni, Nik Patek-Sargeant, Paramita Punnwong, Jan Risberg, Gunhild Rosqvist, Stephen Rucina, Annemiek Schrijver, Lisa Schüler, Lauren Shotter, Daryl Stump, Senna Thornton-Barnett, Geert van der Plas, Dirk Verschuren, Lars-Ove Westerberg, and many others acknowledged in the accompanying data set. We thank Philippa Ascough, Michelle Chaput, Louise Iles, Dirk Seidensticker, and Alayne Street-Perrott for insightful discussions; and the participants of the PAGES-supported LandCover6k meeting hosted by the British Institute in Eastern Africa, Nairobi, Kenya (22–23 October 2015), and at the African Quaternary Association meeting (AFQUA 15–19 July 2018), National Museum of Kenya, Nairobi. CC-M was supported through a European Commission grant to PL and RM (FP7-PEOPLE-2013-ITN 606879); the "Adaptation and Resilience to Climate Change (ARCC)" project awarded to PL, LM, and RM, and administered by the Department of Archaeology and Ancient History, Uppsala University, Sweden, and The Nelson Mandela-African Institution of Science and Technology (NM-AIST), Tanzania, under the Sustainability and Resilience—tackling climate and environmental changes programme funded by the through Swedish Research Council (Vetenskapsrådet), Sida and Formas (2016-06355); and Geoecology, University of Basel. This study benefited from Past Global Changes (PAGES) Global Paleofire (GPWG2),

LandCover6k, and Planetary Boundaries working groups; funded by the Swiss Academy of Sciences and Chinese Academy of Sciences and supported in-kind by the University of Bern, Switzerland. The British Institute in Eastern Africa (BIEA) and National Museum of Kenya are thanked for their continued support.

NOTE

1. This fund was established jointly by the Society of Africanist Archaeologists, the British Institute in Eastern Africa, and the PanAfrican Archaeological Association. In addition to providing a source of funds for African scholars, the Fund has also published guidelines on sample collection, approaches to calibration and reporting radiocarbon dates. See https://safarchaeology.org/African-Chronometric-Dating-Fund (accessed 11 March 2025).

REFERENCES

Andama, M., and Lejju, J., 2012, *Holocene Climate and Environmental History of Lake Victoria Basin: Use of Biogeochemical Proxies*, (London: LAP Lambert Academic Publishing).

Ascough, P., Bompard, N., Garnett, M.H., Gulliver, P., Murray, C., Newton, J.A. and Taylor, C., 2024, ^{14}C measurement of samples for environmental science applications at the National Environmental Isotope Facility (NEIF) Radiocarbon Laboratory, SUERC, UK. *Radiocarbon*, **66**(5), pp. 1020–1031.

Ashley, G.M., Deocampo, D.M., Kahmann-Robinson, J., and Driese, S.G., 2013, Groundwater-fed wetland sediments and paleosols: It's all about water table. *New Frontiers in Paleopedology and Terrestrial Paleoclimatology: Paleosols and Soil Surface Analog Systems: SEPM, Special Publication*, **104**, pp. 47–61.

Ashley, G.M., Maitima Mworia, J., Muasya, A.M., Owen, R.B., Driese, S.G., Hover, V.C., Renaut, R.W., Goman, M.F., Mathai, S. and Blatt, S.H., 2004, Sedimentation and recent history of a freshwater wetland in a semi-arid environment: Loboi Swamp, Kenya, East Africa. *Sedimentology*, **51**(6), pp. 1301–1321.

Attenbrow, V. and Hiscock, P., 2015, Dates and demography: are radiometric dates a robust proxy for long-term prehistoric demographic change? *Archaeology in Oceania*, **50**, pp. 30–36.

Barendsen, G.W., Deevey, E.S. and Gralenski, L.J., 1957, Yale natural radiocarbon measurements III. *Science*, **126**(3279), pp. 908–919.

Barker, H. and Mackey, C., 1960, British Museum natural radiocarbon measurements II. *Radiocarbon*, **2**, pp. 26–30.

Beuning, K.R., Talbot, M.R. and Kelts, K., 1997, A revised 30,000-year paleoclimatic and paleohydrologic history of Lake Albert, East Africa. *Palaeogeography, Palaeoclimatology, Palaeoecology*, **136**(1–4), pp. 259–279.

Bird, D., Miranda, L., Vander Linden, M., Robinson, E., Bocinsky, R.K., Nicholson, C., Capriles, J.M., Finley, J.B., Gayo, E.M., Gil, A., d'Alpoim Guedes, J., *et al.*, 2022, p3k14c, a synthetic global database of archaeological radiocarbon dates. *Scientific Data*, **9**(1), 27.

Blaauw, M. and Heegaard, E., 2012, Estimation of Age-Depth Relationships. In *Tracking Environmental Change Using Lake Sediments* edited by Birks, H., Lotter, A., Juggins, S., Smol, J. Developments in Paleoenvironmental Research, vol. 5, (Dordrecht: Springer). pp. 379–413.

Blaauw, M., van Geel, B., Kristen, I., Plessen, B., Lyaruu, A., Engstrom, D.R., van der Plicht, J., and Verschuren, D., 2011, High-resolution ^{14}C dating of a 25,000-year lake-sediment record from equatorial East Africa. *Quaternary Science Reviews*, **30**, pp. 3043–3059.

Blois, J.L., Williams, J.W.J., Grimm, E.C., Jackson, S.T. and Graham, R.W., 2011, A methodological framework for assessing and reducing temporal uncertainty in paleovegetation mapping from late-Quaternary pollen records. *Quaternary Science Reviews*, **30**(15–16), pp. 1926–1939.

Boisard, S. and Ben Arous, E., 2024, A critical inventory and associated chronology of the Middle Stone Age and Later Stone Age in Northwest Africa. *Journal of Open Archaeology Data*, **12**, p. 5.

Bonnefille, R. and Riollet, G., 1988, The Kashiru pollen sequence (Burundi) palaeoclimatic implications for the last 40,000 yr B.P. in tropical Africa. *Quaternary Research*, **30**, pp. 19–18.

Bower, J. and Chadderdon, T.J., 1986, Further excavations of Pastoral Neolithic sites in Serengeti. *Azania*, **21**(1), pp. 129–133.

Bower, J.R., Mabulla, A.Z. and Kobusiewicz, M., 2012, Loiyangalani: a cultural isolate in the Middle Stone Age of Northern Tanzania. In *Prehistory of Northeastern Africa New Ideas and Discoveries*, edited by Kabaciński, J., Chłodnicki, M., Kobusiewicz, M. Studies in African Archaeology 11, (Poznań: Archaeological Museum), pp. 479–490.

Brachi, R.M., 1960, Excavation of a rock shelter at Hippo Bay, Entebbe. *Uganda Journal*, **24**, pp. 62–70.

Brewer, S., Jackson, S.T. and Williams, J.W., 2012, Paleoecoinformatics: applying geohistorical data to ecological questions. *Trends in Ecology and Evolution*, **27**, pp. 104–112.

Bronk Ramsey, C., 1995, Radiocarbon calibration and analysis of stratigraphy: the OxCal program. *Radiocarbon*, 37(2), pp. 425–430.

Bronk Ramsey, C. and Lee, S., 2013, Recent and planned developments of the program OxCal. *Radiocarbon*, 55(2), pp. 720–730.

Buckland, H.M., Cashman, K.V., Engwell, S.L. and Rust, A.C., 2020, Sources of uncertainty in the Mazama isopachs and the implications for interpreting distal tephra deposits from large magnitude eruptions. *Bulletin of Volcanology*, **82**(3), p. 23.

Casanova, J. and Hillaire-Marcel, C., 1992, Chronology and paleohydrology of late Quaternary high lake levels in the Manyara Basin (Tanzania) from isotopic data (^{18}O, ^{13}C, ^{14}C, Th/U) on fossil stromatolites. *Quaternary Research*, **38**(2), pp. 205–226.

Chaput, M.A. and Gajewski, K., 2016. Radiocarbon dates as estimates of ancient human population size. *Anthropocene*, **15**, pp. 3–12.

Chaput, M.A., Kriesche, B., Betts, M., Martindale, A., Kulik, R., Schmidt, V. and Gajewski, K., 2015, Spatiotemporal distribution of Holocene populations in North America. *Proceedings of the National Academy of Sciences of the United States of America*, **112**, pp. 12127–12132.

Clist, B., Denbow, J., Lanfranchi, R., 2023, Using the radiocarbon dates of Central Africa for studying long-term demographic trends of the last 50,000 years: potential and pitfalls. *Azania*, **58**(2), pp. 235–293.

Colombaroli, D., van der Plas, G., Rucina, S. and Verschuren, D., 2018, Determinants of savanna-fire dynamics in the eastern Lake Victoria catchment (western Kenya) during the last 1200 years. *Quaternary International*, **488**, pp. 67–80.

Cooremans, B. and Mahaney, W.C., 1990, Palynology of a Quaternary paleosol sequence on Mount Kenya, East Africa. *Palaeoecology of Africa*, **21**, pp. 283–294.

Courtney Mustaphi, C., 2016, Radiocarbon dates from eastern Africa in the CARD2.0 format. *Harvard Dataverse*, V5. https://doi.org/10.7910/DVN/NJLNRJ

Courtney Mustaphi, C.J. and Marchant, R., 2016, A database of radiocarbon dates for palaeoenvironmental research in eastern Africa. *Open Quaternary*, **2**, 3. https://doi.org/10.5334/oq.22

Courtney Mustaphi, C.J., Brahney, J., Aquino-López, M.A., Goring, S., Orton, K., Noronha, A., Czaplewski, J., Asena, Q., Paton, S. and Brushworth, J.P., 2019a, Guidelines for reporting and archiving ^{210}Pb sediment chronologies to improve fidelity and extend data lifecycle. *Quaternary Geochronology*, **52**, pp. 77–87.

Courtney Mustaphi, C.J., Capitani, C., Boles, O., Kariuki, R., Newman, R., Munishi, L., Marchant, R., Lane, P., 2019b, Integrating evidence of land use and land cover change for land management policy formulation along the Kenya-Tanzania borderlands. *Anthropocene*, **28**, p. 100228.

Courtney Mustaphi, C.J., Kinyanjui, R., Shoemaker, A., Mumbi, C., Muiruri, V., Marchant, L., Rucina, S., Marchant, R., 2021, A 3000-year record of vegetation changes and fire at a high-elevation wetland on Kilimanjaro, Tanzania. *Quaternary Research*, **99**, pp. 34–62.

Courtney Mustaphi, C.J., Platts, P.J., Willcock, S., Timberlake, J.R., Osborne, J., Matimele, H., Summer Osgood, H., Muiruri, V., Gehrels, M., Bayliss, J., Marchant, R., 2025, Tree demographics and soil charcoal evidence of fire disturbances in an inaccessible forest atop the Mount Lico inselberg. Mozambique. *Plants, People, Planet*, 7(1), pp. 229–244.

Courtney Mustaphi, C.J., Rucina, S.M. and Marchant, R., 2014, Training in emerging palaeoenvironmental approaches to researchers on the dynamics of East African ecosystems. *Frontiers of Biogeography*, **6**, pp. 169–172.

Courtney Mustaphi, C.J., Vos, H.C., Marchant, R., Beale, C., 2022, Charcoal whirlwinds and post-fire observations in Serengeti National Park savannahs. *Tanzania Journal of Science*, **48**(2), pp. 460–473.

Crann, C.A., Patterson, R.T., Macumber, A.L., Galloway, J.M., Roe, H.M., Blaauw, M., Swindles, G.T. and Falck, H., 2015, Sediment accumulation rates in subarctic lakes: Insights into age-depth modeling from 22 dated lake records from the Northwest Territories, Canada. *Quaternary Geochronology*, **27**, pp. 131–144,

De Maret, P., Van Noten, F. and Cahen, D., 1977, radiocarbon dates from west central Africa: a synthesis. *The Journal of African History*, **18**(4), pp. 481–505.

Dunbar, E., Cook, G.T., Naysmith, P., Tripney, B.G. and Xu, S., 2016, AMS ^{14}C dating at the Scottish Universities Environmental Research Centre (SUERC) radiocarbon dating laboratory. *Radiocarbon*, **58**(1), pp. 9–23.

EU CORDIS, 2024, Resilience in East African Landscapes: Identifying critical thresholds and sustainable trajectories – past, present and future. Grant agreement ID: 606879, (Brussels, CORDIS European Commission). Last update: 6 September 2024. https://cordis.europa.eu/project/id/606879

Finch, J., Wooller, M., Marchant, R., 2014, Tracing long-term tropical montane ecosystem change in the Eastern Arc Mountains of Tanzania. *Journal of Quaternary Science*, **29**(3), pp. 269–278.

Gajewski, K., Muñoz, S., Peros, M., Viau, A., Morlan, R. and Betts, M., 2011, The Canadian archaeological radiocarbon database (CARD): archaeological ^{14}C dates in North America and their paleoenvironmental context. *Radiocarbon*, **53**, pp. 371–394.

Gayo, E.M., Latorre, C. and Santoro, C.M., 2015, Timing of occupation and regional settlement patterns revealed by time-series analyses of an archaeological radiocarbon database for the South-Central Andes (16–25 S). *Quaternary International*, **356**, pp. 4–14.

Gillson, L., Seddon, A., Mottl, O., Zhang, K., Kirsten, K., Gell, P., Marchant, R.A., Schwörer, C., Razanatsoa, E., Lane, P.J., Courtney-Mustaphi, C.J., Dearing, J., 2025, Exploring the interface between planetary boundaries and palaeoecology. *Global Change Biology*, **31**(1), p. e70017.

Goldstein, S., Hixon, S., Scott, E., Wolfhagen, J., Iminjili, V., Janzen, A., Chritz, K., Sawchuk, E., Ndiema, E., Sealy, J.C., Stone, A., Zoeller, G., Phelps, L.N., Fernandes, R., 2022, Presenting the AfriArch isotopic database. *Journal of Open Archaeology Data*, **10**, p. 6.

Goring, S., Dawson, A., Simpson, G.L., Ram, K., Graham, R.W., Grimm, E.C. and Williams, J.W., 2015, Neotoma: A programmatic interface to the Neotoma Paleoecological Database. *Open Quaternary*, **1**, p. 2.

Grimm, E.C., 2008, *Neotoma: An Ecosystem Database for the Pliocene, Pleistocene, and Holocene*. Illinois State Museum Scientific Papers E Series 1.

Grimm, E.C., Bradshaw, R.H.W., Brewer, S., Flantua, S., Giesecke, T., Lézine, A.M., Takahara, H., and Williams, J.W., 2013, Databases and their application. In *Encyclopedia of Quaternary Science*, edited by Elias, S.A. and Mock, C.J., (Amsterdam: Elsevier), pp. 831–838.

Grimm, E.C., Maher Jr, L.J. and Nelson, D.M., 2009, The magnitude of error in conventional bulk-sediment radiocarbon dates from central North America. *Quaternary Research*, **72**(2), pp. 301–308.

Grudzinska, I., Courtney-Mustaphi, C., Rey, F., Gobet, E., Tinner, W., Marchetto, A., Heiri, O., 2024, Neolithic human activity caused eutrophication in small central European lakes. *Catena*, **236**, p. 107738.

Hajdas, I., 2008, Radiocarbon dating and its applications in Quaternary studies. *E&G Quaternary Science Journal*, **57**(1/2), pp. 2–24.

Hajdas, I., Ascough, P., Garnett, M.H., Fallon, S.J., Pearson, C.L., Quarta, G., Spalding, K.L., Yamaguchi, H. and Yoneda, M., 2021, Radiocarbon dating. *Nature Reviews Methods Primers*, **1**(1), 62.

Hamilton, A.C. 1982, *Environmental History of East Africa*, (London: Academic Press), p. 328.

Hawthorne, D., Courtney Mustaphi, C.J., Aleman, J.C., Blarquez, O., Colombaroli, D., Daniau, A.L., Marlon, J.R., Power, M., Vannière, B., Han, Y., Hantson, S., Kehrwald, N., Magi, B., Yue, X., Carcaillet, C., Marchant, R., Ayodele, O., Githumbi, E.N., Muriuki, R.M., 2018, Global Modern Charcoal Dataset (GMCD): a tool for exploring proxy-fire linkages and spatial patterns of biomass burning. *Quaternary International*, **488**, pp. 3–17.

Heckmann, M., Muiruri, V., Boom, A., Marchant, R., 2014, Human-environment interactions in an agricultural landscape: a 1400-yr sediment and pollen record from North Pare, NE Tanzania. *Palaeogeography, Palaeoclimatology, Palaeoecology*, **406**, pp. 49–61.

Higgins, L., 2017, Linking lake variability, climate, and human activity in Basotu, Tanzania, PhD thesis, (Stockholm: Department of Physical Geography, Stockholm University).

Hogg, A.G., Heaton, T.J., Hua, Q., Palmer, J.G., Turney, C.S.M., Southon, J., Bayliss, A., *et al.*, 2020, SHCal20 Southern Hemisphere calibration, 0–55,000 years cal BP. *Radiocarbon*, **62**(4), pp. 759–778.

Hua, Q. and Barbetti, M., 2004, Review of tropospheric bomb ^{14}C data for carbon cycle modeling and age calibration purposes. *Radiocarbon*, **46**(3), pp. 1273–1298.

Jacobson, G.L., 2021, A tribute to Eric C. Grimm (1951–2020). *Palynology*, **45**(2), pp. 187–190.

Kendall, R.L., 1969, An ecological history of the Lake Victoria basin. *Ecological Monographs*, **39**(2), pp. 121–176.

King, L., Courtney-Mustaphi, C., Cuenca-Cambronero, M., Wienhues, G., Ngoepe, N., Muschick, M., Temoltzin-Loranca, Y., Vogel, H., Grosjean, M., Tinner, W., Cohen, A., Kishe, M., Heiri, O., Seehausen, O., Matthews, B., 2024, Temporal dynamics of invertebrate community assembly in Lake Victoria since the Late Pleistocene based on chitinous remains. *Freshwater Biology*, **69**(5), pp. 660–678.

Kuhnert, H., Kuhlmann, H., Mohtadi, M., Meggers, H., Baumann, K.H. and Pätzold, J., 2014, Holocene tropical western Indian Ocean sea surface temperatures in covariation with climatic changes in the Indonesian region. *Paleoceanography*, **29**(5), pp. 423–437.

Loftus, E., Mitchell, P.J., Ramsey, C.B., 2019, An archaeological radiocarbon database for southern Africa. *Antiquity*, **93**(370), pp. 870–885.

Lowe, J.J. and Walker, M., 2014, *Reconstructing Quaternary Environments*, (Routledge).

Mahaney, W.C., Slowpoke, R.H., Rutter, N. and Sanmugadas, K., 1991, Stratigraphy and geochemistry of a sequence of Quaternary palaeosols in the Lower Teleki Valley, Mount Kenya, East Africa—Implications for interregional correlations. *Journal of Quaternary Science*, **6**(3), pp. 245–256.

Manyanga, M., 2006, *Resilient landscapes: socio-environmental dynamics in the Shashi-Limpopo Basin, southeastern Zimbabwe c. AD 800 to present*. Studies in Global Archaeology 11, (Uppsala: Department of Archaeology and Ancient History, Uppsala University).

Marchant, R., Richer, S., Boles, O., Capitani, C., Courtney-Mustaphi, C., Lane, P., Prendergast, M., Stump, D., Wynne-Jones, S., Ferro Vázquez, C., Wright, D., Boivin, N., Lang, C., Kay, A., Phelps, L., Fuller, D., Widgren, M., Punwong, P., Lejju, J., Gaillard-Lemdahl, M.-J., Morrison, K.D., Kaplan, J., Benard, J., Crowther, A., Cuní-Sanchez, A., de Cort, G., Deere, N., Ekblom, A., Farmer, J., Finch, J., Gillson, L., Githumbi, E., Kabora, T., Kariuki, R., Kinyanjui, R., Kyazike, E., Muiruri, V., Mumbi, C., Muthoni, R., Muzuka, A., Ndiema, E., Nzabandora, C., Olago, D., Onjala, D., Pas Schrijver, A., Petek, N., Platts, P.J., Rucina, S., Shoemaker, A., Thornton-Barnett, S., van der Plas, G., Watson, L., Williamson, D., 2018, Drivers and trajectories of land cover change in East Africa: human and environmental interactions from 6000 years ago to present. *Earth-Science Reviews*, **178**, pp. 322–378.

Millard, A.R., 2014, Conventions for reporting radiocarbon determinations. *Radiocarbon*, **56**, 555–559.

Mehlman, M.J., 1989, *Late Quaternary Archaeological Sequences in Northern Tanzania*, (Urbana: University of Illinois).

Mes, A., 2024, *Tracking movement: models to reconstruct regional geographic variations of the sub-Saharan Early Iron Age*. Advances in modelling past human ecosystems - bringing together traditional ecological knowledge, archaeological science and computational archaeology. Institute of Archaeology, University of Cologne, Germany, 22–24 May 2024. https://ecosystem-modelling.uni-koeln.de/programme/paper-2-mes

Montade, V., Schüler, L., Hemp, A., Bremond, L., Salamanca Duarte, A. M., and Behling, H., 2018, Late Quaternary ecotone change between sub-alpine and montane forest zone on the leeward northern slope of Mt. Kilimanjaro. *Journal of Vegetation Science*, **29**(3), pp. 459–468.

Morlan, R., 1999, Canadian archaeological radiocarbon database: establishing conventional ages. *Canadian Journal of Archaeology*, **23**, pp. 3–10.

Mottl, O., Grytnes, J.A., Seddon, A.W., Steinbauer, M.J., Bhatta, K.P., Felde, V.A., Flantua, S.G. and Birks, H.J.B., 2021, Rate-of-change analysis in paleoecology revisited: a new approach. *Review of Palaeobotany and Palynology*, **293**, 104483.

Nelson, J.K., Blois, J., Goring, S., Williams, J.W. and Dominguez, S., 2024, User-centered design and evaluation of the neotoma paleoecology open software ecosystem. *International Journal of Digital Earth*, **17**(1), 2378822.

Oslisly, R., Bentaleb, I., Favier, C., Fontugne, M., Gillet, J.F., Morin-Rivat, J., 2013, West Central African peoples: survey of radiocarbon dates over the past 5000 years. *Radiocarbon*, **55**(3), pp. 1377–1382.

Parnell, A.C., Haslett, J., Allen, J.R., Buck, C.E. and Huntley, B., 2008, A flexible approach to assessing synchroneity of past events using Bayesian reconstructions of sedimentation history. *Quaternary Science Reviews*, **27**(19–20), pp. 1872–1885.

Perrott, R.A., 1982, A postglacial pollen record from Mount Satima, Aberdare Range, Kenya. In: American Quaternary Association, Seventh Biennial Conference, Seattle, June 1982. Program and Abstracts, (Seattle: AMQUA), p. 153.

Prendergast, M.E., Grillo, K.M., Mabulla, A.Z. and Wang, H., 2014, New dates for Kansyore and pastoral Neolithic ceramics in the Eyasi Basin, Tanzania. *Journal of African Archaeology*, **12**(1), pp. 89–98.

Reimer, P.J., Austin, W.E.N., Bard, E., Bayliss, A., Blackwell, P.G., Bronk Ramsey, C., Butzin, M., *et al.*, 2020, The IntCal20 Northern Hemisphere radiocarbon age calibration curve (0–55 cal kBP). *Radiocarbon*, **62**(4), pp. 725–757.

Reimer, P.J., Brown, T.A. and Reimer, R.W., 2004, Discussion: reporting and calibration of post-bomb ^{14}C data. *Radiocarbon*, **46**(3), pp. 1299–1304.

Rey, F., Courtney Mustaphi, C., Szidat, S., Gobet, E., Heiri, O., Tinner, W., 2023, Radiocarbon sampling efforts for high-precision lake sediment chronologies. *The Holocene*, **33**(5), pp. 581–591.

Riede, F., 2009, Climate and demography in early prehistory: using calibrated ^{14}C dates as population proxies. *Human Biology*, **81**(3), pp. 309–337.

Rijsdijk, K.F., Hume, J.P., Bunnik, F., Florens, F.V., Baider, C., Shapiro, B., van der Plicht, J., Janoo, A., Griffiths, O., van den Hoek Ostende, L.W. and Cremer, H., 2009, Mid-Holocene vertebrate bone Concentration-Lagerstätte on oceanic island Mauritius provides a window into the ecosystem of the dodo (*Raphus cucullatus*). *Quaternary Science Reviews*, **28**(1), pp. 14–24.

Rippert, N., Baumann, K.H. and Pätzold, J., 2015, Thermocline fluctuations in the western tropical Indian Ocean during the past 35 ka. *Journal of Quaternary Science*, **30**(3), pp. 201–210.
Robertshaw, P., 1986, Engaruka revisited: excavations of 1982. *Azania*, **21**(1), pp. 1–26.
Robinson, E., Zahid, H.J., Codding, B.F., Haas, R. and Kelly, R.L., 2019, Spatiotemporal dynamics of prehistoric human population growth: radiocarbon 'dates as data' and population ecology models. *Journal of Archaeological Science*, **101**, pp. 63–71.
Rubin, M. and Suess, H.E., 1956, US Geological Survey radiocarbon dates III. *Science*, **123**(3194), pp. 442–448.
Russell, T. and Steele, J., 2009, A geo-referenced radiocarbon database for Early Iron Age sites in sub-Saharan Africa: initial analysis. *Southern African Humanities*, 21(1), pp. 327–344.
Sassoon, H., 1966, Engaruka: excavations during 1964. *Azania*, **1**(1), pp. 79–99.
Sassoon, H., 1967, New views on Engaruka, Northern Tanzania: excavations carried out for the Tanzania government in 1964 and 1966. *The Journal of African History*, **8**(2), pp. 201–217.
Shilling, A., 2013, *Characterization of groundwater discharge sites using remote sensing and wetland cores, Lake Eyasi Basin, Tanzania*. MSc thesis, (New Brunswick: Rutgers University).
Schmid, C., Seidensticker, D. and Hinz, M., 2019, c14bazAAR: An R package for downloading and preparing C14 dates from different source databases. *Journal of Open Source Software*, **4**(43), 1914.
Schüler, L., 2012, Studies on late Quaternary environmental dynamics (vegetation, biodiversity, climate, soils, fire and human impact) on Mt Kilimanjaro. PhD Thesis, (Göttingen: Georg-August-Universität Göttingen).
Stager, J.C. and Johnson, T.C., 2000, A 12,400 ^{14}C yr offshore diatom record from east central Lake Victoria, East Africa. *Journal of Paleolimnology*, **23**(4), pp. 373–383.
Stager, J.C., Cumming, B.F. and Meeker, L.D., 2003, A 10,000-year high-resolution diatom record from Pilkington Bay, Lake Victoria, East Africa. *Quaternary Research*, **59**(2), pp. 172–181.
Stuiver, M. and Polach, H.A., 1977, Discussion; reporting of C-14 data. *Radiocarbon* **19**, pp. 355–363.
Stuiver, M., Deevey, E.S. and Gralenski, L.J., 1960, Yale natural radiocarbon measurements V. *Radiocarbon*, **2**, pp. 49–61.
Temoltzin-Loranca, Y., Gobet, E., Vannière, B., van Leeuwen, J.F.N., Wienhues, G., Szidat, S., Courtney-Mustaphi, C., Kishe, M., Muschick, M., Seehausen, O., Grosjean, M., and Tinner, W., 2023, A chronologically reliable record of 17,000 years of biomass burning in the Lake Victoria area. *Quaternary Science Reviews*, **301**, 107915.
Thornton-Barnett, S., 2018, *Persevering with Great Abandon: An Archaeobotanical Investigation of Resilience and Sustainability in Eastern African Irrigated Terrace Agriculture*. PhD thesis, (York: University of York).
Tierney, J.E., Ummenhofer, C. and deMenocal, P.B., 2015, Past and future rainfall in the horn of Africa. *Science Advances*, **1**(9), e1500682.
Tomlinson, R.W., 1974, Preliminary biogeographical studies on the Inyanga Mountains, Rhodesia. *South African Geographical Journal*, **56**(1), pp. 15–26.
Torfing, T., 2015, Layers of assumptions: a reply to Timpson, Manning, and Shennan. *Journal of Archaeological Science*, **63**, pp. 203–205.
Uhen, M.D., Goring, S., Jenkins, J. and Williams, J.W., 2018, Earthlife Consortium: supporting digital paleobiology. *Past Global Changes Magazine*, *26*, pp. 78–79.
University of Arizona, 2025, *RADIOCARBON LABORATORIES*, (Phoenix: University of Arizona), Updated 16 January 2025. Last accessed 10.03.2025. https://radiocarbon.org/laboratories%%%; https://radiocarbon.org/sites/default/files/2025-01/Labs-2025_01_16.pdf
Urrego, D., Bush, M.B., Silman, M., Correa-Metrio, A., Ledru, M., Mayle, F., Valencia, B., 2009, Millennial-scale ecological changes in tropical South America since the last glacial maximum. In *Past Climate Variability from the Last Glacial Maximum to the Holocene in South America and Surrounding Regions*, edited by Vimieux, F., Sylvestre, F. and Khodri, M., (Springer, Paris), pp. 283–300.
Van der Plicht, J. and Hogg, A., 2006, A note on reporting radiocarbon. *Quaternary Geochronology*, **1**, pp. 237–240.
Van der Plicht, J., Bronk Ramsey, C., Heaton, T.J., Scott, E. M., Talamo, S., 2020, Recent developments in calibration for archaeological and environmental samples. *Radiocarbon*, **62**(4), pp. 1095–1117.
van Zinderen Bakker, E.M., 1962, A late-glacial and post-glacial climatic correlation between East Africa and Europe. *Nature*, **194**(4824), pp. 201–203.
Varela, S., González-Hernández, J., Sgarbi, L.F., Marshall, C., Uhen, M.D., Peters, S. and McClennen, M., 2015, paleobioDB: an R package for downloading, visualizing and processing data from the Paleobiology Database. *Ecography*, **38**(4), pp. 419–425.
Vidaña, S.D. and Goring, S.J., 2023, neotoma2: an R package to access data from the Neotoma Paleoecology Database. *Journal of Open Source Software*, **8**(91), 5561.
Vogel, J.C., 1970, Groningen Radiocarbon Dates IX. *Radiocarbon*, **12**(2), pp. 444–471.

Walker, M. and Lowe, J., 2007, Quaternary science 2007: a 50-year retrospective. *Journal of the Geological Society*, **164**(6), pp. 1073–1092.

Wienhues, G., Temoltzin-Loranca, Y., Vogel, H., Morlock, M.A., Anselmetti, F.S., Bernasconi, S.M., Jaggi, M., Tylmann, W., Kishe, M., King, L., Ngoepe, N., Courtney-Mustaphi, C.J., Muschick, M., Matthews, B., Mwaiko, S., Seehausen, O., Tinner, W., Grosjean, M., 2024, From desiccation to wetlands and outflow: rapid re-filling of Lake Victoria during the Late Pleistocene 14 – 13 ka. *Journal of Great Lakes Research*, **50**(3), 102246.

Williams, J.W., Grimm, E.C., Blois, J.L., Charles, D.F., Davis, E.B., Goring, S.J., Graham, R.W., Smith, A.J., Anderson, M., Arroyo-Cabrales, J., and Ashworth, A.C., 2018, The Neotoma Paleoecology Database, a multi-proxy, international, community-curated data resource. *Quaternary Research*, **89**(1), pp. 156–177.

Wood, R., 2015, From revolution to convention: the past, present and future of radiocarbon dating. *Journal of Archaeological Science*, **56**, pp. 61–72.

Wright, D.K., 2017, Accuracy vs. precision: understanding potential errors from radiocarbon dating on African landscapes. *African Archaeological Review*, **34**(3), pp. 303–319.

Zech, M., 2006, Evidence for Late Pleistocene climate changes from buried soils on the southern slopes of Mt. Kilimanjaro, Tanzania. *Palaeogeography, Palaeoclimatology, Palaeoecology*, **242**(3), pp. 303–312.

Impact of Pastoralism on the Vegetation Cover of Koobi Fora Basin, East Turkana, Kenya

11

Husna Mashaka Katambo, Ben N. Nyanchoga, Ephraim W. Wahome, Sylvia N. Wemanya, Venanzio Munyaka, David Braun, Benjamin Davies, Matthew J. Douglass, Emmanuel E. Ndiema and Rahab N. Kinyanjui

INTRODUCTION

Pastoralism spread from the eastern Sahara to central and western Sahara (the Sahel, Sudan, East Africa, and southern Africa) over 8000 years (Gifford-Gonzalez, 2017; Marshall and Hildebrand, 2002; Shahack-Gross *et al.*, 2008). In East Africa, it is reported that early food-producing groups, primarily reliant on domestic livestock, migrated from the north approximately 4,000 years ago, marking a significant emergence of pastoralism, particularly in central and southern Kenya (Marshall, 1986, 1990; Marshall *et al.*, 1984).

The transitions from hunting and gathering to food production in eastern Africa represent a significant shift from hunter-gatherer societies to pastoral modes of food acquisition, primarily driven by animal domestication and mobile settlement patterns (Marshall, 1986, 1990; Marshall *et al.*, 1984). Dongodien (GaJi 4), an archaeological site near Lake Turkana, is dated to around 4,000 years ago and is located in northern Kenya on the Koobi Fora Ridge, within the Holocene Galana Boi Beds (Ashley *et al.*, 2011;

DOI: 10.1201/9781003727781-11

Hildebrand and Grillo, 2012; Marshall *et al.*, 1984; Ndiema *et al.*, 2011). This site provides the earliest reliable evidence of domestic stock in Kenya and is associated with Nderit pottery (Marshall *et al.*, 1984; Ndiema *et al.*, 2011). The site is one of the few well-dated early pastoral sites in eastern Africa, where the community engaged in fishing, hunting, and herding cattle, sheep, and goats (Marshall *et al.*, 1984).

Animal domestication has significantly influenced the African savanna through grazing, trampling, and nutrient inputs, highlighting the need to understand the background effects of traditional pastoralism on vegetation cover (Muchiru *et al.*, 2008, 2009). In the Central Rift Valley, the growth of savannas and the decline of bush and forest are influenced by pastoralists' fire-setting practices to improve grazing and promote grass regrowth, leading to the establishment of fire-climax species such as *Themeda triandra* (Marean *et al.*, 1994). Additionally, overgrazing has been recognized as a significant factor in the maintenance of East African grasslands, primarily through the suppression of grass growth and the removal of vegetation cover from the landscape (Lamprey and Waller, 1990; Marean, 1992; Pratt and Gwynne, 1977). African savannas display significant resilience (Walker and Noy-Meir, 1982). However, there is ongoing debate about whether they are experiencing a lasting decline in productivity due to intensified livestock grazing, permanent settlements, and other human activities, or if climate rather than these factors largely influences their resilience (Behnke *et al.*, 1994; Muchiru *et al.*, 2009; Sullivan and Rohde, 2002; Vetter, 2005).

There is a need to understand the ecological importance of the impact of growing pastoral activities on the African savanna. This is because human impact on vegetation intensified with the emergence of pastoralism (Marean, 1992; Marean *et al.*, 1994). A study conducted by Shahack-Gross *et al.* (2003) involved comparing modern and degraded livestock dung in enclosures at Maasai settlements with control soils nearby that lacked visible dung. Only a limited number of studies have explored the effects of abandoned pastoral homesteads on vegetation dynamics and nutrient succession (Muchiru *et al.*, 2008, 2009). This gap exists because most research has concentrated on natural ecosystems, as seen in the work of Gallagher and Carpenter (1997), rather than on the human-dominated ecosystems that characterize much of Africa. Few studies have explored pastoralism as an integral factor rather than an aberrant one in savanna ecosystems, particularly in Amboseli since 1967 (Muchiru, 1981; Western, 1973). Livestock settlements represent the most visible impact that pastoralists have on the savannas, primarily due to the significant daily concentration of animals (Augustine, 2003; Augustine *et al.*, 2003; Muchiru *et al.*, 2009) and the resulting destruction of woody vegetation for housing and fencing (Ellis *et al.*, 1984; Muchiru *et al.*, 2009). The effect of settlements is temporary and is succeeded by the swift recolonization of vegetation on sites that have been abandoned settlements or bomas (Muchiru *et al.*, 2009; Reid and Ellis, 1995). This process is exceptionally rapid after rainfall, which provides essential moisture for plant growth.

Previous studies have focused on identifying and understanding pastoralist activities through the analysis of dung deposits, utilizing methods such as examining phytoliths, calcium oxalate druse, phosphate compounds, and dung spherulites. Notable contributions to this field include the following: Akeret and Rentzel, 2001; Boschian and Montagnari-Kokelj, 2000; Shahack-Gross *et al.*, 2003, 2008. In the Koobi Fora region, eastern Turkana, a single study has been conducted using phytolith analysis to reconstruct past vegetation cover during the Pleistocene and Holocene periods (Kinyanjui, 2018). Phytoliths are microscopic remnants of plants that form when hydrated silicon dioxide is deposited within and between the cell walls of living plant tissues (Piperno, 2006; Stromberg, 2004). This palaeobotanical study recorded that the vegetation cover and structure changed in response to climatic variability, transitioning from wooded-dominated vegetation cover during the early Holocene to open grasslands toward the late Holocene as climates became increasingly dry, with a subtle indication of the impact of human activities. However, there is limited data on modern vegetation cover, the impact of climate variability, and the changing human adaptation strategies, particularly those related to domestication, on vegetation cover.

Phytoliths have demonstrated their importance as a reliable proxy, along with new methods, in identifying degraded livestock enclosure sediments at archaeological sites (Shahack-Gross *et al.*, 2003, 2008). These biogenic silica bodies can be deposited in various parts of plants, including stems, leaves, fruits, bark, and roots. While plants also produce other minerals, such as calcium oxalate, siliceous phytoliths are highly resistant to degradation and often persist in sediments long after the plant has died (Weiner 2010). As plants decompose, these silica cells are preserved in the soil, becoming essential plant microfossils that can be studied to reconstruct the history of vegetation, even after the vegetation has long gone. Additionally, phytolith indices calculated from specific morphotypes serve as vital proxies for evaluating temporal changes in precipitation, modifications in vegetation structural dynamics, and the reconstruction of historical temperature variations (Alexandre *et al.*, 1997; Bremond *et al.*, 2008; Fredlund and Tieszen, 1994).

Therefore, this study aims to build on existing phytolith data, with a focus on abandoned modern pastoralist (Dassanech community) bomas in the Koobi Fora basin, East Turkana. *Boma* is a Swahili name for homesteads constructed by herding communities, and dung accumulations in livestock enclosures can leave lasting signatures on the landscapes. We utilize phytoliths to investigate the impact of livestock on vegetation cover over time (years) in abandoned bomas. Specifically, we examine the correlation between phytolith assemblages and above-ground vegetation over time, focusing on two major grazing sites, Ileret and Karari.

STUDY SITE

Koobi Fora is located in the northern region of Kenya's Rift Valley, at 3°N, 36°E within Marsabit County. The area extends from Kenya to the Ethiopian border at Ileret and reaches southward to Alia Bay. It is situated within the confines of Sibiloi National Park, which was established as a national park in the 1970s (Figure 11.1). The study sites (abandoned bomas) are also situated in Ileret and Karari in eastern Turkana.

The average annual rainfall in this region ranges from approximately 130 mm to 150 mm. It is characterized by predominantly high temperatures, with an average range of 27°C to 37°C, and a soil temperature at a depth of 25 cm averaging around 37°C (Kinyanjui, 2018; Mbaluka and Brown, 2016). The prolonged rainy season occurs between March and May, while the shorter period falls between September and November (Mbaluka and Brown, 2016). The peak of the rainfall season is typically observed in April and November (Mbaluka and Brown, 2016).

Vegetation

Koobi Fora basin is predominantly covered by *Acacia-Commiphora* grasslands and shrublands, which include herbaceous vegetation, semi-desert grasslands with perennial and annual bushes, riverine forests, woodlands, and semi-desert shrublands (Mbaluka and Brown, 2016). Mbaluka and Brown (2016) asserted that the unique flora diversification in the Koobi Fora basin was influenced by soil/rock type, topography, and the drainage system.

Abandoned settlements, or bomas, exhibit a remarkable diversity in plant species, with a range of 21 to 27 species identified in these areas. The predominant grass species found within the abandoned bomas include *Sporobolus* sp., *Stipagrostis* sp., *Eragrostis* sp., *Leptothrium senegalense, Chloris* sp., *Dinebra retroflexa, Dactyloctenium aegyptium, Aristida* sp., and *Cynodon dactylon*. Additionally, a variety of herbs and forbs are present, comprising *Solanum incanum, Aerva javanica, Tribulus* sp., *Amaranthus, Sueda monoica, Duosperma longicalyx, Sericocomopsis* sp., *Kendrotis gijef, Barleria* sp., *Chenopodiaceae* sp., and *Heliotropium* sp.

Woody species, including *Salvadora persica, Boscia coriacea, Maerua* sp., *Prosopis juliflora, Acacia* spp., *Commiphora edulis, Cadaba rotundifolia, Balanites rotundifolia, Grewia* sp., and *Dobera gabra*, primarily dominate older bomas. This ecological differentiation highlights the unique plant community structures present both within and adjacent to the abandoned bomas.

Bomas

Bomas are circular thorn enclosures, marked by a thorn fence around their perimeter. Family huts are situated along the inner fence. The domestic species identified in the study area include cattle (*Bos taurus*), goats (*Capra hircus*), sheep (*Ovis aries*), and donkeys (*Equus*). At night, adult cattle are housed within the boma, while calves, sheep, and goats are secured in smaller compounds at the center of the settlement. The Dassanech community relies heavily on the production of these domesticated animals. The construction and upkeep of these fences, as well as the need for firewood, place considerable strain on the surrounding trees.

The Dassanech pastoralist community in the eastern Turkana basin, particularly in Karari and Ileret (Figure 11.1), tends to move frequently between locations, utilizing grazing grounds during both the dry and wet seasons to accommodate their cattle to have access to pasture and water sources (Mashaka, 2022). They ultimately use both semi-permanent base camps in the villages and seasonal fly/satellite camps (*foras*) (Kinyanjui *et al., in prep*). During the wet seasons, they reside in the floodplains and near the rivers that are now flowing. During this season, they take advantage of the grasses, herbs, and new leaf shoots from trees and shrubs. During the dry season, they migrate to the ridges and lake shores, where their resources persist longer due to the availability of hydrological input (Mashaka, 2022). This is due to climate change, and bimodal rainfall each year has made it increasingly challenging for this pastoralist group to remain in one place. This mobility has a significant impact on the vegetation changes within the landscape (Mashaka, 2022).

Once the bomas are abandoned after a few years of seasonal occupation, especially after a rainy season, due to the high uric acid and nitrogen from livestock dung and urine (Muchiru, 1981; Muchiru *et al.*, 2009; Vuorio *et al.*, 2014), new plant species colonize the area, particularly within cattle enclosures. Eventually, the bomas host a high concentration of new plant species that grow across the boma enclosures (Mashaka, 2022; Muchiru *et al.*, 2008, 2009; Vuorio *et al.*, 2014). Species abundance and diversity differ between occupied areas with varying ages (inside abandoned bomas) due to differences in water chemistry, soil structure, and sediment composition. Consequently, all evaluations in this study focus on the effects within the abandoned boma (termed "nside"). The timeframe of boma abandonment is a crucial factor in explaining the variation in vegetation cover diversity, abundance, and density *(Kinyanjui in press)*. The longer the boma has been abandoned, the denser the succeeding vegetation cover (Muchiru *et al.*, 2009; Vuorio *et al.*, 2014).

METHODS AND MATERIALS

Boma Ages and Vegetation Identification

A total of five abandoned bomas from Ileret (N = 1) and Karari (N = 4) were selected, with abandonment periods ranging from one to ten years. We identified and located these sites with the assistance of Dassanech informants and through observations made during field seasons. The pastoralists tend to reoccupy the abandoned bomas after a period of time (Muchiru *et al.*, 2009). It is noteworthy that the Dassanech community often reoccupies these bomas; therefore, for this study, we excluded any reoccupied bomas to eliminate the potential confounding effects of multiple occupations of the same site. Vegetation data were recorded within (inside) abandoned settlements (Muchiru *et al.*, 2009; Vuorio *et al.*, 2014) to assess species abundance and diversity. Plant species were identified, tallied, and grouped using plant identification guides from various East African plant species books, including grasses (Martins, 2022), herbs/forbs (Blundell, 1992; Powys, 2019), trees, and shrubs (Dharani, 2019).

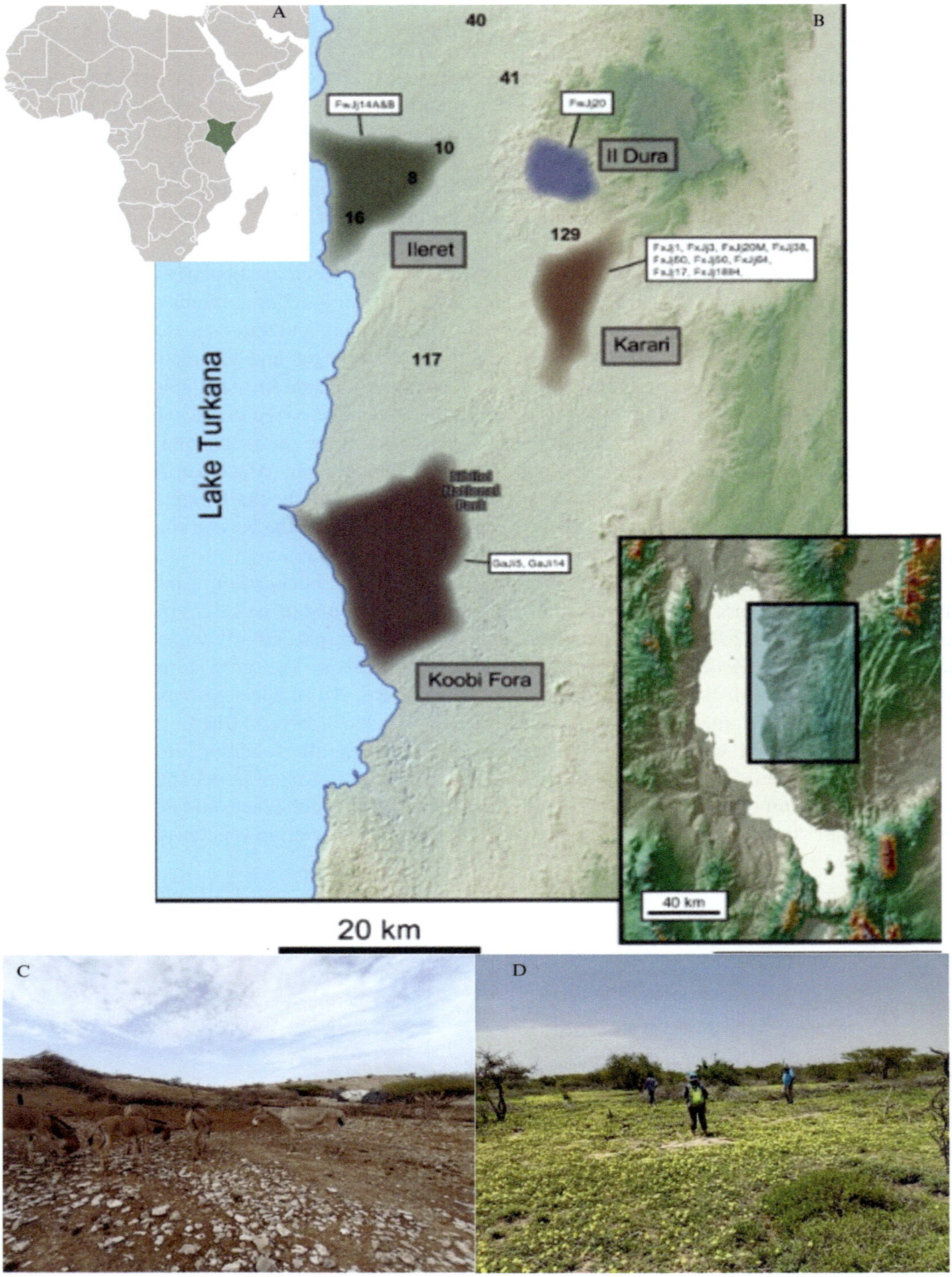

FIGURE 11.1 Location map of East Turkana. (a) Location of Kenya in Africa. (b) Location of abandoned bomas. (c) The boma before it was abandoned, where Dassanech people reside with domesticated animals. (d) The boma after being abandoned by the Dassanech.

•hytolith Analyses from Soils

. total of ten soil samples for phytolith analysis were collected from inside abandoned bomas to deter-ıine the correlation between the standing vegetation cover from inside the boma and phytolith concen-ation. Six samples were collected from the boma in Area 8, El-Kimere, and River Loborana at Ileret, nd four samples from Area 117 and 129 bomas in Karari. First, we scraped the surface using a trowel 3undala, 2024), collected the soil while wearing gloves, placed each sample in a bag, and labeled it. The ɔil samples were assigned field numbers according to a standard convention: for example, I1 denoting the eret sample 1, and K1 denoting the Karari sample 1.

•hytolith Sample Preparation

'he soil samples collected from abandoned bomas were extracted for phytoliths using a rapid procedure s highlighted in the study by Katz *et al.* (2010). The sediment sample was screened to eliminate any articles larger than 0.5 mm. Approximately 10 mg of sediment was treated with 50 μL hydrochloric acid HCl) at 10% for carbonate and carbonated hydroxylapatite removal, including small bone fragments. 'hen, 50 μL of 10% nitric acid was added to the sample to oxidize the organic material. Between each cid treatment, the samples were placed in hot water for ten minutes to accelerate the process and then entrifuged at 3,000 rpm for five minutes. The supernatant was decanted and washed twice with distilled 'ater (dH_2O) and then centrifuged at 3,000 rpm for five minutes to remove any remaining acids. 50 μL of odium Polytungstate Solution (SPT) at a specific gravity of 2.3 was added for the heavy liquid flotation rocess and centrifuged at 3,000 rpm for five minutes. The flotation fraction was carefully transferred) different vials, as this is the phytolith-containing fraction, and washed twice with distilled water to emove the salt that remained during the process.

The final solutions of each sample, approximately 1 mg, were mounted on 22 mm × 22 mm frosted ıicroscopic slides using Entellan mounting media and covered with 11 mm × 11 mm coverslips. hytoliths were systematically identified and counted to characterize assemblages within abandoned omas. Phytolith counting was conducted using a Zeiss Axioskop at 400x magnification. One slide was ounted for rich extracts, and a second slide was mounted for those with fewer morphotypes.

•hytoliths Morphological Classification

hytoliths in this study were identified according to their morphotype categories and taxonomic affilia-on (parent plant species). The general morphological classification of phytoliths follows the published ıternational Code of Phytolith Nomenclature (Madella *et al.*, 2005; Neumann *et al.*, 2019). For this study, ɔecific plant taxa with published studies are considered, such as Palms (Albert *et al.*, 2015a, 2015b), edges (Albert *et al.*, 2006), and grasses (Alexandre *et al.*, 1997; Rossouw, 2009; Stromberg, 2004; Twiss, 992; Twiss *et al.*, 1969). In addition, other studies have focused on East-Central Africa as a modern ref-rence, and vegetation reconstruction has been included (Bremond *et al.*, 2005b; Neumann *et al.*, 2009; ovello and Barboni, 2015).

Morphotypes were grouped into phytolith plant functional types: Grass silica short cell phytolith 3SSCP) includes Panicoideae (Bilobate, Cross-shaped, and Polylobate), and Chloridoideae (Saddle-ıaped) (Madella *et al.*, 2005). Panicoideae and Chloridoideae grasses are commonly C4 tropical grasses 3remond *et al.*, 2008, 2005a; Stromberg, 2004; Strömberg *et al.*, 2024). The short silica cells help dif-erentiate between C3 and C4 grasses and are good indicators of climatic conditions and plant habitats Rossouw, 2009). Chloridoideae grasses utilize C4 photosynthesis and are typically dominant in dry envi-ɔnments at low elevations and latitudes. Conversely, Panicoideae grasses also primarily employ C4 pho-ɔsynthesis and are commonly found in moist environments within similar low-elevation and low-latitude regions (Bremond *et al.*, 2008). Forest/woody indicator (spheroid ornate, spheroid psilate, polyhedra epidermis, tracheid, sclereid, and block) (Mercader *et al.*, 2009; Piperno, 2006; Strömberg *et al.*, 2024 Sedges include an achene and a hat-shaped (Piperno, 2006). Other grasses include bulliform, trichom Scutiform, elongate sinuate, and elongate dendritic (Strömberg *et al.*, 2024). Herbs and others includ epidermal appendages (Piperno, 2006), elongate facetate, and other unknown morphotypes (Stromber 2003; Strömberg *et al.*, 2024).

Statistical Analysis

The vegetation abundance data and phytoliths derived from soil samples were analyzed using R versic 4.5.1 (R Core Team, 2025) to assess species abundance and diversity within bomas. For the standin vegetation, rarefaction curves were generated using the iNEXT package (Hsieh *et al.*, 2016) to evaluat vegetation sample completeness and species richness within each abandoned boma. The relative abu dance and dominance of species were analyzed based on their proportional representation within eac site. Additionally, the Shannon diversity index was calculated and standardized to quantify the evenne of individual species distribution (Magurran, 2013; Spellerberg and Fedor, 2003). These indices we bootstrapped (N = 10,000) to derive 95% confidence intervals, providing robust estimates of variability

The phytolith assemblages, classified by functional types, were visualized through a pie chart. T explore the underlying patterns and variations within these assemblages, Principal Component Analys (PCA) was subsequently performed. Phytolith data were analyzed to create indices that determine whether the local environment was arid or humid and whether it was primarily composed of grasslan or forested ecosystems. To infer climate conditions and vegetation structure, providing insight into th region's ecological dynamics. Two indices were employed:

Aridity index (Iph): The Iph index, a reliable measure of aridity, quantifies the ratio of Chloridoideae to the sum of Chloridoideae and Panicoideae short-cell phytoliths (Bremond *et al.*, 2008). It accurately determines the presence of short and tall C4 grasses (Alexandre *et al.*, 1997; Bremond *et al.*, 2005a, 2008; Fredlund and Tieszen, 1994), offering precise information about aridity-humidity habitats at low altitudes. High Iph values (>20–40%) indicate that grasslands are dominated by Chloridoideae, which is typical of dry, warm conditions. Low values (<20–40%) indicate the dominance of Panicoideae, suggesting moist, warm environments (Bremond *et al.*, 2005a).

Dicotyledon versus Poaceae index (D/P^0): The D/P0 index indicates tree cover density for tropical areas of low altitude. It is calculated by determining the percentage of phytoliths identified as tropical woody dicots versus short-cell phytoliths, which are identified as grasses (Bremond *et al.*, 2005a, 2008). When the D/P0 index is high, close to one, it suggests close-forest vegetation with grass dominance (Bremond *et al.*, 2008).

RESULTS

Species Composition

A total of 34 plant species were recorded in the boma, comprising 11 grasses, 13 herbs and forbs, and 1 trees and shrubs. Grasses exhibit high abundance as bomas grow older, except for E-Kimere, which showe the lowest abundance of grasses, despite having been abandoned for ten years. The species compositic within bomas was primarily represented by *Stipagrostis* sp. and *Aristida* sp., which swiftly colonized a of the abandoned sites. Herbs/forbs exhibit high abundance over time across the bomas, while two bomas Area 117- 2 years and El-Kimere) revealed low abundance, even though they differ in age.

The species is shown to have a high composition in the four boma, *Amaranthus* sp., compared to other herbaceous and forb species. Trees and shrubs have the lowest abundance across all bomas, regardless of boma age. However, the El-Kimere boma, which has been abandoned for ten years, exhibits a slightly higher abundance of trees and shrubs, primarily among the smaller species, *Acacia* spp.

Species Richness

After abandonment, plant species began to colonize the interiors of bomas (Muchiru *et al.*, 2009; Vuorio *et al.*, 2014) (fig. 11.1). The total number of trees and shrubs shows low richness across all species within all boma ages ($p < 0.01$), indicating a significant decrease corresponding to the age of the boma. However, this decrease was only weakly significant ($p < 0.5$) during the initial two to five years. Remarkably, tree and shrub species peaked at ten years in the El-Kimere abandoned boma. *Acacia* species such as *Acacia tortilis*, noted

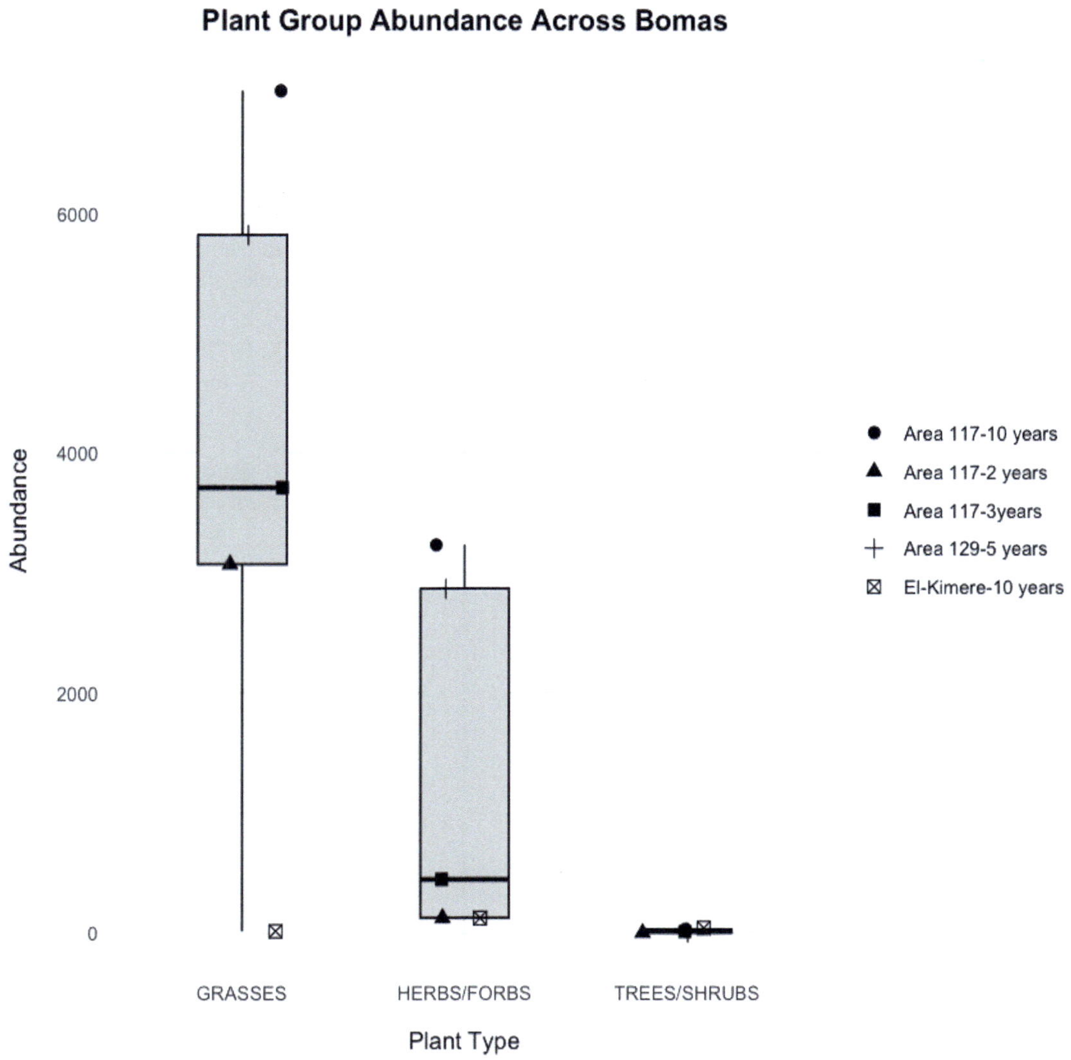

FIGURE 11.2 Vegetation composition across inside bomas of all ages of abandonment.

TABLE 11.1 Bomas Comparing Species Richness and Evenness Using (Magurran, 2013; Spellerberg and Fedor, 2003).

BOMAS	*SHANNON INDEX (H')*	*SIMPSON INDEX (1/D)*	*RICHNESS*	*EVENNESS*
Area 117-2years	1,602,803	0.7422021	17	0.5657192
Area 117 -3years	1,887,367	0.7817081	16	0.6807239
Area 129-5years	1,908,970	0.7981724	16	0.6885154
El-Kimere-10 years	2,394,693	0.8718927	16	0.8637028
Area 117–10 years	1,862,133	0.8167239	20	0.6215954

for their slow growth and longevity (Andersen and Krzywinski, 2007; Atkin *et al.*, 1999; Muchiru *et al.* 2009), displayed an increased correlation with age, particularly evident at the ten-year mark in El-Kimere. Other woody species, including *Prosopis* sp., *Maerua crassifolia*, *Cadaba* sp., and *Commiphora edulis* showed slightly lower biomass levels compared to *Acacia* sp. at the same location. Furthermore, the biomass of other species remained stable throughout the two to ten-year abandonment period.

The Shannon and Simpson diversity index show a relatively high and evenly distributed value across all bomas. The grass and herb/forb species richness increases strongly with age from two to ten years (Table 11.1), except for Boma in El-Kimere, which has the lowest mean value. The mean of woody species increases weakly for the first 2–3 years and slightly in the 5–10 years after abandonment, especially in the El-Kimere old boma (fig. 11.2).

The total number of species in all bomas increased over 5–10 years compared to the first years. The rarefied species richness across bomas indicated that abandoned bomas, two and ten years old, have the highest richness, compared to those three to five years old, which have equal richness. The rarefaction curve (fig. 11.3) for the plant species within each abandoned boma demonstrates that the recorded plant counts are sufficient to represent the species richness in these environments.

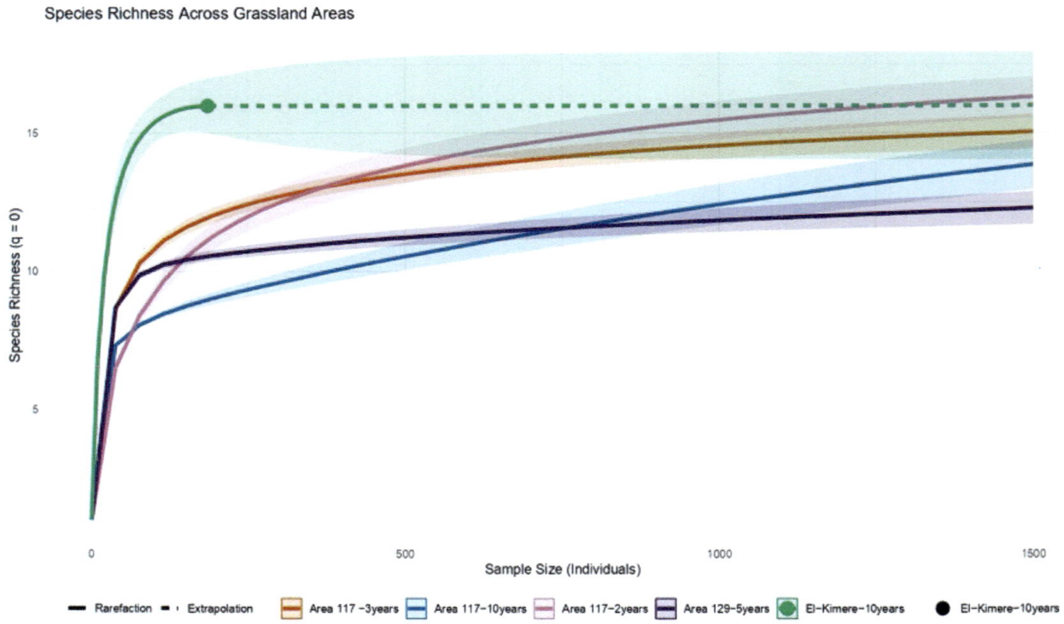

FIGURE 11.3 Bomas vegetation richness rarefaction curve. (a) Rarefaction curves for Area 117–2 years (pink) Area 117-3 years (orange), Area 129–5 years (purple), El-Kimere-10 years (green), Area 117–10 years (blue) Shading represents the 95% bootstrap confidence limits, including both the upper and lower bounds.

hytolith Analysis

f the ten samples collected, both contained well-preserved phytoliths suitable for quantitative anal-sis. All counted morphotypes exhibit similar diagnostic phytolith plant types, including GSSCP Chloridoideae and Panicoideae), forest indicator phytoliths, and wetland plants (such as sedges). Non-diagnostic, including herbaceous type and other grasses. The assemblages include Saddle C4-Chloridoideae) and Bilobate, Cross, and Polylobate (C4-Panicoideae) (Figure 11.4). The concentra-ons of phytoliths within the bomas (fig. 11.5) are expressed as the phytolith function types, including Chloridoideae and Panicoideae, which show low concentrations across all bomas (figs. 11.4 and 11.5). Notably, a high concentration was observed in one boma located along the River Loborana (aban-oned for one year), with Chloridoidea exhibiting high abundance, and at El-Kimere (abandoned for ten ears). The Panicoideae group remains consistently low in all bomas.

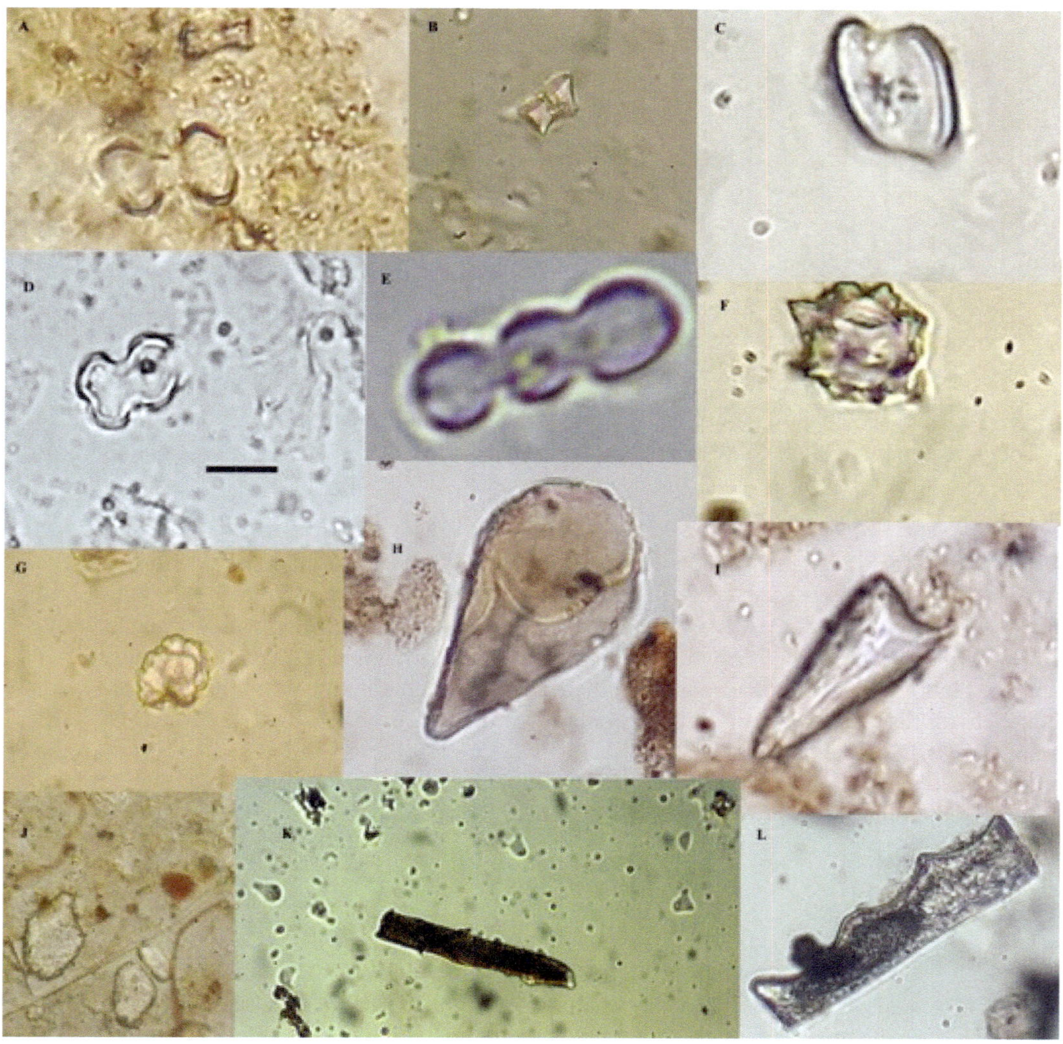

IGURE 11.4 (a) Bilobate, (b) Rondel, (c) Saddle, (d) Cross, (e) Polylobate, (f) Spheroid echinate, (g) Verrucate pheroid, (h) Trichome, (i) Scutiform, (j) Bulliforms, (k) Smooth elongate, and (l) Elongate echinate.

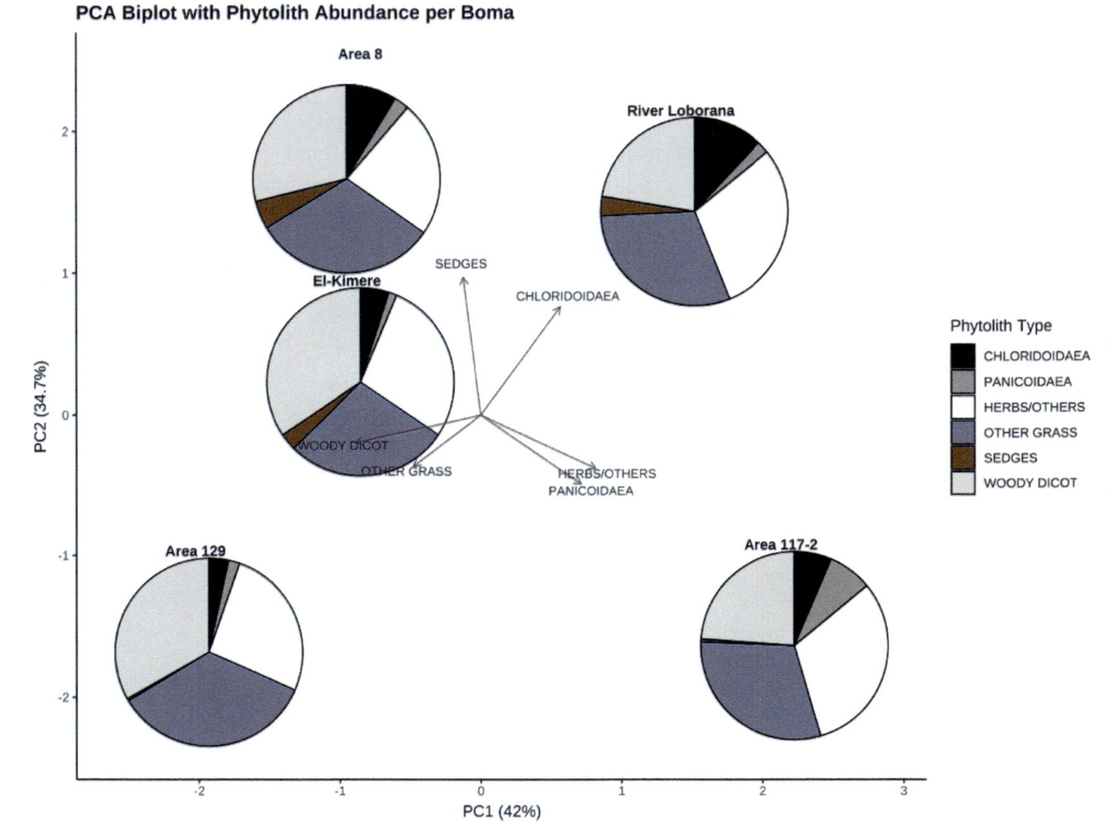

FIGURE 11.5 Principal component analysis (PCA) of phytolith assemblages represented in a pie chart c the relative abundance of phytolith plant function type. Arrows show phytolith-based functional grou variables.

Herbaceous and other grass habits exhibit high phytolith concentrations across all bomas, rangin from 23% to 34.88%. In contrast, sedge phytoliths are generally low in most bomas, except those adjace to riverine environments such as River Loborana and El-Kimere, where slightly higher concentrations ar observed. Woody plant phytolith remains consistently high throughout all bomas. Overall, the concer trations of phytoliths from herbaceous and other grasses are notably higher within habitat-type boma whereas phytolith function group diversity remains limited. The two main components, PC1 and PC account for 76.7% of the variance (fig. 11.5). On PC1, Chloridoideae and Panicoideae are positioned i opposite directions. Variables strongly associated with PC2 include woody dicots and sedges. Meanwhil herbs/others and other grass groups tend to cluster near the center.

Phytolith Indices

All bomas, irrespective of age, exhibited high Iph values ranging from > 20% to 40%, indicating a dom nance of Chloridoideae C4 grasses. Despite the presence of Panicoid grasses, the Iph values suggest th Chloridoideae are the most abundant. Moreover, all bomas exhibited a D/P ratio of less than 1, characte izing dominance of grasses.

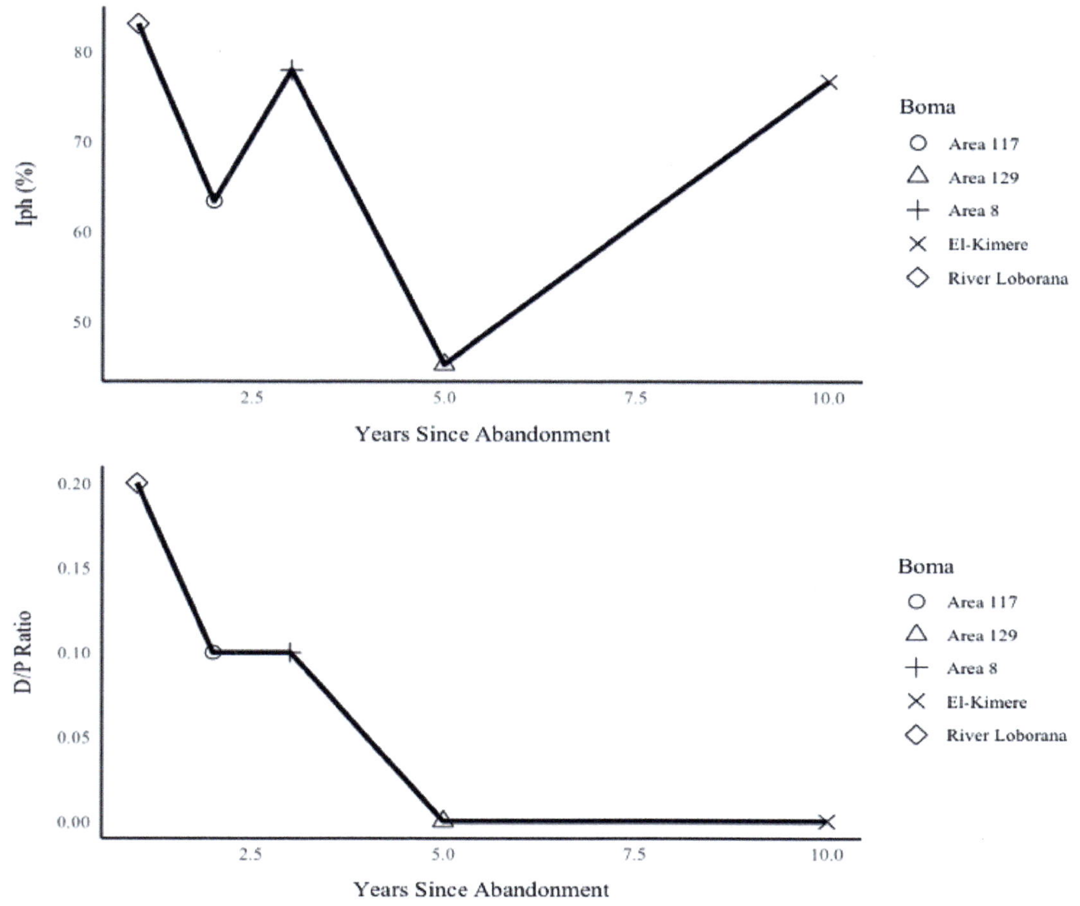

FIGURE 11.6 Phytolith indices Iph (Chloridoideae phytoliths (saddle-shaped)/(Chloridoideae + Panicoideae phytoliths (saddle + cross + bilobate + polylobate) and D/P ratio calculated for the abandoned boma in relation to site age of abandonment.

DISCUSSION

Plant Succession in Dassanech Abandoned Bomas

Pastoral communities that heavily graze in savanna ecosystems can induce substantial ecological shifts, frequently resulting in the transition from grassland biomes to woodland or shrubland habitats (Tobler *et al.*, 2003; Vuorio *et al.*, 2014). Pastoralists frequently herd their livestock in proximity to settlements, resulting in modifications that are often patchy (Muchiru *et al.*, 2008; Reid and Ellis, 1995; Vuorio *et al.*, 2014). The nutrients within abandoned bomas enhance the vegetation composition. Herbaceous species colonize the area, especially within cattle enclosures, due to the high levels of uric acid and nitrogen from livestock dung and urine, which begin to integrate with the soil (Augustine, 2003; Muchiru *et al.*, 2008, 2009; Vuorio *et al.*, 2014). As the boma aged, the herbaceous biomass increased within the abandoned bomas, especially soon after rain (fig. 11.6).

The herbaceous biomass exhibited a rapid increase ten years after abandonment, reflecting the species richness within these bomas. These results suggest that abandoned bomas in East Turkana (Ileret and

Karari) possess a more diverse vegetation composition across all bomas. The observed decline in herbaceous species, coupled with an increase in woody species in El-Kimere, may be attributed to the shading effects of trees and the specific location of the boma (Muchiru *et al.*, 2009). The boma in El-Kimere is in a riverine environment. Erosion created small canals within parts of the boma, facilitating the transport of manure and dung through water. Additionally, the high-water table promotes the growth of more trees in the area. Shadwell and February (2017) suggested that trees located along ephemeral rivers access deep groundwater reserves during periods of increased rainfall. Conversely, during drought conditions, trees downstream of boreholes—where groundwater extraction occurs—exhibit physiological stress and increasingly depend on isotopically enriched soil water. Importantly, despite the persistent aridity, dominant species such as *Acacia erioloba* maintain their canopy health in regions where groundwater remains accessible. These findings suggest that the dominance of trees in arid riparian zones is strongly correlated with groundwater availability and that groundwater extraction has the potential to disrupt this ecological equilibrium.

Dassannech Pastoralist bomas are comparable to those in Amboseli; during the construction of bomas, pastoralists tend to clear fewer trees within a 150-meter radius. Following abandonment, the site gradually becomes reforested, with *Acacia tortilis*, eventually establishing biomass dominance over grasses and forbs in older settlements. This phenomenon is particularly evident in East Turkana, where most *Acacia* sp. tend to colonize abandoned bomas after three to ten years.

The high composition of herbs/forbs and woody plants over time reflects the high nutrient content in the bomas and local livestock feeding behaviors, notably that of goats and camels. Goats and camels consume *Barleria* and *Duosperma* species during dry and wet seasons. When the environmental conditions are dry, goats and camels primarily consume *Acacia species*. Muchiru *et al.* (2009) and Reid and Ellis (1995) observed that goats ingest pods and seeds, subsequently excreting them within the boma during nighttime. Also, herders collected pods and seeds to nourish young goats near settlements during dry seasons. In Turkana, Reid and Ellis (1995) noted that the pods of *Acacia tortilis* are a nutritional resource for goats.

Most bomas, particularly around the age of ten years, exhibit a high concentration of trees and shrub species such as *Balanite* spp. These species are frequently utilized as fencing material for pastoralist bomas, especially on the cattle side, to prevent carnivores such as hyenas from attacking the cattle. Additionally, species like *Boscia* spp., *Cadaba* spp., and *Salvadora* spp. are often found near bomas, where they provide shade and are sometimes used as toothbrushes. The high concentration of woody species over time is attributable to their ability to accumulate nutrients in the absence of disturbance, particularly when grazers prefer herbaceous vegetation (Archer *et al.*, 2017).

The widespread presence of *Aristida* spp. and Stipagrostis spp., across all bomas underscores the remarkable adaptive strategies these species employ to survive and thrive within Africa's challenging arid and semi-arid environments. These grasses have evolved specific traits, such as drought tolerance and efficient water use, which enable them to withstand harsh climatic conditions, making them crucial components of arid and semi-arid environments (Ghasemkhani *et al.*, 2008). Here, we corroborate this by observing nutrient-rich boma sites sequentially colonized by grasses and forbs over time.

Phytolith Morphologies Concentration

The observed phytolith assemblages indicate the mixed vegetation of woodland and open habitat. Groups are relatively scarce across all bomas but tend to thrive in arid and semi-arid environments. The notable dominance of Chloridoideae at River Loborana and El-Kimere may be attributed to the seasonal availability of moisture or the presence of temporary water during the rainy season. Additionally, the marginally higher concentration of Panicoideae at Area 129 suggests that the soil in this boma was well-drained, likely due to recent rainfall.

Shahack-Gross *et al.* (2004) noted that dung from grazers during humid seasons has higher grass phytolith concentrations than dung from the same grazers during dry seasons. Although dung is better reserved under moist conditions, it also disintegrates during wet seasons, facilitating the release of nutrints and humus into the soil. The marginally increased presence of Chloridoideae phytoliths, predomiantly found in dung within the enclosure, is likely attributable to accumulation during drier periods haracterized by low grass phytolith concentrations. This is consistent with the fact that the dominant lant species in these periods are shrubs and woody plants, which are typically browsed rather than razed by herbivores. Furthermore, the dung within the boma appears to have accumulated following a ainy season, as evidenced by the heightened presence of Panicoideae phytoliths.

The woody taxa phytoliths in the boma are low compared to herbaceous and non-grass taxa. Woody nd shrub species generally produce low phytoliths, which decrease further during dry conditions than rasses (Shahack-Gross *et al.*, 2003, 2004). The observed phytolith morphologies do not exhibit signifiant temporal variation; however, the concentration remains relatively consistent throughout the samples. hytoliths, for example, remain stable within a pH range of five to nine. Given the alkaline conditions of ast African soils (Shahack-Gross *et al.*, 2003, 2004), opal phytoliths are expected to last for extended eriods (Berna *et al.*, 2004; Shahack-Gross *et al.*, 2004). Therefore, the relative concentration of phytoths may be attributable to limited mechanical mixing resulting from root activity and bioturbation.

hytolith Indices

he Iph index is employed in this study to quantify the relative dominance of short C4 grasses, preominantly belonging to the Chloridoideae subfamily, in comparison to tall C4 grasses, primarily from ne Panicoideae subfamily, within a specified vegetation zone (Bremond *et al.*, 2005a, 2008). Given that hloridoideae species exhibit greater tolerance to extreme temperature conditions and aridity relative to ther grasses, the Iph index is also influenced by the balance between humidity and aridity (Bremond *et al.*, 005a). The Iph indices, ranging from >45% across all bomas, indicate the dominance of Chloridoideae, hich aligns with the phytolith composition, as the Chloridoideae plant function type has a higher perentage compared to Panicoideae.

The low value of D/P^0 indicates the low component of woody taxa compared to grasses. All samples om Bomas indicate high aridity and low woody taxa components. The phytolith indices suggest that erbaceous grasslands characterize the bomas of all ages.

CONCLUSION

he Dassanech population has been increasing, which has also led to a corresponding rise in the livestock opulation. The community has been seasonally migrating across the landscape to ensure that grazing nds are utilized during the optimal season and are not overgrazed. They have "mapped" pastures for dry nd rainy seasons for grazing purposes. This socio-cultural practice has significantly influenced vegetaon cover dynamics, particularly within abandoned settlements. This impact is mainly due to the differntial nitrogen input from the dung that decomposes, adding to the soil nutrients in the cattle enclosures nd thereby creating microhabitats characterized by the abundance and diversity of herbaceous species, s well as the reduction or absence of other species. During the rainy season, various plant species can be nore easily observed compared to the dry season.

Specifically, our findings indicate that enclosures demonstrate higher herbaceous species richness, reater abundance of desirable species, and increased biomass compared to areas outside these settlement ites. These settlements are typically utilized asynchronously by traditional pastoralists and encompass broad spectrum of ages and successional stages. It is hypothesized that settlement activities, in connction with successional processes, substantially influence the development of mosaic patterns of bush nd grass cover, as well as the patch dynamics characteristic of savanna ecosystems. This suggests that pastoral settlement patterns are a significant factor in the redistribution of nutrients within East Turkana. The study concludes that phytolith assemblages correlate with vegetation and remain unchanged over time. The results suggest that phytolith data can identify past archaeological occupations and are crucial for reconstructing paleoenvironments, indicating changes in vegetation and human activities.

ACKNOWLEDGMENT

This study is supported by the Paleontological Scientific Trust (PAST) and the Baldwin Fellowship from the Leakey Foundation, both of which covered all of my living and educational expenses while I was studying at the University of Nairobi in Kenya, as well as the lab costs for the samples from Koobi Fora for my research. I would like to thank the Koobi Fora Field School (KFFS) for allowing me to join the research team and for their assistance with the logistics of sampling. The Kenyan Immigration office approved my application for a study permit—the National Museums of Kenya, Palynology and Palaeobotany Lab for affiliation recognition.

REFERENCES

Akeret, Ö., and Rentzel, P. (2001). Micromorphology and plant macrofossil analysis of cattle dung from the Neolithic lake shore settlement of Arbon Bleiche 3. *Geoarchaeology*, *16*(6), 687–700.

Albert, R. M., Bamford, M. K., and Cabanes, D. (2006). Taphonomy of phytoliths and macroplants in different soils from Olduvai Gorge (Tanzania) and the application to Plio-Pleistocene palaeoanthropological samples. *Quaternary International*, *148*(1), 78–94.

Albert, R. M., Bamford, M. K., and Esteban, I. (2015a). Reconstruction of ancient palm vegetation landscapes using a phytolith approach. *Quaternary International*, *369*, 51–66.

Albert, R. M., Bamford, M. K., Stanistreet, I., Stollhofen, H., Rivera-Rondón, C., and Rodríguez-Cintas, A. (2015b). Vegetation landscape at DK locality, Olduvai Gorge, Tanzania. Palaeogeography, Palaeoclimatology, *Palaeoecology*, 426, 34–45.

Alexandre, A., Meunier, J.-D., Lézine, A.-M., Vincens, A., and Schwartz, D. (1997). Phytoliths: Indicators of grassland dynamics during the late Holocene in intertropical Africa. *Palaeogeography, Palaeoclimatology, Palaeoecology*, *136*(1–4), 213–229.

Andersen, G. L., and Krzywinski, K. (2007). Longevity and growth of Acacia tortilis; insights from 14C content and anatomy of wood. *BMC Ecology*, *7*(1), 4.

Archer, S. R., Andersen, E. M., Predick, K. I., Schwinning, S., Steidl, R. J., and Woods, S. R. (2017). Woody plant encroachment: Causes and consequences. In D. D. Briske (Ed.), *Rangeland Systems: Processes, Management and Challenges*, pp. 25–84. Springer International Publishing.

Ashley, G. M., Ndiema, E. K., Spencer, J. Q. G., Harris, J. W. K., & Kiura, P. W. (2011). Paleoenvironmental context of archaeological sites: Implications for subsistence strategies under Holocene climate change, northern Kenya. Geoarchaeology, 26(6), 809–837. https://doi.org/10.1002/gea.20374

Atkin, O. K., Schortemeyer, M., McFarlane, N., and Evans, J. R. (1999). The response of fast-and slow-growing Acacia species to elevated atmospheric CO2: An analysis of the underlying components of relative growth rate. *Oecologia*, *120*(4), 544–554.

Augustine, D. J. (2003). Long-term, livestock-mediated redistribution of nitrogen and phosphorus in an East African savanna. *Journal of Applied Ecology*, *40*(1), 137–149.

Augustine, D. J., McNaughton, S. J., and Frank, D. A. (2003). Feedback between soil nutrients and large herbivores in a managed Savanna ecosystem. *Ecological Applications*, *13*(5), 1325–1337.

Behnke, R., Scoones, I., and Kerven, C. (1994). Range ecology at disequilibrium: New models of natural variability and pastoral adaptation in African Savannas. *Africa: Journal of the International African Institute*, *64*.

Berna, F., Matthews, A., and Weiner, S. (2004). Solubilities of bone mineral from archaeological sites: The recrystallization window. *Journal of Archaeological Science*, *31*(7), 867–882.

Blundell, M. (1992). *Wild Flowers of East Africa*. Harper Collins Publishers.

Boschian, G., and Montagnari-Kokelj, E. (2000). Prehistoric shepherds and caves in the Trieste Karst (Northeastern Italy). *Geoarchaeology, 15*(4), 331–371.

Bremond, L., Alexandre, A., Hély, C., and Guiot, J. (2005a). A phytolith index as a proxy of tree cover density in tropical areas: Calibration with Leaf Area Index along a forest–savanna transect in southeastern Cameroon. *Global and Planetary Change, 45*(4), 277–293.

Bremond, L., Alexandre, A., Peyron, O., and Guiot, J. (2005b). Grass water stress estimated from phytoliths in West Africa. *Journal of Biogeography, 32*(2), 311–327.

Bremond, L., Alexandre, A., Peyron, O., and Guiot, J. (2008). Definition of grassland biomes from phytoliths in West Africa. *Journal of Biogeography, 35*(11), 2039–2048.

Bundala, M. J. (2024). *Middle Pleistocene paleoenvironmental reconstruction through phytolith analysis at the Manyara Beds, northern Tanzania* [PhD Thesis]. University of Calgary.

Dharani, N. (2019). *Field Guide to Common Trees & Shrubs of East Africa*. Struik Nature.

Ellis, J. E., Coppock, D. L., McCabe, J. T., Galvin, K., and Wienpahl, J. (1984). Aspects of Energy Consumption in a Pastoral Ecosystem: Wood Use by the South Turkana. Wood, Energy and Households: Perspectives on Rural Kenya. Beijer Institute and the Scandinavian Institute of African Studies.

Fredlund, G. G., and Tieszen, L. T. (1994). Modern phytolith assemblages from the North American Great Plains. *Journal of Biogeography, 21*(3), 321–335.

Gallagher, R., and Carpenter, B. (1997). Human-dominated ecosystems. *Science, 277*(5325), 485–485.

Ghasemkhani, M., Akhani, H., Sahebi, J., and Scholz, H. (2008). The genera Aristida and Stipagrostis (Poaceae) in Iran. *Willdenowia - Annals of the Botanic Garden and Botanical Museum Berlin-Dahlem, 38*, 135–148.

Gifford-Gonzalez, D. (2017). Pastoralism in sub-Saharan Africa: Emergence and ramifications. In U. Albarella, M. Rizzetto, H. Russ, K. Vickers, and S. Viner-Daniels (Eds.), *The Oxford Handbook of Zooarchaeology*. Oxford University Press.

Hildebrand, E. A., and Grillo, K. M. (2012). Early herders and monumental sites in Eastern Africa: Dating and interpretation. *Antiquity, 86*(332), 338–352.

Hsieh, T. C., Ma, K. H., and Chao, A. (2016). iNEXT: An R package for rarefaction and extrapolation of species diversity (Hill numbers). *Methods in Ecology and Evolution, 7*(12), 1451–1456.

Katz, O., Cabanes, D., Weiner, S., Maeir, A. M., Boaretto, E., and Shahack-Gross, R. (2010). Rapid phytolith extraction for analysis of phytolith concentrations and assemblages during an excavation: An application at Tell es-Safi/Gath, Israel. *Journal of Archaeological Science, 37*(7), 1557–1563.

Kinyanjui, R. N. (2018). *Phytolith analysed to compare changes in vegetation structure of Koobi Fora and Olorgesailie basins through the mid-pleistocene-holocene periods* [PhD Thesis]. University of Witwatersrand.

Lamprey, R., and Waller, R. (1990). The Loita-Mara region in historical times: Patterns of subsistence, settlement and ecological change. *Early Pastoralists of South-Western Kenya, 3*, 16–35.

Madella, M., Alexandre, A., and Ball, T. (2005). International Code for Phytolith Nomenclature 1.0. *Annals of Botany, 96*(2), 253–260.

Magurran, A. E. (2013). *Measuring Biological Diversity*. John Wiley & Sons.

Marean, C. W. (1992). Implications of late Quaternary mammalian fauna from Lukenya Hill (south-central Kenya) for paleoenvironmental change and faunal extinctions. *Quaternary Research, 37*(2), 239–255.

Marean, C. W., Mudida, N., and Reed, K. E. (1994). Holocene paleoenvironmental change in the Kenyan Central Rift as indicated by micromammals from Enkapune Ya Muto Rockshelter. *Quaternary Research, 41*(3), 376–389.

Marshall, F. B. (1986). *Aspects of the Advent of Pastoral Economies in East Africa (Neolithic, Fauna, Kenya)*. University of California, Berkeley.

Marshall, F. B. (1990). Origins of specialized pastoral production in East Africa. *American Anthropologist, 92*(4), 873–894.

Marshall, F. B., and Hildebrand, E. (2002). Cattle before crops: The beginnings of food production in Africa. *Journal of World Prehistory, 16*(2), 99–143.

Marshall, F. B., Stewart, K., and Barthelme, J. (1984). Early domestic stock at Dongodien in Northern Kenya. *Azania: Archaeological Research in Africa, 19*(1), 120–127.

Martins, D. J. (2022). *Grasses of East Africa*. Struik Nature.

Mashaka, H. (2022). *Palaeo Environmental Reconstruction of Holocene Vegetation in East Turkana, Kenya* [MA Thesis]. University of Nairobi.

Mbaluka, J. K., and Brown, F. H. (2016). Vegetation of the Koobi Fora Region Northeast of Lake Turkana, Marsabit County, Northern Kenya. *Journal of East African Natural History, 105*(1), 21–50.

Mercader, J., Bennett, T., Esselmont, C., Simpson, S., and Walde, D. (2009). Phytoliths in woody plants from the Miombo woodlands of Mozambique. *Annals of Botany, 104*(1), 91–113.

Muchiru, A. N. (1981). *The ecological impact of abandoned Maasai settlements on savanna vegetation and its herbivores in the Amboseli ecosystem* [PhD Thesis]. University of Nairobi.

Muchiru, A. N., Western, D. J., and Reid, R. S. (2008). The role of abandoned pastoral settlements in the dynamics of African large herbivore communities. *Journal of Arid Environments*, *72*(6), 940–952.

Muchiru, A. N., Western, D., and Reid, R. S. (2009). The impact of abandoned pastoral settlements on plant and nutrient succession in an African savanna ecosystem. *Journal of Arid Environments*, *73*(3), 322–331.

Ndiema, K. E., Dillian, C. D., Braun, D. R., Harris, J. W. K., and Kiura, P. W. (2011). Transport and subsistence patterns at the transition to pastoralism, Koobi Fora, Kenya. *Archaeometry*, *53*(6), 1085–1098.

Neumann, K., Fahmy, A., Lespez, L., Ballouche, A., and Huysecom, E. (2009). The early holocene palaeoenvironment of Ounjougou (Mali): Phytoliths in a multiproxy context. *Palaeogeography, Palaeoclimatology, Palaeoecology*, *276*(1), 87–106.

Neumann, K., Strömberg, C. A. E., Ball, T., Albert, R. M., Vrydaghs, L., and Cummings, L. S. (2019). International Code for Phytolith Nomenclature (ICPN) 2.0. *Annals of Botany*, *124*(2), 189–199.

Novello, A., and Barboni, D. (2015). Grass inflorescence phytoliths of useful species and wild cereals from sub-Saharan Africa. *Journal of Archaeological Science*, *59*, 10–22.

Piperno, D. R. (2006). *Phytoliths: A Comprehensive Guide for Archaeologists and Paleoecologists.* Bloomsbury Publishing PLC.

Piperno, D. R., and Pearsall, D. M. (1998). The silica bodies of tropical American grasses: Morphology, taxonomy, and implications for grass systematics and fossil phytolith identification. *Smithsonian Contributions to Botany*.

Powys, A. (2019). *Wildflowers of Kenya and Northern Tanzania*. Struik Nature.

Pratt, D. J., and Gwynne, M. D. (1977). *Rangeland Management and Ecology in East Africa.*

R Core Team. (2025). *R: The R Project for Statistical Computing.*

Reid, R. S., and Ellis, J. E. (1995). Impacts of pastoralists on woodlands in South Turkana, Kenya: Livestock-mediated tree recruitment. *Ecological Applications*, *5*(4), 978–992.

Rossouw, L. (2009). *The application of fossil grass-phytolith analysis in the reconstruction of Cainozoic environments in the South African interior* [PhD Thesis]. University of the Free State.

Shadwell, E., and February, E. (2017). Effects of groundwater abstraction on two keystone tree species in an arid savanna national park. *PeerJ*, *5*, e2923.

Shahack-Gross, R., Marshall, F., Ryan, K., and Weiner, S. (2004). Reconstruction of spatial organization in abandoned Maasai settlements: Implications for site structure in the Pastoral Neolithic of East Africa. *Journal of Archaeological Science*, *31*(10), 1395–1411.

Shahack-Gross, R., Marshall, F., and Weiner, S. (2003). Geo-ethnoarchaeology of pastoral sites: The identification of livestock enclosures in abandoned Maasai settlements. *Journal of Archaeological Science*, *30*(4), 439–459.

Shahack-Gross, R., Simons, A., and Ambrose, S. H. (2008). Identification of pastoral sites using stable nitrogen and carbon isotopes from bulk sediment samples: A case study in modern and archaeological pastoral settlements in Kenya. *Journal of Archaeological Science*, *35*(4), 983–990.

Spellerberg, I., and Fedor, P. (2003). A tribute to Claude Shannon (1916–2001) and a plea for more rigorous use of species richness, species diversity and the 'Shannon–Wiener' Index. *Global Ecology & Biogeography*, *12*, 177–179.

Stromberg, C. (2004). Using phytolith assemblages to reconstruct the origin and spread of grass-dominated habitats in the Great Plains of North America during the late Eocene to early Miocene. *Palaeogeography, Palaeoclimatology, Palaeoecology*, *207*(3–4), 239–275.

Stromberg, C. A. (2003). *The origin and spread of grass-dominated ecosystems during the Tertiary of North American and how it relates to the evolution of hypsodonty in equids.* University of California, Berkeley.

Strömberg, C. A. E., Saylor, B. Z., Engelman, R. K., Catena, A. M., Hembree, D. I., Anaya, F., and Croft, D. A. (2024). The flora, fauna, and paleoenvironment of the late Middle Miocene Quebrada Honda Basin, Bolivia (Eastern Cordillera, Central Andes). *Palaeogeography, Palaeoclimatology, Palaeoecology*, *656*, 112518.

Sullivan, S., and Rohde, R. (2002). On non-equilibrium in arid and semi-arid grazing systems. *Journal of Biogeography*, *29*(12), 1595–1618.

Tobler, M. W., Cochard, R., and Edwards, P. J. (2003). The impact of cattle ranching on large-scale vegetation patterns in a coastal savanna in Tanzania. *Journal of Applied Ecology*, *40*(3), 430–444.

Twiss, P. C. (1992). Predicted world distribution of C3 and C4 grass phytoliths. In *Phytolith Systematics: Emerging Issues*, pp. 113–128. Boston, MA: Springer.

Twiss, P. C., Suess, E., and Smith, R. M. (1969). Morphological classification of grass phytoliths. *Soil Science Society of America Journal*, *33*(1), 109–115.

Vetter, S. (2005). Rangelands at equilibrium and non-equilibrium: Recent developments in the debate. *Journal of Arid Environments*, *62*(2), 321–341.

uorio, V., Muchiru, A., Reid, R. S., and Ogutu, J. O. (2014). How pastoralism changes savanna vegetation: Impact of old pastoral settlements on plant diversity and abundance in south-western Kenya. *Biodiversity and Conservation*, *23*(13), 3219–3240.

/alker, B. H., and Noy-Meir, I. (1982). Aspects of the Stability and Resilience of Savanna Ecosystems. In B. J. Huntley and B. H. Walker (Eds.), *Ecology of Tropical Savannas*, pp. 556–590.

/einer, S. (2010). *Microarchaeology: Beyond the visible archaeological record.* Cambridge University Press.

/estern, D. (1973). *The structure, dynamics and changes of the Amboseli ecosystem* [PhD Thesis]. University of Nairobi.

Index

For Product Safety Concerns and Information please contact our
EU representative GPSR@taylorandfrancis.com Taylor & Francis
Verlag GmbH, Kaufingerstraße 24, 80331 München, Germany